The Physical Chemistry

of Steelmaking

TECHNOLOGY PRESS BOOKS IN SCIENCE AND ENGINEERING

John F. Elliott, Editor

The Physical Chemistry
of Steelmaking

Proceedings of the Conference
THE PHYSICAL CHEMISTRY OF IRON AND STEELMAKING
Endicott House, Dedham, Massachusetts
28 May to 3 June, 1956

Sponsored by

DEPARTMENT OF METALLURGY

MASSACHUSETTS INSTITUTE OF TECHNOLOGY

The M.I.T. Press
Massachusetts Institute of Technology
Cambridge, Massachusetts

SECOND PRINTING, JUNE, 1959

Library of Congress Catalog Card Number: 58–12709
Printed in the United States of America

CONFERENCE CHAIRMAN

Professor John Chipman

Head, Department of Metallurgy

Massachusetts Institute of Technology

TECHNICAL ARRANGEMENTS

John F. Elliott

Associate Professor of Metallurgy

Massachusetts Institute of Technology

GENERAL ARRANGEMENTS

Thomas B. King

Associate Professor of Metallurgy

Massachusetts Institute of Technology

CONFEREES

Harry L. Bishop—*Melting Advisor*
Allegheny Ludlum Steel Corporation
Brackenridge, Pennsylvania
(Formerly: Jones and Laughlin
Steel Corporation)

John Chipman—*Professor*
Department of Metallurgy
Massachusetts Institute of Technology
Cambridge, Massachusetts

L. S. Darken—*Associate Director*
U. S. Steel Corporation
Fundamental Research Laboratory
Monroeville, Pennsylvania

G. Derge—*Jones and Laughlin Professor*
Carnegie Institute of Technology
Schenley Park
Pittsburgh, Pennsylvania

John F. Elliott—*Associate Professor*
Department of Metallurgy
Massachusetts Institute of Technology
Cambridge, Massachusetts

Harold B. Emerick—*Director of Technical Services*
Jones and Laughlin Steel Corporation
3 Gateway Center
Pittsburgh, Pennsylvania

Karl L. Fetters—*Assistant Vice President*
Youngstown Sheet and Tube Company
Youngstown, Ohio

W. A. Fischer—*Professor*
Max Planck Institute
Düsseldorf, Germany

James C. Fulton—*Associate Director of Research*
Allegheny Ludlum Steel Corporation
Brackenridge, Pennsylvania

Daniel J. Girardi
>Steel and Tube Division
>Timken Roller Bearing Company
>Canton, Ohio

Nev A. Gokcen—*Associate Professor*
>*University of Pennsylvania*
>*Philadelphia, Pennsylvania*

N. J. Grant—*Professor*
>*Department of Metallurgy*
>*Massachusetts Institute of Technology*
>*Cambridge, Massachusetts*

Donald C. Hilty—*Manager Research Information*
>*Metals Research Laboratories*
>*Electro Metallurgical Company*
>*Niagara Falls, New York*

Ulf Kalling—*Manager Steel Division*
>*Stora Kopparberg Corporation*
>*230 Park Avenue*
>*New York, New York*

T. B. King—*Associate Professor*
>*Department of Metallurgy*
>*Massachusetts Institute of Technology*
>*Cambridge, Massachusetts*

H. Kosmider—*Dr. Ing.*
>*Klückner Hüttenwerk Haspe*
>*Hagen-Haspe*
>*Westphalia, Germany*

P. P. Kozakevitch—*Director of Chemistry Division*
>*Institut de Recherches de la Sidérurgie*
>*Saint-Germain-en-Laye, France*

William A. Krivsky
>*Electro Metallurgical Company*
>*Niagara Falls, New York*

Frederick C. Langenberg—*Supervisor*
>*Pyrometallurgical Research*
>*Crucible Steel Company of America*
>*Pittsburgh, Pennsylvania*

B. M. Larsen—*Assistant Director*
>*Edgar C. Bain Research Laboratory*
>*U. S. Steel Research Center*
>*Monroeville, Pennsylvania*

Hugo R. Larson—*Research Metallurgist*
>*American Brake Shoe Company*
>*Mahwah, New Jersey*

S. Matoba—*Professor*
>*Tôhoku University*
>*Sendai, Japan*

W. H. Mayo—*Manager Process Control*
>*U. S. Steel Corporation*
>*Pittsburgh, Pennsylvania*

C. Law McCabe—*Associate Professor*
>*Carnegie Institute of Technology*
>*Pittsburgh, Pennsylvania*

Joseph P. Morris
>*U. S. Bureau of Mines*
>*4800 Forbes Street*
>*Pittsburgh, Pennsylvania*

Arnulf Muan—*Assistant Professor*
>*Pennsylvania State University*
>*University Park, Pennsylvania*

D. W. Murphy—*Technical Adviser*
>*Bethlehem Steel Company*
>*Bethlehem, Pennsylvania*

Michael Olette
>*Institut de Recherches de la Sidérurgie*
>*Saint-Germain-en-Laye, France*

J. Pearson
>*British Iron and Steel Research Association*
>*Hoyle Street*
>*Sheffield, England*

W. O. Philbrook—*Professor*
>*Department of Metallurgical Engineering*
>*Carnegie Institute of Technology*
>*Pittsburgh, Pennsylvania*

Bernard R. Queneau—*Assistant Manager, Metallurgy*
>*Inspection and Research*
>*Tennessee Coal and Iron Division*
>*U. S. Steel Corporation*
>*Fairfield, Alabama*
>*(Formerly: U. S. Steel Corporation Duquesne, Pennsylvania)*

F. D. Richardson—*Professor*
>*Imperial College of Science and Technology*
>*University of London*
>*Royal School of Mines*
>*London, England*

George R. St. Pierre—*Assistant Professor*
>*Ohio State University*
>*Columbus, Ohio*

C. E. A. Shanahan
>*The R. T. S. C. Laboratories*
>*The Firs*
>*Whitchurch, Aylesbury*
>*Buckingham, England*

Charles W. Sherman
>*Jones and Laughlin Steel Corporation*
>*3 Gateway Center*
>*Pittsburgh, Pennsylvania*
>*(Formerly: Director of Research Latrobe Steel Company)*

Charles R. Taylor
>*Research Laboratories*
>*Armco Steel Corporation*
>*Middletown, Ohio*

Michael Tenenbaum—*Superintendent*
>*Quality Control Department*
>*Inland Steel Company*
>*East Chicago, Indiana*

G. Trömel—*Professor*
>*Max Planck Institute*
>*Düsseldorf, Germany*

Georges Urbain
>*Institut de Recherches de la Sidérurgie*
>*Saint-Germain-en-Laye, France*

Pierre Vallet
> *Institut de Recherches de la Sidérurgie*
> *Saint-Germain-en-Laye, France*

John H. Walsh—*Scientific Officer*
> *Mines Branch*
> *Department of Mines and Technical*
> *Surveys*
> *Ottawa, Ontario, Canada*

Carl W. Wagner—*Director*
> *Max Planck Institut für Physikalische*
> *Chemie*
> *Göttingen, West Germany*
> *(Formerly: Professor*
> *Massachusetts Institute of Technology)*

Others Attending

C. M. Adams—*Assistant Professor*
> *Department of Metallurgy*
> *Massachusetts Institute of Technology*
> *Cambridge, Massachusetts*

W. D. Kingery—*Associate Professor*
> *Department of Metallurgy*
> *Massachusetts Institute of Technology*
> *Cambridge, Massachusetts*

J. Bruce Wagner—*Assistant Professor*
> *Pennsylvania State University*
> *University Park*
> *Pennsylvania*
> *(Formerly: Research Associate*
> *Massachusetts Institute of Technology)*

Massachusetts Institute of Technology Students Attending

T. Floridas

T. Fuwa

J. Humbert

P. Koros

Preface

This volume contains the papers presented at the Conference on The Physical Chemistry of Iron and Steelmaking which was held late in May and early in June of 1956. For brevity, the pertinent points of the discussions have been summarized, and these summaries follow the papers. The aim of the conference was to bring together metallurgists who were actively working in the field and who were qualified by background and experience to contribute significantly both formally and informally. The program was designed to include papers on the latest research in the field and also to stimulate the conferees to think about how these fundamental data can be used to obtain a better understanding of actual steelmaking systems.

The idea of this conference developed in the summer of 1955 between Professor John Chipman and several of his friends in the field of metallurgy. The conference was to differ from other meetings and conferences which had been held in the past in that it would be held in a secluded place where those attending would not only discuss metallurgy together but would also have ample opportunity to enjoy each other's company while they lived together. The Endicott Estate in Dedham, Massachusetts, was a suitable site because it can house comfortably approximately fifty men, has adequate dining and meeting facilities for the group, and is sufficiently far from Boston to avoid the intrusion of the activities of a large city.

In retrospect it appears that the aim of Professor Chipman and his planning committee was realized. A free exchange of views prevailed during the technical sessions. The discussions of the day were continued on a more informal basis during meals and in the late afternoons and evenings. In a short time the conferees came to know each other personally. Small groups took advantage of the pleasant surroundings and went for walks in the woods among the late spring flora. The weather was pleasant and warmed sufficiently so that many enjoyed a short swim in the pool on Friday.

Although the program allowed some time for discussion after each paper, in many cases it was necessary to move on to the next paper before all had their say. These discussions were renewed on a very informal basis over a cup of tea, during dinner, or on the patio. It is felt that one of the primary benefits of the conference was derived from the associations made and the viewpoints exchanged during these informal discussions. Unfortunately, these benefits can neither be chronicled nor transmitted on the printed page. Although the papers in themselves are important contributions to the technical literature, it is hoped that the reader might also gain an appreciation of the atmosphere of the conference as he reads the discussion summaries and the commentaries.

JOHN F. ELLIOTT

Cambridge, Mass.
February 6, 1958

Contents

SECTION 4

Slag-Metal Equilibria in Blast-Furnace and Steelmaking Furnace Systems

SECTION 5

Kinetics and Slag-Metal Reactions

SECTION 6

Reaction Rates in Iron and Steelmaking Processes

SECTION 7

Application of Fundamental Data to Process Development and Metallurgical Problems in the Steel Industry

SECTION 8

Solidification of Castings and Ingots

SECTION 9

Research Planning

SPECIAL LECTURE

N. J. Grant
C. E. A. Shanahan
Presiding

Section 1

Liquid Metals
and Properties of Solutes
in Liquid Iron and Steel

by A. Rist
and J. Chipman

Activity of Carbon

in Liquid Iron-Carbon Solutions

INTRODUCTION

The early physical chemistry of steelmaking was content with applying the law of mass action in its original form to solutes in liquid iron. Much was written then on how to express the concentration of carbon. Was carbon dissolved as atoms or as molecules such as Fe_3C? Precarious phase-diagram and thermal data could not help to settle the issue.[1] As metallurgical chemistry was gradually approached from a more thermodynamical point of view, non-ideal solutions gained recognition, and the necessity to establish activity-concentration relationships was realized. The nature of the solution, now known to be interstitial in the case of carbon,[2] could be ignored for that purpose.

The attention given by metallurgists to the reaction of carbon and oxygen in the open-hearth bath greatly delayed the study of the simple binary iron-carbon system. The equilibrium of laboratory melts with $CO-CO_2$ mixtures, defining fixed carbon and oxygen potentials, was used by a number of workers [3, 4, 5, 6, 7] to study the equilibrium value of the product [C]·[O]. Such experiments may give information on the activity of carbon if accurate gas analyses are obtained. In fact, Phragmén and Kalling[6] did compute an activity coefficient for Henry's law from their data which ranged below 0.1% carbon. They remarked that the value which they found had to increase very fast with concentration if the solubility limit was to be accounted for. Marshall and Chipman[7] reached carbon contents as high as 2.0% by operating under pressure. They found that the activity coefficient of carbon may be regarded as constant up to 1% and that it increases thereafter. Later work was not to confirm this view.

In 1953, Richardson and Dennis[8] contributed the first study devoted primarily to the determination of the carbon activity in liquid iron. Melts with carbon contents between 0.1 and 1.1% were equilibrated with controlled $CO-CO_2$ mixtures at 1560, 1660, and 1760° C. The experiments were carried out with extreme care, and the data are very consistent. They point to an appreciable deviation from Henry's law down to the lowest carbon investigated.

The work of Richardson and Dennis is authoritative and covers most of the range of interest in steelmaking. Nevertheless, it seemed desirable, for the sake of completeness as well as to provide data for ironmaking, to explore the entire liquid field. The present work, although it met with limited success, was undertaken with this purpose.

Concentrated iron-carbon solutions have already been studied by Esin and Gavrilov[9] and by Sanbongi and Ohtani.[10] These authors built electrochemical cells of the type:

Based on a thesis submitted in partial fulfillment of the requirements for the degree of Doctor of Science at the Massachusetts Institute of Technology. Dr. Chipman is Professor of Metallurgy, Massachusetts Institute of Technology. Dr. Rist is now with the Institut de Recherches de la Sidérurgie, Saint-Germain-en-Laye, France.

The authors wish to thank Professor C. Wagner for many helpful discussions, T. Fuwa for assistance in the equilibrium measurements, and D. L. Guernsey for carbon determinations. Financial support was received from the Institut de Recherches de la Sidérurgie, the Allegheny Ludlum Steel Corporation, and the Office of Naval Research.

| Liquid iron-carbon alloy (sat.) | Carbide slag | Liquid iron-carbon alloy |

and measured their electromotive forces at various concentrations of the alloy on the right-hand side.

STATEMENT OF THE PROBLEM AND METHOD

It was proposed to measure the activity of carbon dissolved in liquid iron as a function of concentration and temperature through the study of the equilibrium:

$$CO_2 \text{ (g)} + [C] = 2CO \text{ (g)} \tag{1}$$

where [C] represents carbon dissolved in liquid iron. The equilibrium constant for the reaction at temperature T is

$$K_1 = \frac{(p_{CO})^2}{p_{CO_2} \cdot a_C}$$

where p's represent partial pressures in the gas phase and a_C is the activity of carbon in solution.

Temperature and the "gas ratio" $(p_{CO})^2/p_{CO_2}$ are taken as the two independent variables. The activity is proportional to the gas ratio, the proportionality factor $1/K_1$ being determined at each temperature by the choice of a standard state for carbon. In the experiments, gas ratio and temperature are maintained constant, and the metal which is exposed to the flowing gas adjusts its composition to the carbon activity imposed by the gas phase. The study is thus designed to yield the activity-composition relationship.

The main experimental problems are: the control and measurement of the gas composition, the control and measurement of temperature, and the analysis of the metal.

In order to cover the range of high carbon concentration, high gas ratios must be attained with mixtures extremely dilute with respect to carbon dioxide. A situation thereby arises in which composition control in the gas is difficult (and, in fact, becomes impossible when carbon deposition steps in), the reaction is slow, and side reactions between melt and crucible are favored.

APPARATUS AND EXPERIMENTAL PROCEDURE

Preparation of the Gas Mixtures

Ternary gas mixtures of CO_2, CO, and argon were used, argon being added in order to benefit by the effects of an increased total flow rate. The sketch of Fig. 1 shows clearly the three gas lines.

Argon was purified from water vapor, carbon dioxide, and oxygen in columns containing anhydrone, ascarite, and magnesium turnings at 590° C. Carbon dioxide was dried over anhydrone and purified from oxygen over copper at 450° C. Carbon monoxide was manufactured by passing dry carbon dioxide over graphite at 1100° C and then purified from residual carbon dioxide over a concentrated potassium hydroxide solution and ascarite.

The flow rates of the component gases were controlled by adjusting the pressure drops across capillary flow meters. Through most of the experiments the flow rates of CO and argon were kept constant

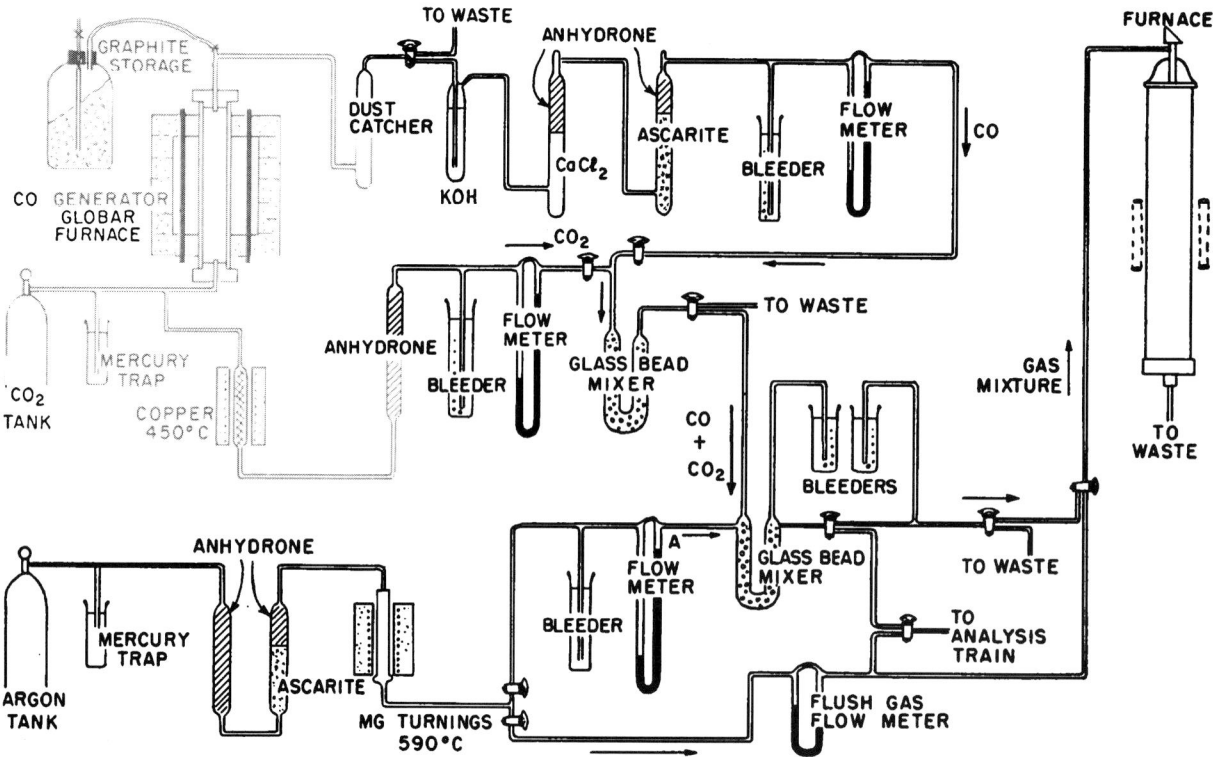

Fig. 1. Diagram of gas system.

and each equal to about 300 ml/min while the flow rate of CO_2 was varied to obtain the required gas ratios. The entire range of gas ratios (100 to 4300) was covered with CO_2 flow rates ranging roughly from 2 to 0.03 ml/min.

Gas Analysis

The apparatus was equipped with facilities for the gravimetric analysis of CO_2 by absorption on ascarite and of CO by conversion to CO_2 in a cupric-oxide furnace followed by absorption on ascarite. Analysis was used to establish or check the calibration of the capillary flow meters. In the case of argon, a volumetric method was used.

The analysis of CO_2 required special care in view of the small quantities involved. Over 4 hr were necessary to collect about 15 mg at the lowest flow rate. Two ascarite bulbs were put in series, the first one being capable of absorbing over 99% of the incoming CO_2. Argon had first been used as a flushing gas; later, hydrogen was substituted for it to minimize the weight fluctuations of the enclosed gas, and a dummy bulb was used to suppress buoyancy corrections.

Furnace Design

The metal was contained in alumina crucibles and was heated and stirred by high-frequency in-

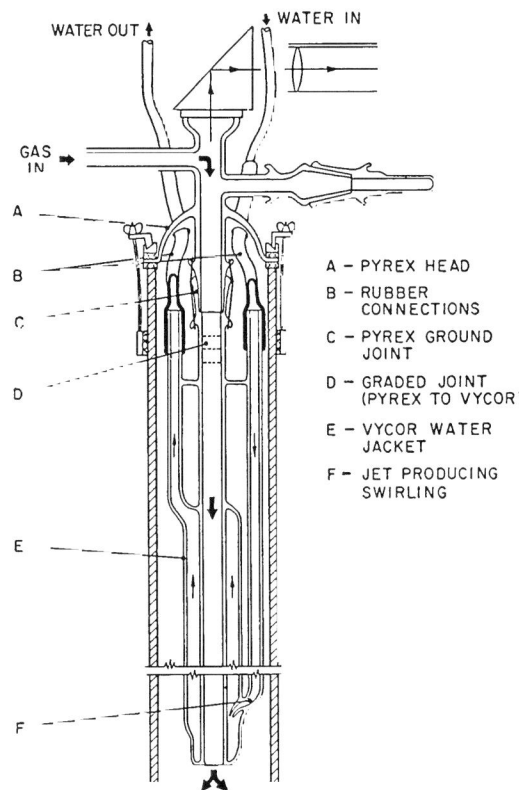

Fig. 3. Furnace head mounted with watercooled Vycor tube.

A - PYREX HEAD
B - RUBBER CONNECTIONS
C - PYREX GROUND JOINT
D - GRADED JOINT (PYREX TO VYCOR)
E - VYCOR WATER JACKET
F - JET PRODUCING SWIRLING

duction. The furnace, as first designed and mounted, is shown in Fig. 2. An Alundum tube, 13 mm i.d., led the gas flow downward to the melt surface. The crucible was surrounded by an annular graphite susceptor in order to delay cooling of the gases as they left and thereby to delay carbon deposition. The furnace enclosure was a glazed silica tube equipped with a sight glass and prism at the top to permit optical temperature readings.

The above version of the furnace failed at gas ratios higher than 1150 when carbon deposition began to appear in the Alundum inlet tube. A new inlet tube was installed, made of Vycor and water-cooled all the way down to its mouth above the melt (see Fig. 3). That second version, which was successful in preventing carbon deposition at the lower temperatures used (1360 and 1260° C) introduced other errors to be discussed later.

Temperature Measurement

Temperature was measured with a disappearing-filament pyrometer. Previous work[11] gave information on the emissivity of pure iron and its variation with temperature, thus permitting calibration of the instrument at the melting point of iron and providing an optical temperature scale over a range of temperatures. The validity of the calibration has been extended to iron-carbon alloys at lower temperatures by using the eutectic point (1153° C) as a reference in conjunction with a

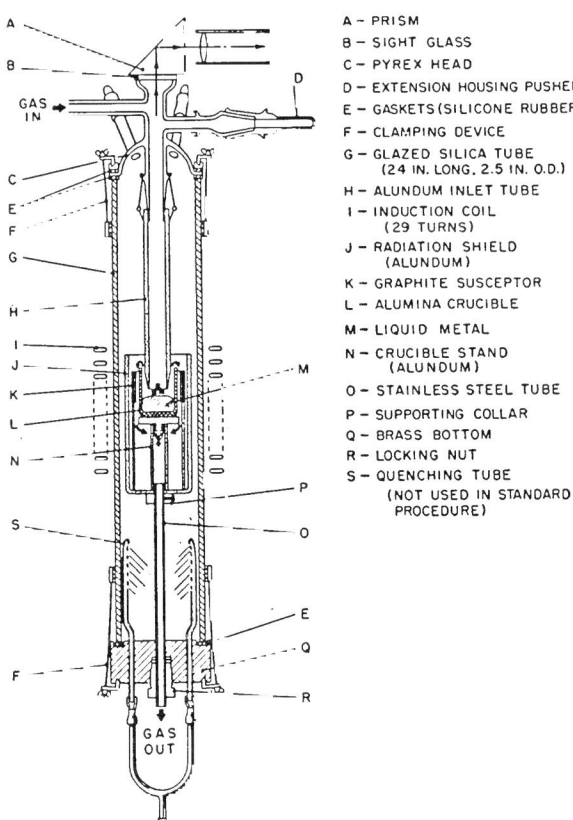

A - PRISM
B - SIGHT GLASS
C - PYREX HEAD
D - EXTENSION HOUSING PUSHER
E - GASKETS (SILICONE RUBBER)
F - CLAMPING DEVICE
G - GLAZED SILICA TUBE (24 IN. LONG. 2.5 IN. O.D.)
H - ALUNDUM INLET TUBE
I - INDUCTION COIL (29 TURNS)
J - RADIATION SHIELD (ALUNDUM)
K - GRAPHITE SUSCEPTOR
L - ALUMINA CRUCIBLE
M - LIQUID METAL
N - CRUCIBLE STAND (ALUNDUM)
O - STAINLESS STEEL TUBE
P - SUPPORTING COLLAR
Q - BRASS BOTTOM
R - LOCKING NUT
S - QUENCHING TUBE (NOT USED IN STANDARD PROCEDURE)

Fig. 2. Furnace arrangement.

linear extrapolation of the emissivity curve for pure iron. Agreement was found within two degrees by observing the solidification of alloys of slightly hypoeutectic composition.

Running Procedure

For each run a temperature and a gas ratio were selected. A 30-gram charge was prepared from electrolytic iron and a very pure grade of graphite. Air was flushed out of the furnace with argon, and the charge was heated under argon. Melting was completed under the ternary gas mixture to avoid excessive reaction between metal and crucible, and temperature was stabilized at the assigned value after 15 min of heating.

The heat was held at temperature for times which varied between a few minutes in recovery runs and a maximum of 31 hr (see Tables 1 and 2). Temperature was controlled manually through the power output of the high-frequency converter unit. Fluctuations in temperature were normally less than ±10 degrees.

At the end of the run, argon was substituted for the gas mixture, the power turned off, and the melt cooled under argon. The heats containing less than 2.0% carbon were killed with aluminum. Quenching had been planned originally and was to be effected between two helium jets at the bottom of the furnace. It was abandoned, however, to suppress opportunities for scraping or shaking loose any carbon deposited on the exit path of the gas.

Metal Analysis

The metal analysis was performed by the conventional combustion method. One-gram samples were taken from milling chips representing one-half of the solidified ingot and thoroughly mixed. The analysis was thus made insensitive to segregation if any was present. When solidification had produced grey or mottled iron, a certain amount of graphite powder was present with the chips. It was carefully screened out and weighed, and proportional amounts of powder and chips were taken for each *analytical sample. The spread of duplicate carbon determination was constant at all carbon levels and equal to* ±0.01% carbon.

EXPERIMENTAL RESULTS

The experiments reported are divided into series A and series B according to the furnace setup used. As a rule, only the heats corresponding to the closest approach to equilibrium are listed.

Series A. The heats were made with Alundum inlet tubes, and results were obtained free of any recognized systematic error at 1560 and 1460° C with gas ratios up to 1150. The data are recorded

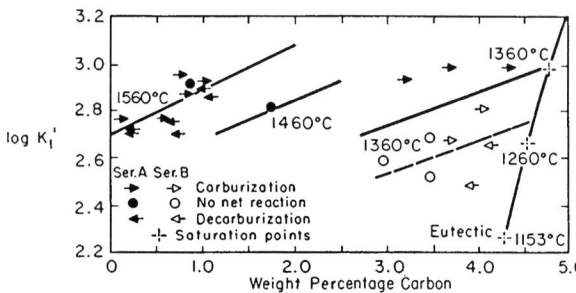

Fig. 4. Experimental data; $K_1' = p^2_{CO}/(p_{CO_2} \cdot \% C)$. The solid lines represent equilibrium; broken line at 1360° represents series B, known to be subject to errors of thermal diffusion and incomplete equilibrium.

in Table 1 and plotted in Fig. 4. In the fifth column of Table 1, the "initial % C" of a heat was

Table 1. Experimental Results, Series A

Heat No.	Temperature, °C	Gas Ratio	Time, hr	% C Initial	% C Final	Δ% C	log K'*
53	1560	104	4.0	0.16	0.19	+0.03	2.750
53	1560	103	6.8	0.31	0.20	−0.11	2.710
54	1560	102	3.0	0.20	0.19	−0.01	2.725
58	1560	336	4.0	0.55	0.57	+0.02	2.770
59	1560	325	6.0	0.64	0.62	−0.02	2.720
61	1560	325	6.0	0.59	0.56	−0.03	2.765
72	1560	1045	6.0	1.12	1.14	+0.02	2.965
68	1560	1030	5.0	1.17	1.17	0.00	2.945
63	1560	990	6.0	1.18	1.19	+0.01	2.920
66	1560	1030	5.25	1.29	1.27	−0.02	2.910
65	1560	1035	6.0	1.23	1.22	−0.01	2.930
75	1560	1150	6.0	1.28	1.29	+0.01	2.950
133	1460	1140	6.0	1.76	1.76	0.00	2.810
81	1360	2750	6.0	3.19	3.21	+0.02	2.930
82	1360	3705	6.0	3.78	3.81	+0.03	2.995
83	1360	4290	6.0	4.37	4.38	+0.01	2.990

$$* K' = \frac{(p_{CO})^2}{p_{CO_2} \cdot \% C}.$$

calculated after recovery runs showing that over 99% of the carbon charged was recovered. In the seventh column, Δ% C is the difference between final and initial % C. The ranges of temperatures and gas ratios which could be investigated were limited by the occurrence of carbon deposition, which is discussed later. Justification for quoting heats 81, 82, and 83 is also given later.

Series B. The heats were made with the Vycor

Table 2. Experimental Results, Series B

Heat No.	Temperature, °C	Gas Ratio	Time, hr	% C Initial	% C Final	Δ% C	log K'†
195	1360	1160	9.0	2.98	2.98	0.00	2.590
194	1360	1165	10.0	3.48	3.48	0.00	2.525
196	1360	1160	10.3	3.88	3.86	−0.02	2.480
201*	1360	1700	30.25	3.48	3.48	0.00	2.690
188	1360	1820	10.5	3.73	3.75	+0.02	2.685
204	1360	1840	31	4.08	4.05	−0.03	2.660
187	1360	2720	10.5	4.17	4.18	+0.01	2.815
203	1260	1150	27.2	4.17	4.15	−0.02	2.450
186	1260	1475	10.5	4.07	4.08	+0.01	2.560
182	1260	1810	10.0	4.07	4.08	+0.01	2.645

* In heat 201, helium was substituted for argon.

$$† K' = \frac{(p_{CO})^2}{p_{CO_2} \cdot \% C}$$

water-cooled tube mostly at 1360° C. The results have been recognized to be affected by a large systematic error, other than carbon deposition, and are discussed later. They are reported, however, in Table 2 since they suggest some interesting comments. For the sake of clarity, only the 1360° C heats are shown in Fig. 4.

The co-ordinates selected to plot the data on Fig. 4 are % C as abscissa and $\log K_1' = \log ((p_{CO})^2 / p_{CO_2} \cdot \% C)$ as ordinate. They are well suited to the case where the standard state for carbon is defined by the condition that its activity should become equal to its weight % at infinite dilution. The plot yields readily:

(a) The logarithm of the equilibrium constant, $\log K_{1(T)}$, by extrapolation to zero % carbon of the isotherm T.

(b) The logarithm of the activity coefficient, $\log f_C$ (where $f_C = a_C / \% C$), at any concentration by reading off the plot $\log f_C = \log K_{1(T)}' - \log K_{1(T)}$.

Similar plots will be presented where the mole fraction N_C is used as a unit of concentration.

The full lines on Fig. 4 have been drawn according to the treatment given below. At 1360° C the broken line is drawn through the experimental points and parallel to the full line.

DISCUSSION OF THE MAIN SOURCES OF ERROR

Carbon Deposition

Carbon deposition is the reaction $2CO \rightarrow CO_2 + C$ (amorphous). Its effect is to lower the carbon potential in the gas and correspondingly in the metal. Carbon deposition could affect the measurements only if it occurred:

(a) During preheating in the inlet tube so as to alter the composition of the fresh gas.

(b) During cooling on the exit path of the gas at such a short distance from the melt as to permit mixing of used and fresh gas (see Fig. 5). Con-

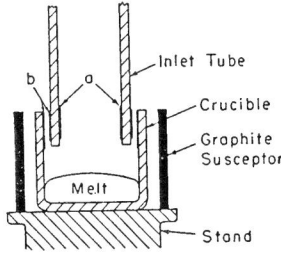

Fig. 5. Most harmful locations for carbon deposition.

densed iron was especially efficient in catalyzing the formation of such a deposit.

Heats retained in series A were free of both types. Type b could be suppressed by heating the crucible externally with the graphite susceptor. Type a ap-

peared only at temperatures above 1600° C or at gas ratios higher than 1150. All heats made under the latter conditions were discarded, except heats 81, 82, and 83, which were run at 1360° with gas ratio much above 1150. In spite of visible carbon deposition, they still showed carburization, and therefore, they set lower limits of the equilibrium concentrations.

In the heats of series B, carbon deposition of type a was suppressed since the gas was kept cold in the inlet tube. Type b deposits tended to be heavier because of increased iron condensation on cold surfaces. Heats were retained at 1360 and 1260° C when carbon deposition was not visible or when it was light, provided its level on the crucible wall did not reach below the tube mouth. More serious errors were to affect series B heats which will be discussed presently.

Thermal Diffusion

Thermal diffusion in the gas phase, if it is appreciable, will result in an excessive carburization of the melt since CO, the lighter gas, tends to diffuse toward the hot surface. Other workers in gas-metal equilibrium studies[12, 13] have resorted to full preheating of the gas to suppress the temperature gradient in the vicinity of the melt. This could not be done here because of carbon deposition. Addition of a heavy inert gas, which was found beneficial by the same authors to preserve the ratio p_i/p_j of gases i and j in a mixture, is slightly detrimental when it comes to preserving the ratio $(p_{CO})^2/p_{CO_2}$. Comparison with the experiments of Dastur and Chipman[12] on thermal diffusion in H_2—H_2O mixtures under very similar conditions bears out the fact that, in series A, the error on $\log K_1'$ ($\Delta \log K'$) is less than 0.03 at 1560° C. In fact, the agreement of the present data with those of Richardson and Dennis obtained at the same temperature in a resistance furnace confirms that no large error was introduced in series A heats by thermal diffusion.

In the heats of series B, there is no other basis for estimating the error than comparison with measurements of thermal diffusion at equilibrium in similar gas mixtures, although such an equilibrium is not likely to be reached in the fast-flowing system under consideration. Gillespie's equation,[14] when tested on the available data,[15] is found to exaggerate the thermal separation of CO and CO_2. If applied to the maximum temperature gradient found here, it gives:

$$-\Delta \log K_1' \leqslant 0.13$$

Although every step of the calculation exaggerates the estimate of the error, an even larger error is found, approaching 0.2. This is evidenced by the discrepancy between the known saturation points (equilibrium of graphite with CO and CO_2) and the extrapolation to saturation of series B data at 1360° C (see Fig. 4).

A few heats in which conditions were identical in series A and B show a displacement of the points of the same order. These facts suggest the existence of another large error affecting the measurements in the same direction as thermal diffusion, which, according to the authors, is lack of thermal equilibrium.

Lack of Thermal Equilibrium

The heat transfer from the hot metal to the cold gas is not instantaneous and, for short retention times, the gas at the interface will contain "cold" molecules (i.e., the average stored energy is less than the average at thermal equilibrium). Fewer molecules will reach the activated state required for them to react, and reaction rates will be slower. Chemical equilibrium, which is a balance between the rates of two opposite reactions, may be displaced if one of them is slowed down more than the other. This may happen in two ways:

(a) The reactants being equally "cold" in both, one reaction requires more activation energy than the other, or

(b) Activation energies being equal, the reactants for one reaction are "colder" than for the other.

Short of any better working hypothesis, the mechanism proposed by Doehlemann[16] for carburization and decarburization of austenite is applied to liquid iron:

$$CO_2 \underset{\text{carb.}}{\overset{\text{decarb.}}{\rightleftarrows}} CO + O \text{ (adsorbed)} \qquad \text{(step I)} \quad \text{slow}$$

$$O \text{ (adsorbed)} + C \underset{\text{carb.}}{\overset{\text{decarb.}}{\rightleftarrows}} CO \qquad \text{(step II)} \quad \text{fast}$$

Step I is rate controlling. The difference between the heats of activation for the forward and the backward reactions is equal to $\Delta H_{(I)}$, the heat of reaction (I), a low estimate of which may be obtained by the standard heat of the reaction:

$$CO_2 \text{ (g)} + Fe \text{ (l)} = CO \text{ (g)} + FeO \text{ (l)}$$

$$\Delta H° = 7500 \text{ cal}$$

One may, therefore, write:

$$\Delta H^*_{\text{I(forward)}} > \Delta H^*_{\text{I(backward)}} + 7500 \text{ cal}$$

If it is assumed that the reactants be equally "cold," decarburization is, therefore, slowed down more than carburization.

Had the activation energies turned out to be equal, the same conclusion could be reached by arguing that the CO_2 molecules (reactants in decarburization) which have more degrees of freedom may be expected to stay "colder" than CO molecules (reactants in carburization). In all cases, therefore, if the mechanism is correct, the total effect is a displacement of equilibrium towards higher carbon content, this is indeed found by experiment.

A quantitative evaluation of the error introduced by lack of thermal equilibrium is not possible. The large systematic error which steps in when going from series A to series B (i.e., when cooling of the gas is substituted for natural preheating) can be interpreted merely as the joint contribution of thermal diffusion and lack of thermal equilibrium, without it being possible to determine how much each contributes. One may only show that both are independent of gas composition so that, for a given temperature, all the equilibrium points are displaced the same distance parallel to the ordinate axis on Fig. 4, the slope of the line being preserved.

Other Sources of Error

When no such large errors as have just been discussed are present, minor errors become of interest to assess the precision of the measurements.

In series A, errors on log K_1' due to carbon analysis, inlet gas ratio (impurities, flow measurement), and temperature simultaneously were such that:

$$\Delta \log K_1' \leqslant 0.033$$

At high-carbon contents, the reaction was so slow that equilibrium, or "pseudo-equilibrium," could not be approached closely. Analytical errors could theoretically result in the wrong interpretation of the sign of the concentration changes when those were smaller than 0.03% carbon. Consistency was, however, obtained when they were trusted as low as 0.005%.

All the impurities in the metal which might affect the activity coefficient of carbon, except aluminum and oxygen, were controlled by selecting pure charge materials and properly purifying the gases. Oxygen was controlled through the equilibrium:

$$C + [O] = CO$$

and its concentration, according to Marshall and Chipman,[7] was always lower than 0.01 weight percentage. Aluminum was controlled through the reaction of the melt with the crucible. Aluminum was analyzed and found to be always less than 0.01% in the range of the data presented here. It increased rapidly with temperature (0.11 at 1760° C in a 1.2% carbon melt), and the evidence that the reaction reached equilibrium was spectacular at 1760, 1660, or even 1560° C at high carbon. Alumina particles on the melt surface could be formed or suppressed at will with temperature fluctuations in a 20° interval. Independently of carbon deposition, the crucible reaction sets a limit to this study at high temperatures.

INTERPRETATION OF THE DATA AND THERMODYNAMIC CALCULATIONS

At 1560° C, three equilibrium points have been established with good accuracy. They are also in good agreement with those of Richardson and Dennis at the same temperature. At 1460° C, a single heat, showing no net reaction under conditions where reaction rates were high, is taken as defining equilibrium within the accuracy of the method. This point fits the temperature dependence of the equilibrium found by Richardson and Dennis at higher temperatures.

At 1360 and 1260° C, the equilibrium lines could not be determined in the present work. Reliable data are limited at the present time to the solubility limit and the equilibrium of graphite with CO and CO_2. Successful experimental work is still needed between 2% carbon and saturation.

In view of the modest contribution of this work, it seems desirable to propose a joint interpretation of all the data available. All of the experimental points of Richardson and Dennis and of the authors have been plotted in Fig. 6. The choice of $(1 -$

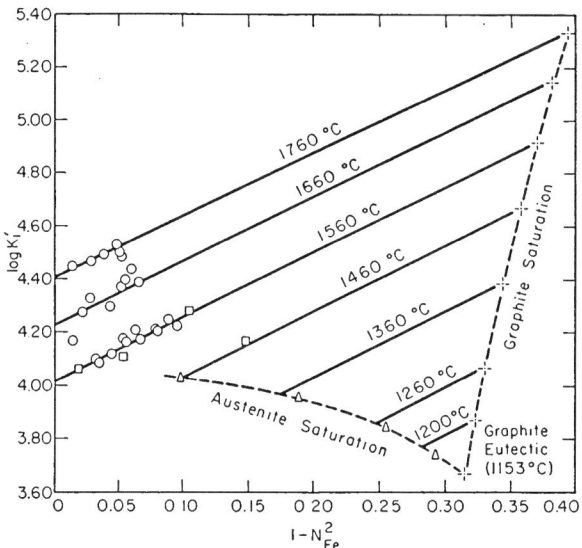

Fig. 6. Equilibrium of carbon with CO-CO₂ mixtures. $K_1' = p^2_{CO}/(p_{CO_2} \cdot N_C)$.

$N_{Fe}^2)$ as abscissa permits a linear extrapolation of the 1560° C data to a point determined by the known carbon content and gas ratio of the graphite-saturated melt. The relative position of the lines for other temperatures will be discussed presently.

The isotherms have equations of the type:

$$\log K_1' = \log K_1 + \log \gamma_C$$

$$K_1' = \frac{(p_{CO})^2}{p_{CO_2} \cdot N_C}$$

The 1560° C isotherm is the best defined experimentally and may be represented by the equation:

$$\log K_1' = 4.02 + 2.43(1 - N_{Fe}^2)$$

where 4.02 is the value of $\log K_1$ determined by the intercept and the last term represents $\log \gamma_C$ at 1560° C over the entire range of liquid compositions.

To proceed further, two assumptions are made:

(a) The intercepts of the isotherms (i.e., values of $\log K_1$) are a linear function of $1/T$ which, in view of the relationship:

$$\frac{d \log K_1}{d(1/T)} = -\frac{\Delta H_1^\circ}{2.3R}$$

is equivalent to assuming that the standard heat of reaction 1 is independent of temperature.

(b) The slopes of the isotherms are proportional to $1/T$, following the treatment of the iron-carbon system by Darken and Gurry.[17] These authors assume the relationship:

$$\log \gamma_C = -\frac{A}{T}(1 - N_{Fe}^2)$$

where A is a constant. Hence:

$$\frac{\partial \log K_1'}{\partial(1 - N_{Fe}^2)} = \frac{\partial \log \gamma_C}{\partial(1 - N_{Fe}^2)} = -\frac{A}{T}$$

All slopes may, therefore, be calculated from the 1560° C isotherm, the value of A being $A = -4450$. A tentative general expression for the activity coefficient is therefore

$$\log \gamma_C = \frac{4450}{T}(1 - N_{Fe}^2)$$

The data at 1760° C are used along with the previous equation for $\log \gamma_C$ to establish the temperature dependence of K_1 with the following result:

$$\log K_1 = \frac{-7280}{T} + 7.98$$

These equations reproduce the data of Richardson and Dennis and of the authors at carbon concentrations below 2%. The expression for the activity coefficient, however, is not valid at high carbon concentrations at temperatures other than 1560° C, and slight modification is required to conform with what is known about high-carbon solutions. The equilibrium constant of the producer-gas reaction is known from thermodynamic data[18] and may be represented by the equation:

$$CO_2 \text{ (g)} + C \text{ (graphite)} = 2CO \text{ (g)};$$

$$\log K = \frac{-8460}{T} + 8.85$$

The solubility of graphite according to Chipman and co-workers[19] is:

$$\% \, C = 1.34 + 2.54 \times 10^{-3} \, t \, (°C)$$

From these equations, values of $\log K_1'$ were calculated and are shown along the line of saturation in Fig. 6. The lines are fitted to those points by a correction whereby A is made a function of temperature. The expression of $\log K_1$ remaining unchanged, the best fit is obtained when $\log \gamma_C$ is written:

$$\log \gamma_C = \frac{4350}{T}[1 + 4 \times 10^{-4}(T - 1770)](1 - N_{Fe}^2)$$

The lines of Fig. 6 are drawn to conform to this equation, and values of $\log \gamma_C$ are shown in Fig. 7.

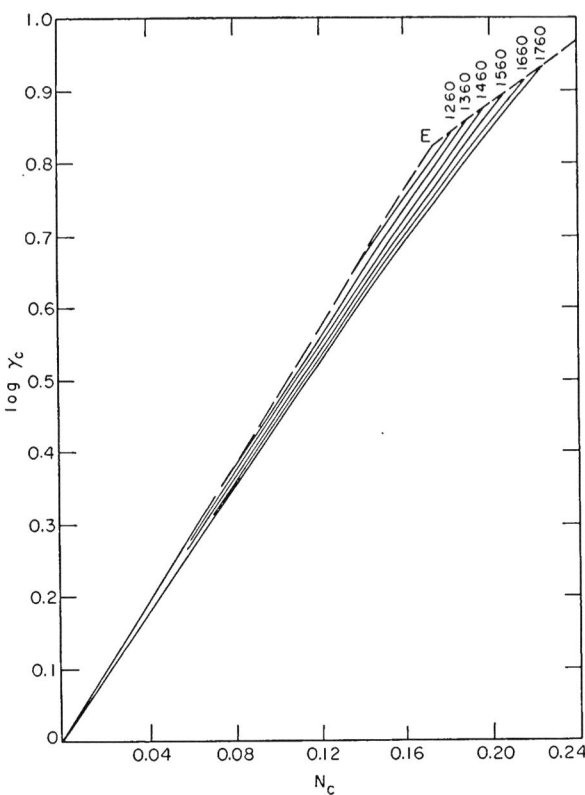

Fig. 7. Activity coefficient of carbon in liquid iron (mole fraction basis).

Figure 8 is a translation of Fig. 7 on the weight-percentage basis, and the line earlier proposed by Chipman[20] is shown on the same graph.

Comparison with Data on Austenite

The data of Smith[21] on the equilibrium of carbon in austenite with $CO-CO_2$ mixtures may be extrapolated across the two-phase field where austenite is in equilibrium with liquid alloys. The points placed on the liquidus line on Fig. 6 have been calculated in a manner to be described here. First the liquidus and solidus lines of the iron-carbon diagram were redrawn on the following basis: the eutectic was taken at 1153° C[17] and 4.27% carbon[19] and the peritectic at 1499° C and 0.53% carbon.[17] The experimental points of several investigators,[22, 23, 24, 25] when corrected to fit the above end points, define the liquidus used here. The agreement with the line proposed by Darken and Gurry[17] is very close. The end points of the solidus are taken at 1499° C, 0.16% carbon, and at 1153° C, 2.01%. Short of any justified choice among the widely scattered experimental determinations of the solidus, a straight line was drawn between the two end points.

Second, the data of Smith were extrapolated to the solidus concentrations at 1200° C (the highest experimental temperature), 1260, 1360, and 1460° C to yield the corresponding gas ratios. In doing so, the equilibrium isotherms were drawn as a set of parallel straight lines on a plot of

$$\log\left(\frac{p_{CO}^2}{p_{CO_2}} \cdot \frac{N_{Fe}}{N_C}\right) \text{ versus } \frac{N_C}{N_{Fe}}$$

and spaced on the assumption that the heat of transfer of carbon from gas to metal is independent of temperature.

Third, the gas ratios obtained were applied to the liquid alloys of the liquidus line at the same temperatures, and $\log K_1'$ was calculated.

The agreement with the equilibrium lines of Fig. 6 is fair. It could be improved by selecting a solidus line slightly concave downward, since the location of the final points is rather sensitive to the choice of the solidus. In fact, the uncertainty regarding both solidus and liquidus is such that Darken and Gurry[17] preferred to calculate those lines from activity data.

Thermodynamic Summary

The experimental data and the thermodynamic implications of the above treatment regarding the reactions of $CO-CO_2$ mixtures with carbon in solution or as graphite and the various solution and

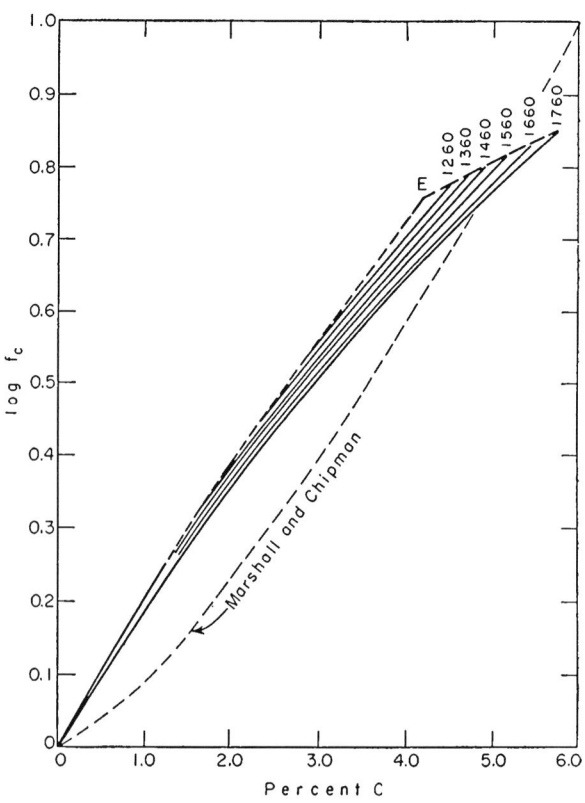

Fig. 8. Activity coefficient of carbon in liquid iron (weight-percentage basis). Broken line based on data of Marshall and Chipman.[20]

dilution processes for liquid iron-carbon alloys are summarized* in the following statements and equations. In particular, expressions are given for the activity of carbon with respect to both graphite and the infinitely dilute solution as standard states. The free-energy equations are well established because they follow directly from the experimental data. The heat terms in equations 4 to 7 that follow should not be considered as accurate since they are very sensitive to small errors in the temperature coefficients of free-energy terms.

$$CO_2 \text{ (g)} + \text{[C] (inf. dil.)} = 2CO \text{ (g)} \qquad (1)$$

$$\Delta F_1^\circ = 33{,}300 - 36.5T$$

$$\log K_1 = \frac{-7280}{T} + 7.98$$

The standard state is defined by $a_C/N_C = 1$ when $N_C = 0$. The enthalpy term, $\Delta H_1^\circ = +33{,}300$ cal is an average for the experimental range and is assumed to be independent of T.

When it is desired to express carbon concentration in weight per cent, making $a_C/[\% \ C] = 1$ when $[\% \ C] = 0$, the equation becomes

$$\Delta F_{1a}^\circ = 33{,}300 - 30.40T$$

$$\log K_{1a} = \frac{-7280}{T} + 6.65$$

$$CO_2 \text{ (g)} + C \text{ (graphite)} = 2CO \text{ (g)} \qquad (2)$$

$$\Delta F_2^\circ = 38{,}700 - 40.5T$$

$$\log K_2 = \frac{-8460}{T} + 8.85$$

The enthalpy term is an average between 39,700 cal at 1150° C and 37,900 cal at 2030° C, based on National Bureau of Standards' data.

$$C \text{ (graphite)} = \text{[C] (inf. dil.)} \qquad (3)$$

$$\Delta F_3^\circ = 5400 - 4.00T$$

$$\log K_3 = \frac{-1180}{T} + 0.87$$

The enthalpy term $\Delta H_3^\circ = 5400$ cal is the heat of solution of graphite in the infinitely dilute solution.

$$\text{[C] (inf. dil.)} = \text{[C]} (N_C) \qquad (4)$$

$$\Delta F_4 = RT \ln N_C + \Delta F_4^x$$

$$\Delta F_4^x = +19{,}900[1 + 4 \times 10^{-4}(T - 1770)]$$
$$(1 - N_{Fe}^2)$$

$$\Delta H_4 = \overline{H}_C - \overline{H}_C^\circ \text{ (inf. dil.)} = 5810(1 - N_{Fe}^2)$$

The excess partial molar free energy of carbon ΔF_4^x and its relative partial molar enthalpy ΔH_4 are obtained directly from the equation for activity coefficient:

* See Ref. 26.

$$\log \gamma_C = \frac{4350}{T}[1 + 4 \times 10^{-4}(T - 1770)](1 - N_{Fe}^2)$$

$$C \text{ (graphite)} = \text{[C]} (N_C) \qquad (5)$$

$$\Delta F_5 = \overline{F}_C - F_C^\circ \text{ (graphite)} = \Delta F_4 + \Delta F_3^\circ$$

$$\Delta H_5 = \overline{H}_C - H_C^\circ \text{ (graphite)}$$

$$= 5400 + 5810(1 - N_{Fe}^2)$$

These equations give the free-energy and enthalpy change for dissolving graphite in a solution of mole fractions N_C and N_{Fe}. The activity of carbon in the graphite-saturated solution follows from 3:

$$\log a_C \text{ (sat.)} = \frac{-1180}{T} + 0.87$$

If a_C' and γ_C' are the activity and activity coefficient referred to graphite as the standard state, then a_C' (sat.) $= 1$ and

$$\log a_C' = \log N_C + \log \gamma_C'$$

$$\log \gamma_C' = \log \gamma_C + \frac{1180}{T} - 0.87$$

$$C \text{ (graphite)} = \text{[C] (sat.)} \qquad (6)$$

$$\Delta F_6 = 0$$

$$\Delta H_6 = \overline{H}_C \text{ (sat.)} - H^\circ \text{ (graphite)}$$

$$= 5400 + 5810[1 - N_{Fe}^2 \text{ (sat.)}]$$

For a mean temperature of 1500° corresponding to N_C (sat.) $= 0.2$, the heat of solution of graphite is 7500 cal.

$$C(N_{C_1}) = C(N_{C_2}) \qquad (7)$$

$$\Delta F_7 = RT \ln \frac{a_2}{a_1}$$

$$\Delta H_7 = 5810(N_{Fe_1}^2 - N_{Fe_2}^2)$$

The last is a general expression for the heat of dilution.

References

1. J. Chipman, *Trans. Am. Soc. Metals*, **22**, 385 (1934).
2. N. J. Petch, *J. Iron Steel Inst.*, **145**, 111 (1942).
3. H. C. Vacher and E. H. Hamilton, *Trans. A.I.M.E.*, **95**, 124 (1931).
4. H. C. Vacher, *J. Research Natl. Bur. Standards*, **11**, 541 (1933).
5. S. Matoba, *Honda Anniv. Vol.*, 548 (1936).
6. G. Phragmén and B. Kalling, *Jernkontorets Ann.*, **123**, 199 (1939).
7. S. Marshall and J. Chipman, *Trans. Am. Soc. Metals*, **30**, 695 (1942).
8. F. D. Richardson and W. E. Dennis, *Trans. Faraday Soc.*, **49**, 171 (1953).
9. O. A. Esin and L. K. Gavrilov, *Izves. Akad. Nauk SSSR, Otdel. Khim. Nauk*, No. 7, 1040 (1950).
10. K. Sanbongi and M. Ohtani, *Sci. Rep., Res. Inst. Tohoku Univ., A*, **5**, 263 (1953).

11. M. N. Dastur and N. A. Gokcen, *Trans. A.I.M.E.*, **185**, 665 (1949).
12. M. N. Dastur and J. Chipman, *Discussions Faraday Soc.*, No. 4, 100 (1948).
13. C. W. Sherman, H. I. Elvander, and J. Chipman, *Trans. A.I.M.E.*, **188**, 334 (1950).
14. L. J. Gillespie, *J. Chem. Phys.*, **7**, 530 (1939).
15. H. Kitagawa and M. Wakoo, *J. Chem. Soc. Japan*, **62**, 100 (1941).
16. E. Doehlemann, *Z. Elektrochem.*, **42**, 561 (1936).
17. L. S. Darken and R. W. Gurry, *Physical Chemistry of Metals*, McGraw-Hill Book Co., New York, 1953.
18. "Selected Values of Chemical Thermodynamic Properties," U. S. Natl. Bur. Standards, Series III, 1948-1949.
19. J. Chipman, R. M. Alfred, L. W. Gott, R. B. Small, D. M. Wilson, C. N. Thomson, D. L. Guernsey, and J. C. Fulton, *Trans. Am. Soc. Metals*, **44**, 1215 (1952).
20. *Basic Open Hearth Steelmaking*, Second Edition, Physical Chemistry of Steelmaking Committee, A.I.M.E., New York, 1951.
21. R. P. Smith, *J. Am. Chem. Soc.*, **68**, 1163 (1946).
22. H. C. H. Carpenter and B. F. E. Keeling, *J. Iron Steel Inst.*, **65**, 224 (1904).
23. R. Ruer and R. Klesper, *Ferrum*, **11**, 258 (1913-14).
24. R. Ruer and F. Goerens, *Ferrum*, **14**, 161 (1916-17).
25. J. H. Andrew and D. Binnie, *J. Iron Steel Inst.*, **119**, 309 (1929).
26. A. Rist and J. Chipman, *Rev. mét.*, **53**, 796 (1956).

Discussion

DARKEN pointed out that in his laboratory they had made similar calculations several years ago. They concluded that the eutectic point provided the most reliable source of information. The activities at point E of Fig. 7 could be calculated from the thermodynamic properties of iron, the temperature and composition of the eutectic, and the assumption that the activity coefficients approach unity at high temperatures.

CHIPMAN said that their calculations did not include the thermodynamic functions of iron, but that $\log K_1'$ at the eutectic was calculated from the graphite-gas equilibrium and the eutectic composition. The result was reasonably close to Darken's previous calculation. The discrepancy may be due in part to the values taken for the eutectic composition. Darken used 4.24%, while Chipman and Rist used 4.27%.[19]

RICHARDSON discussed briefly some of the experimental difficulties inherent in a system where heating is by induction and a cold gas impinges on the surface of the melt. These conditions could be expected to have an influence on the constant K_1' and on the temperature measurement. CHIPMAN and RICHARDSON agreed that the general effect would tend to lead to too-low results, as shown in the experimental points on the right side of Fig. 4.

WAGNER brought out the fact that in a gas like CO_2 there are several degrees of freedom. Equilibrium involving the vibrational energy is probably not readily attained. Therefore, it seems likely that CO_2 is not in equilibrium with the melt, and on Richardson's question, he surmised that this effect would tend to shift K_1' as shown in Fig. 4.

PEARSON described a program of research being conducted in the Chemical Laboratories of B.I.S.R.A. in London. Initially the interest was in measuring the activity of carbon in iron-carbon alloys, but in reviewing the picture, it was decided that although the CO/CO_2 ratio in equilibrium with liquid iron did not give trouble at low carbon levels, it would in the high carbon ranges. As a result, the H_2/CH_4 reaction was tried. A mixture of hydrogen and methane was circulated over a liquid iron-carbon alloy contained in a lime crucible. The gas was recirculated continuously and was analyzed in an infrared gas analyzer. Equilibrium was established in approximately two hours. It was run for another two hours to be sure. Subsequently, the methane-hydrogen gas was replaced with argon, the sample was quenched, and analyzed for carbon. The general accuracy of the method was checked by using iron contained in graphite crucibles. Their results checked with the existing data on methane. A quick calculation from data sent recently from England showed that the results agreed reasonably well with Fig. 7. Up to 0.16 atom fraction of carbon, the agreement is good. At $N_C = 0.18$, they begin to drop a little bit below the line given.

MORRIS inquired whether the presence of the water-cooled Vycor tube influenced the temperature measurement by changing the emissivity of the iron surface. CHIPMAN said that it did not, as the emissivity values are for cold surroundings.

by J. A. Cordier
and P. P. Kozakevitch

Activity of Carbon in Molten Fe—C Alloys at High Carbon Content—A Progress Report

Because of the recent interest in the activity of carbon in molten high-carbon Fe—C alloys, it was considered to be of interest to this Conference to review the current research on this subject being conducted at IRSID. The work is aimed at measuring the activity of carbon at concentrations where difficulties are encountered with carbon deposition when using gas mixtures containing CO and CO_2 alone.

Equilibrium is established between the metal and a gas mixture of hydrogen, methane, and carbon monoxide. The metal samples consist of drops which are cooled under purified argon and then analyzed in total for carbon. The gas mixture is prepared by passing wet hydrogen through a furnace containing graphite powder and iron as a catalyst. Its CO content is controlled by the humidity level in the hydrogen, and the CH_4 content by the temperature of the graphite. The composition is calculated as follows:

(a) $H_2O + C \rightarrow CO + H_2$;

$$p_{CO} = p_{H_2O} \quad \text{(in the incoming gas)} \quad (1)$$

$$C_{(graphite)} + 2H_2 = CH_4;$$

$$K_2 = \frac{p_{CH_4}}{p_{H_2}^2} \quad \text{(at temperature } T_1) \quad (2)$$

and K_2 is calculated from extrapolated thermo-

dynamic data of good accuracy. This gas is stable upon cooling and does not react with alumina since the P_{CO} is high enough (10^{-3} atm) to prevent reaction 3 from taking place.

(b) $3CH_4 + 2Al_2O_3 \rightarrow Al_4O_4C + 6H_2 + 2CO \quad (3)$

The CO gas does not disturb reaction 4 at temperature T_2.

$$[C]_{(dissolved)} + 2H_2 = CH_4 \quad (4)$$

$$K_4 = \frac{p_{CH_4}}{p_{H_2}^2} \frac{1}{a_C}; \qquad a_C = \frac{K_1}{K_2}$$

if graphite is chosen as the standard state.

As $T_1 - T_2$ is small, the ΔH (and ΔS) for reactions 4 and 2 are almost equal. So that, with a good approximation:

$$\log a_C = \log K_2 - \log K_4 = \frac{\Delta H}{R} \frac{T_1 - T_2}{T_1 \cdot T_2}$$

PRELIMINARY RESULTS

(a) Saturation runs

Some preliminary experiments have been made which show that the hydrogen reaches equilibrium with the graphite to form CH_4, that the rate of transfer of carbon between the melt and the gas is fast enough (experiments of 10 to 20 hours are sufficient), and that pure alumina is a suitable refractory for this study.

J. A. Cordier is a Member of the Staff and P. P. Kozakevitch is Head, Physical Chemistry Department, Institut de Recherches de la Sidérurgie, Saint-Germain-en-Laye, France.

Table 1. Carbon Content of Carbon-Saturated Iron

T_2,°C	T_1,°C	$T_1 - T_2$	% C at Equilibrium	Initial % C		
				First Drop	Second Drop	Third Drop
1302	1167	−135	4.54	4.6	1.6	0
1260	1250	−10	4.52	5.6	2.8	1.6
1562	1552	−10	4.71	5.6	2.8	1.6

(b) Activity measurements

To permit comparison of data obtained at different temperatures, it is interesting to compare the results by means of the co-ordinates ϵ and Γ : $\epsilon =$ % C_e/% C_s, i.e., the ratio between carbon concentrations at equilibrium and saturation. $\Gamma = a_C/\epsilon$, i.e., the ratio beteen carbon activity and ϵ.

Table 2. Activities of Carbon

$N°$	T_2,°C	T_1,°C	$(T_1 - T_2)$,° C	a_C	$%C_s$	ϵ	Γ	$-\log \Gamma$
34	1185	1250	+65	0 725	<4.2	0.96	0.755	0.122
36	1190	1300	+110	0.59	3.9	0.89	0.66	0.18
38	1270	1372	+102	0.64	3.90	0.855	0.75	0.125

SUMMARY

For these runs the gas mixture was prepared by bubbling hydrogen through liquid iron saturated with graphite. The hydrogen was dry. The temperatures were not controlled closely enough. Equilibrium was not reached from both sides in runs 34 and 36. Carbon analysis (by volumetric method) was not accurate enough. Therefore, these results are approximate and cannot be relied upon, but they do indicate that the method can be fruitful.

It has been assumed that of the hydrocarbons only CH_4 is present in the gas in important quantities (with CO and H_2). There are no thermodynamic data available on other hydrocarbons than C_2H_2. According to these data, C_2H_2 is present in very small amounts in the gas (10^{-7} atm). Some runs have been made to study the kinetics of decarburization of Fe—C alloys by pure hydrogen. The results are in good agreement with the calculations based on the fact that the only decarburizing reaction taking place is

$$2H_2 + [C] = CH_4 \qquad (5)$$

This tends to prove that the assumption stated above is justified.

The final apparatus where all the necessary conditions will be met is being built.

The same method is used for studying the influence of alloying elements (which do not form volatile hydrides) on the activity of carbon.

Discussion

RICHARDSON asked Pearson to say a little more about his technique for the H_2–CH_4 equilibration, particularly with regard to the possible reactions with refractories.

PEARSON said that circulating dry hydrogen through the system gave a reaction between the methane, the hydrogen, and the alumina boat on which the lime crucible rested. Water vapor was produced by this reaction. If the water-vapor content was increased too much, they found the water oxidized methane. However, a trap cooled with solid CO_2, which was inserted in the circulating system, maintained the water-vapor content at the point where it prevented reaction with the alumina but also was not high enough to oxidize the methane.

by G. Derge

The Distribution of Alloying Elements between Molten Iron and Silver

INTRODUCTION

Much has been learned about liquid metal behavior by the study of metal-gas reactions, since activities of components in the metal can be related to the relatively well understood and easily controlled gas equilibria. The method is limited by the fact that many elements of interest, particularly the metals, do not have compounds of suitable volatility or stability at the required temperatures. Metal-slag reactions have provided another vast body of information. However, the interpretation of these data is often limited by lack of suitable knowledge of slag constitution.

Another approach is to study the distribution of alloying elements between two immiscible metals. For iron this often has the advantage of providing a bridge to data obtained in other systems which are easier to handle experimentally because of lower temperature or less sensitivity to oxidation. A good example is Fe—Ag, which has the added feature of negligible solubility of carbon in silver. This allows observations on the interactions of alloying elements with carbon in iron.

EXPERIMENTAL

The general features of the furnace assembly and experimental techniques were conventional. The working crucible was MgO, in an induction-heated graphite holder. The entire system was contained in a quartz tube with head assembly providing for vacuum or neutral gas, optical temperature measurement calibrated by thermocouple, and an externally operated sampling tube. It was found that the iron layer samples were subject to segregation from top to bottom during freezing, and a great deal of attention was devoted to establish correct procedures for each alloy studied. These details, though important, are too numerous to be included in this digest.

Selected alloy distributions will be shown to illustrate the types of behavior found. All examples are for the liquid iron-liquid silver system.

A. Distribution of Copper

Copper was selected as an element which might be expected to show no interaction with carbon in iron. The distribution ratio is represented as:

$$K_{Cu} = \frac{N_{Cu}}{N_{Cu}'} = \frac{\text{mole fraction of Cu in Fe}}{\text{mole fraction of Cu in Ag}}$$

It is found that the ratio varies with the carbon content when Fe, C, and Cu are considered as independent species in the iron layer, as shown in Table 1.

Table 1. The Distribution of Copper

% C in Fe	K_{Cu}*	$K_{Cu}^{Fe_3C}$
0	0.234	0.234
1.72	0.177	0.228
3.57	0.128	0.230

* K_{Cu} is an average of three or more separate runs.

No temperature dependence was found in the range 1550° C to 1650° C.

Based on the Doctoral Thesis of Yuan Hsi Chou, Department of Metallurgical Engineering, Carnegie Institute of Technology, February, 1947. G. Derge is Professor of Metallurgy, Carnegie Institute of Technology, Pittsburgh, Pennsylvania.

If one makes the simplest possible assumption that the carbon is present as undissociated Fe_3C, the corresponding distribution ratio $K_{Cu}^{Fe_3C}$ is found to be independent of carbon content, as seen in the third column of Table 1. The same result is obtained by using the more complicated expressions for carbon suggested by either Chipman[1] or Darken.[2]

If the Ag—Cu system is assumed to obey the regular solution relation of Hildebrand:[3]

$$RT \ln \gamma_{Cu \text{ in } Ag} = \lambda N_{Ag}^2$$

where γ = the activity coefficient
N = mole fraction
λ = a constant independent of temperature and composition

it is possible to calculate the activity of Cu in Ag from the phase diagram. It is found that $\gamma_{Cu \text{ in } Ag} = 2.14$ at 1600° C (referred to liquid Cu as the standard state). This permits the calculation of the activity of Cu in iron from the distribution data

$$a_{Cu \text{ in } Fe} = \gamma_{Cu \text{ in } Ag} N_{Cu}' = \frac{2.14}{0.234} N_{Cu} = 9.12 N_{Cu}$$

If carbon can be considered as Fe_3C and the concentration of copper is low, the total moles in 100 grams of melt is

$$N = 1.79 - 0.184 \, (\% \, C)$$

and

$$a_{Cu \text{ in } Fe} = 9.12 \, (\% \, Cu)/63.57[1.79 - 0.184 \, (\% \, C)]$$
$$= \frac{\% \, Cu}{12.5 - 1.29 \, (\% \, C)}$$

B. Distribution of Mn

Manganese was selected as an alloying element which might be expected to show strong interaction with carbon in iron. Again, no temperature coefficient was observed for the distribution ratio K_{Mn}, but a marked dependence on carbon in iron is shown by Table 2.

Table 2.   The Distribution of Manganese

% C in Fe	K_{Mn}	$K_{Mn}^{Fe_3C}$	$K_{Mn}^{(Fe_{3-n}, Mn_n)C}$
0	0.47	0.47	—
1.18	0.52	0.62	13.4
2.62	0.54	0.82	10.1
4.19	0.83	1.72	13.8

If, as in the example of copper alloys, all of the carbon is considered present as undissociated Fe_3C, the ratio is even more dependent on carbon, as shown by $K_{Mn}^{Fe_3C}$ in Table 2. The most consistent values for the distribution data were found by assuming the carbon present as $(Fe_{3-n}, Mn_n)C$ with the best average value of $n = 1.25$ for the composi-

tion range 0.4 to 2% Mn and 0.9 to 4.2% C, as shown by $K_{Mn}^{(Fe_{3-n}, Mn_n)C}$ of Table 2. The artificiality of such treatments is apparent. Another approach is to consider the Fe—Mn system as ideal and regard the variation of K_{Mn} in Table 2 as a measure of the influence of carbon on the activity of manganese in molten iron. The experimental results from distribution data may then be compared with Darken's[4] use of the relations for regular ternary solutions. This is done by Table 3, in which values of log a_{Mn}/N_{Mn} are shown as a function of carbon for both examples. The agreement is excellent.

Table 3.   The Influence of Carbon on Manganese Activity

% C_{Fe}	Log $\dfrac{a_{Mn}}{N_{Mn}}$	
	Distribution Data	Darken[4]
0.89	−0.032	−0.031
1.53	−0.051	−0.061
3.33	−0.151	−0.147
4.19	−0.251	−0.193

C. Distribution of Sulfur

A good example of a distribution ratio with a measurable temperature coefficient is that of sulfur. This example will be limited to the carbon-free system. Average values of the distribution ratio are shown in Table 4.

Table 4.   The Distribution of Sulfur

Temperature, °C	$K_S' = \dfrac{\% \, S \text{ in Fe}}{\% \, S \text{ in Ag}}$	K_S
1540	15.5	8.25
1600	11.1	5.92
1660	7.9	4.18

If the sulfur is considered to be present in the iron as FeS and in the silver as Ag_2S, two homogeneous reactions can be postulated:

$$FeS \xrightleftharpoons{\text{in Fe}} Fe + S \qquad (1)$$

$$Ag_2S \xrightleftharpoons{\text{in Ag}} 2Ag + S \qquad (2)$$

If the distribution data are represented on the corresponding molar basis, and simplifications are made because the sulfur in the silver layer is small, one can evaluate the equilibrium constant, K_S, of the heterogeneous reaction:

$$Ag_2S + Fe \rightleftharpoons FeS + 2Ag \qquad (3)$$

as shown in Table 4.

The corresponding plot of log K_S versus $1/T$ is linear and represented by the equation:

$$\log K_S = \frac{8670}{T} - 3.86$$

and therefore, for reaction 3,

$$\Delta F^\circ = -39{,}660 + 17.66T$$

The influence of manganese on the sulfur distribution was also observed and is illustrated by the variation of K_S' at 1600° C, as shown in Table 5.

Table 5. The Combined Distributions of Manganese and Sulfur at 1600° C

% Mn$_{Fe}$	% Mn$_{Ag}$	$K_S' = \dfrac{\% \text{ S in Fe}}{\% \text{ S in Ag}}$
0.0	0.0	11.1
0.21	0.23	9.1
0.46	0.52	8.2
0.69	0.77	7.3
1.08	1.17	6.1
1.63	1.74	4.7

SUMMARY

The distributions of Cu, Mn, and S between the nearly immiscible liquids Fe and Ag illustrate an interesting variety of problems on liquid-metal interactions which can be examined by the experiments described. This pair is particularly useful for studying the influence of carbon in iron. Interpretation of the data can be elaborated to any desired degree and will become increasingly useful as our knowledge of any particular alloy system increases.

References

1. J. Chipman, *Trans. Am. Soc. Metals*, 30, 817 (1942).
2. L. S. Darken, *Trans. A.I.M.E.*, 140, 204 (1940).
3. J. H. Hildebrand and R. L. Scott, *Solubility of Non-electrolytes*, Third Edition, Reinhold Publishing Corp., New York, 1950.
4. L. S. Darken, Private communication.

Discussion

RICHARDSON inquired of Derge whether the authors had coupled their work with that of Morris and established the behavior of copper in the copper-silver system. In the ensuing discussion DERGE said that he had not worked it out in detail, but rough calculations indicated that the liquid copper-silver system was close to ideality in its behavior. RICHARDSON noted that manganese appears to be quite ideal in both iron and silver, in accordance with the authors' last figure.

CHIPMAN remarked that a study had been made at M.I.T. on the distribution of copper between iron and silver by Mr. Peter Koros,* who was present at the meeting. KOROS reported that his work indicated that $\log \gamma_{Cu}$ in the infinitely dilute solution of copper and iron was 8.0, which was in very good agreement with that found by Morris and Zellers. To answer Richardson's question, KOROS pointed out that his work indicated that the activity coefficient of infinitely dilute solutions of

* Graduate Student, Department of Metallurgy, M.I.T.

copper in silver was 1.9 and on the other side it was 2.1. The calculations were based on Hanson's phase diagram, with the assumption that the activity of the solvent in the solid-solution phase was equal to its mole fraction.

DERGE noted that the distribution runs were made in graphite crucibles, which meant that the metals were saturated with carbon. To compensate for this effect, they calculated that carbon was present as iron carbide in the iron alloy, in accordance with equations proposed by both Chipman and Darken. The merits of this type of assumption were discussed, and CHIPMAN felt that there was little value in such assumptions in the light of present-day understanding.

CHIPMAN noted that he was glad to see this work of Chou's brought out for publication. In the past, it has provided considerable stimulation to students both at Carnegie Institute of Technology and at the Massachusetts Institute of Technology in this field. He noted that there have been several places where the unpublished work of Chou has been given as a reference. It has stimulated a number of people to do considerable work in the area and thereby has rendered excellent service.

An extended discussion of difficulties of this experimental technique ensued. GOKCEN stated that a quenching technique would be superior to that of sampling the liquid melt for certain metals. He also noted that the iron-bismuth system could be studied at 1600° C, but the iron-antimony system could not be because of the vapor pressure of antimony. He proposed a design for sampling the liquid melts in which a side duct brought in at an angle communicated only with the more dense metal which lay underneath. The less dense metal would float up to the top and, as a consequence, could be sampled directly. He obtained a very contorted surface between the two liquid phases of iron and bismuth if the bismuth and iron were melted together. The iron should be added after the bismuth has been brought up close to the temperature at which the experiment was to be run. DERGE pointed out that quenching the whole sample gave them a distribution that was independent of the equilibration temperature of the melt. Their assumption was that they were actually obtaining a distribution which pertained at the melting point of iron. Chou found that they could not get a good separation between the two liquid-metal layers when graphite crucibles were used, and as a result a magnesia liner was inserted to avoid this. KOZAKEVITCH said that they had encountered a similar difficulty in equilibrating silver and iron in graphite crucibles. DERGE stated that Chou did not encounter this problem when using refractory inserts. In his work with the copper distribution between iron and silver, KOROS observed no emulsification when the copper level was below about 3 per cent in the silver. KOROS also found that the best technique for sampling the lower layer was to use a Vycor tube with the tip almost closed in order that entrainment of the upper layer be eliminated.

KRIVSKY told of a technique that had been developed in his laboratory for sampling. The lower alloy layer was sampled by means of a tube which was brought out to a fine tip and closed. This tip was then submerged and broken on the bottom of the crucible, thus insuring that the lower liquid layer be sampled. The top layer was sampled by means of a tube which was attached to a vacuum line and a manometer. The displacement of the liquid level in the manometer indicated when the tip of the tube touched the top layer of the metal.

by M. Olette

An Adiabatic Dropping Calorimeter for Enthalpy Measurements at High Temperatures. The Heat Content of Silicon from 1200 to 1550°C

INTRODUCTION

The study of the thermodynamic properties of the elements and of inorganic compounds has been the object of numerous experimental investigations especially during the last fifty years. In the United States, in particular, important work has been carried out in the laboratory of Dr. K. K. Kelley to whom we also owe the publication of wide critical selections[1] of the best experimental data. Such compilations have become the major tool of all thermodynamic calculations concerning metallurgical processes. Careful examination, however, shows how incomplete our knowledge of thermodynamic constants still is, notably concerning enthalpies and specific heats of substances entering iron, steel, blast-furnace, and steelmaking slags in the range of metallurgical temperatures.

Rules of additivity may give an idea[2] of the specific heat of technical substances when the individual specific heats of the constituents are known. However, their limited applicability, to-

gether with the scarcity of measurements on such substances, points to the necessity of further work in this field, as was recently stressed by Pattison.[3]

Furthermore, the knowledge of the heats of fusion of the elements and their compounds is essential to the undertaking of any thermodynamic calculation concerning a metallurgical process in which one of the products or reactants is in the liquid state. Here again, it is important to fill many gaps.

It is with the purpose of contributing to these various fields of knowledge that a calorimeter has been built which we believe presents some new features.

Critical Survey of the Various Experimental Methods

The methods for experimentally determining the enthalpy or the specific heat of a substance belong to two separate groups according to whether the measurement is aimed to yield directly the one or the other property.[4]

Direct Measurement of the True Specific Heat. When the temperature of the sample is varied slightly around a given value, the determination of the heat involved gives the specific heat of the substance at that temperature. The enthalpy variation between two temperatures may then be calculated by integrating the specific heat/temperature

Dr. Olette is a Member of the Staff, Physical Chemistry Department, Institut de Recherches de la Sidérurgie, Saint-Germain-en-Laye, France.

The author is indebted to Mrs. M. F. Ancey-Moret and Mr. A. Ferrier for participation in the experimental work, to Mr. C. Tuppin for advice on electronic problems, and to Dr. G. Pomey for X-ray analysis. Continued encouragement and interest from Dr. P. P. Kozakevitch and valuable discussions with Dr. F. D. Richardson are gratefully acknowledged.

function between the two given limits. This method, first applied to metals as early as 1910–1912 by Corbino[5] and von Pirani[6] and to silica in 1918 by Perrier and Roux,[7] was later improved by numerous authors. Recently, Pallister[8] in the study of steels and Winckler[9] in the study of refractory materials have built calorimeters which seem to satisfy jointly a number of experimental requirements.

Methods involving the direct measurement of the true specific heat have in principle the great advantage of following closely the definition of specific heat. They also make it possible to measure with good accuracy small heat effects such as heats of transformation between allotropic varieties of a substance. However, they do not seem to be capable of high accuracy: Winckler's calorimeter, for example, gives data with errors of 3 to 5%. The error, which is due primarily to the difficulty of estimating the heat lost by radiation, increases with temperature.

Direct Measurement of the Heat Content. For the measurement of the heat content above room temperature, Regnault's "method of mixtures" or "dropping method" has proved to be the most simple and the most reliable. A high degree of precision has been imparted to it, mainly after the work of White.[10] In principle, it consists in dropping a specimen of accurately known weight and temperature into a previously calibrated calorimeter and in measuring the latter's rise in temperature. During cooling of the sample to the vicinity of room temperature, an amount of heat, equal to the enthalpy variation between final and initial temperatures, is transferred to the calorimeter.

The numerous experimental designs described in the literature vary essentially according to the nature of the medium in which the sample is dropped. In the early calorimeters of the Regnault type, the sample was dropped into a liquid, usually water. An elaborate version of such a calorimeter has been described by White.[10] The sample was heated in a furnace consisting of a platinum-wound Alundum tube insulated with calcined magnesia. The furnace was independent of the calorimeter proper. The temperature of the water calorimeter was measured with a series of ten Copper-Constantan thermocouples connected to a high-precision potentiometer, thus warranting ±0.0003° C.

In another type of calorimeter, the sample falls into a high-conductivity metal block (copper or aluminum). This type is derived from the calorimeter used by Nernst, Koref, and Lindemann[11] for thermodynamic measurements at low temperatures. Magnus,[12] Jaeger and Rosenbohm,[13] and more recently Southard[14] have contributed numerous improvements. In Southard's apparatus, the furnace and the calorimeter were connected. A water-cooled gate protected the calorimeter from furnace radiation, and the whole space was filled with a CO_2 atmosphere. The calorimeter proper was immersed in an oil bath maintained at 25.00 ± 0.01° C.

Accurate data have also been obtained with Bunsen's ice calorimeter, a modern version of which was built by Ginnings and Corruccini[15] to measure the enthalpy of extremely pure alumina, considered to be a calorimetric standard at elevated temperatures. This method is delicate in its application and requires a careful degassing of the water. The apparatus is fragile and must be manipulated with much care. It seems that it ought to be reserved for high-precision measurements.

An interesting improvement proposed by Roth and Bertram[16] can be applied to certain dropping calorimeters: it consists in coating the inside walls of the calorimeter with a low-melting-point material (a eutectic mixture of sodium and potassium nitrate which melts at 218° C). When the hot sample comes into contact with the coating, the latter melts and thereby secures excellent thermal contact with the calorimeter.

The main advantage of the dropping method is high precision: 0.1 to 0.5%. Conversely it permits only a rough estimation of small heat effects such as heats of transformation.

Determination of Heats of Fusion. Latent heats of fusion as well as heats of transformation can be measured directly in calorimeters of the first type, although with considerable difficulties for most substances of interest in steelmaking.

In the "method of mixtures," the heat of fusion may be obtained as the difference between the enthalpies of the liquid and of the solid at the melting point. Those values result from an extrapolation of the enthalpy/temperature curve.

A third method has often been used when experimental data were lacking: namely, the application of the Van't Hoff equation to the equilibrium diagram of a binary system. An exact calculation requires the knowledge of activity coefficients. In most cases, such coefficients are unknown, and the solutions are assumed to be ideal; consequently only a rough estimation of the heat of fusion can be made.

Choice of an Experimental Method

In setting up heat balances of industrial furnaces, the knowledge of the enthalpy rather than that of specific heat is of interest. In order to minimize the error on the quantity in question, it is obviously better to select a method yielding the enthalpy directly.

The water calorimeter was ruled out in spite of its simplicity and of the reproductibility of its calibration mainly because of the difficulties encountered in preventing losses, which may be large, due to sudden vaporization of water when the sample is dropped.

Finally, the metal-block calorimeter has been selected. The apparatus is robust and is capable of being operated under a vacuum or with a controlled atmosphere. The relatively complex mecha-

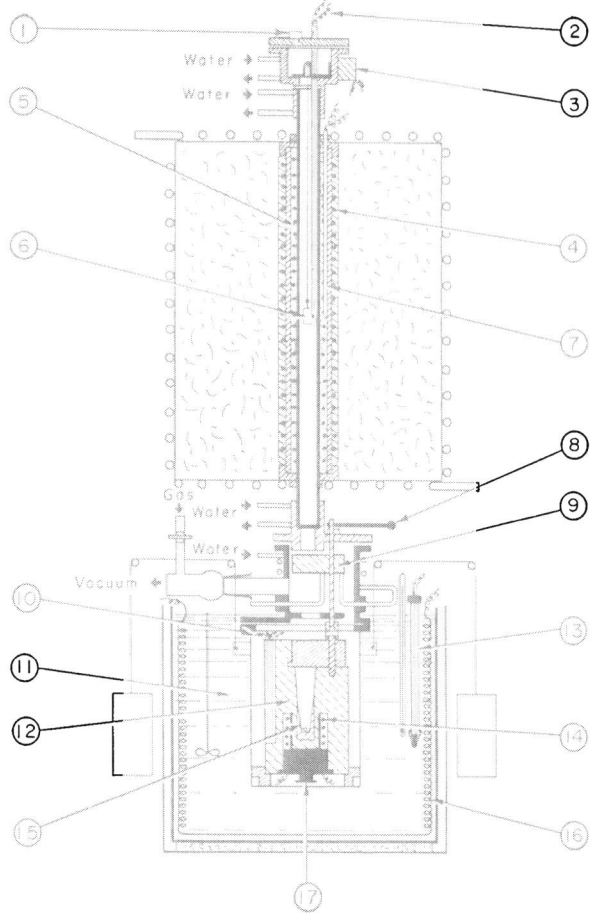

Fig 1. Schematic cross section of the apparatus. 1, Sight glass; 2, 13% Rh platinum-rhodium thermocouple for temperature measurement in the furnace; 3, electromagnet operating the suspension clamp; 4, kanthal winding; 5, 20% Rh platinum-rhodium wire, wound on a Morgan mullite tube; 6, capsule containing the substance under investigation; 7, 13% Rh platinum-rhodium regulation thermocouple; 8, handle operating the gates; 9, water-cooled gate; 10, hole containing the thermistors for measurement and control; 11, water bath with controlled temperature; 12, O.F.H.C. copper calorimetric block; 13, control thermistor; 14, calibration resistance; 15, copper cup containing chips of Wood's metal; 16, heating resistances; 17, orifice for the introduction of helium.

nism of heat transfer within a metal block is no serious objection, provided the metal is a very good conductor and the calorimeter operates adiabatically.

DETAILED DESCRIPTION OF THE APPARATUS

The apparatus built at IRSID (see Figs. 1 and 2) shows the same general design as Southard's.[14] It consists of a furnace in which the samples are brought up to the desired temperature and of a copper-block calorimeter in which they are dropped.

Its essential features are:

(a) The temperature of the substance under investigation must be known with precision and must be the same at all points; the furnace was therefore designed so as to secure an isothermal zone as long as possible.

(b) To reduce to a minimum the heat transfer between the calorimeter block and the outside, an adiabatic method was selected. For this purpose, an automatic device including an electronic regulator* controls the temperature of a water bath surrounding the calorimeter, to be equal at any time to that of a point located in the metal block close to its outer surface.

(c) To reduce to a minimum the heat transfer between the furnace and the calorimeter, three particular features were included:

1. The outer jacket of the furnace is water-cooled to prevent warming up of the air which might transfer heat to the calorimeter.
2. The whole furnace and calorimeter space is gastight. It can be evacuated to pressures of 10^{-3}—10^{-4} mm Hg or maintain a rigorously controlled atmosphere. In most cases, a low pressure of argon (20 mm Hg) was used.
3. The connection between the furnace and the calorimeter is maintained at a temperature equal to that of the calorimeter block through a circulation of water drawn from the surrounding bath.

(d) The sample as it is dropped from the furnace to the calorimeter travels in a space with reduced pressure; its cooling is therefore minimum.

(e) The sensitivity and simplicity of the measurements have been increased by use of a thermistor† which permits easy determination of the block temperature with an accuracy of a thousandth of a degree C.

(f) The whole apparatus was placed in an air-conditioned room whose temperature was maintained at 19 ±1° C.

The Furnace

The furnace contains two windings (see Fig. 1). The inner one, 5, is made of 20% Rh platinum-rhodium wire (0.6 mm in diam) and is divided into three sections fed independently. It is wound with constant spacing on a gastight mullite Morgan tube (i.d. 28 mm, length 600 mm). Powdered alumina contained in a larger coaxial alumina tube is used as a calorifuge. The outer winding, 4, is made of Kanthal, wound with variable spacing (closer at the ends) on a porous alumina (RA 98) Norton tube and imbedded in aluminous cement. It is surrounded by porous alumina bricks placed in a sheet-metal cylinder on which a water-cooled copper coil

* The electronic regulator was built owing to the kind collaboration of C. Tuppin, Member of the Staff, Electronics Department, IRSID.

† The thermistors are semiconductors which follow Ohm's law and whose temperature coefficient is negative and 10 times higher than that of usual metals (−0.04 to −0.05 at 25° C, instead of +0.0039 for platinum).

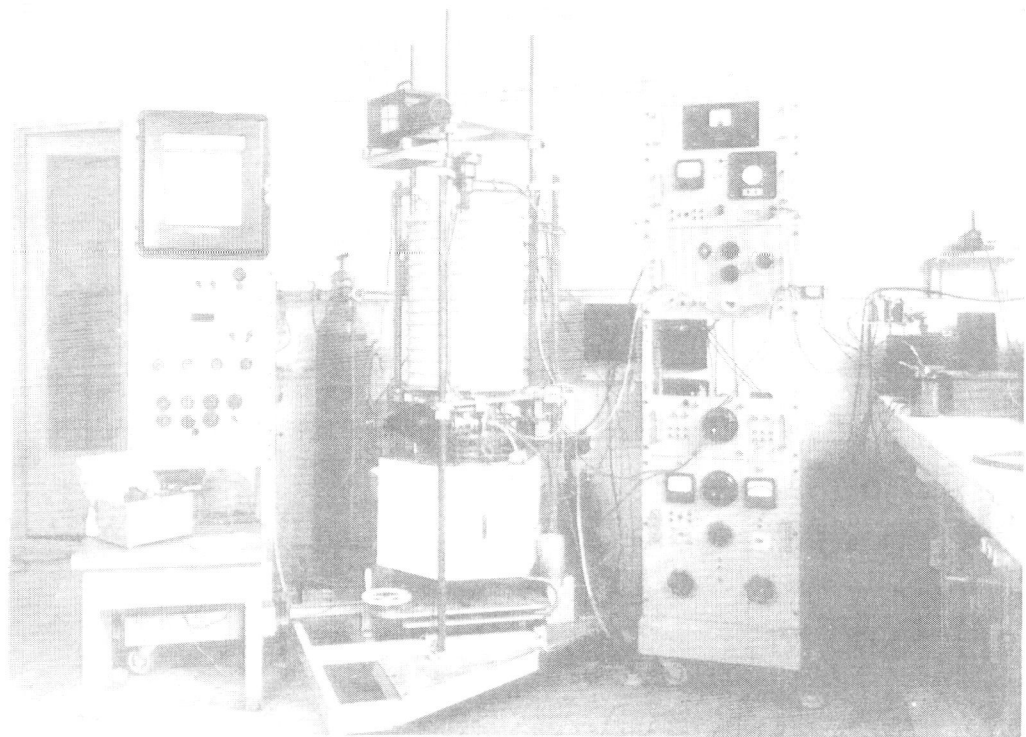

Fig. 2. General view of the apparatus.

is welded. The coaxial tubes are centered by means of Asbestolite flanges. The length of the hot zone is over 400 mm. The windings are fed through continuously adjustable autotransformers. The total power consumed is about 2.5 kw at 1550° C.

The temperature of the furnace is controlled through a 13% Rh platinum-rhodium thermocouple, 7, located against the inner winding, 5, and connected to a MECI electronic regulator (1-mv full-scale deviation). The regulator operates a relay which puts in or cuts out of the circuit an adjustable resistance connected to that winding.

The design presents these advantages:

1. The energy radiated by the platinum winding is only a fraction of the total required, hence a more economical utilization of the precious metal than in a single-winding furnace. In fact, the furnace has been operated for nearly a year without repairs.
2. The heating up of the furnace is faster.
3. Adequate distribution of the power between the various circuits secures a large isothermal zone. Temperature profiles show that up to 1550—1580° C, temperature is constant with time over a length of 80 to 120 mm, fluctuations being less than ±1° C.

The Calorimeter

The calorimeter proper consists of a massive high-conductivity copper block, 12, held in a stainless steel container which in turn is immersed in a 70 l water bath. The water, whose temperature is con-trolled by that of the block, is heated by resistors of low thermal inertia and is stirred mechanically. In the copper block, the cup which receives the sample is surrounded by a hollow annular space which may be filled with helium, by 17, and in which the manganin calibration coil, 14, is fitted.

A hole, 10, is drilled to accommodate two thermistors, one for measuring temperature and the other to control the water-bath temperature. A small quantity of diffusion pump oil (low vapor pressure) is poured into the hole to improve thermal contacts. The sample falls into a cylindrical copper cup, 15, which is fitted in the block. Chips of Wood's metal (about 2 grams) are laid on the bottom of the cup to prevent breakage of the capsule and to facilitate heat transfer once melted.

Two gates, one made of solid copper and closing the block entrance, the other cooled by water drawn from the surrounding water bath may rotate around a common axis and are operated from outside by a handle, 8. When open, they permit the dropping of the samples; when closed, they provide insulation between the calorimeter and the furnace. The block temperature drifts by only 1 or 2 thousandths of a degree per hour.

Measurement of Temperatures

In the Furnace. The furnace temperature is measured with a 13% Rh platinum-rhodium thermocouple connected to a MECI potentiometer (ESPM type) accurate to 5 μv, that is, ±0.2° C. Frequent calibrations are necessary, especially above

1400° C; they are run in the furnace itself at the gold and palladium melting points. By checking the homogeneity of the wires near the hot junction, the length to be cut off after contamination is determined. In practice, each measurement above 1450° C was preceded and followed either by a palladium melting point or by a homogeneity check.

In the Calorimeter. The temperature of the calorimetric block was measured by means of a needle-shaped thermistor* used as one arm of a precision d-c Wheatstone bridge. The unbalanced emf of the bridge is applied to a MECI electronic recording potentiometer (1-mv full-scale deviation). The thermal evolution of the block is thus recorded as a function of time. The thermistor resistance is about 29,000 ohms at 25° C and varies by 1.2 ohms for a temperature variation of 0.001° C. This corresponds to a distance traveled by the recorder pen of about 1.5 mm. To avoid temperature fluctuations of the thermistor through Joule's effect, it is essential that its power supply be controlled with accuracy, for instance, by keeping constant the voltage applied to the Wheatstone bridge. The exponential-like variation law of the thermistor resistance as a function of temperature was determined by comparison with two mercury precision thermometers calibrated at the National Physical Laboratory. With adequate co-ordinates, a linear function could be obtained, and the mean two straight regression lines led to the following relationship:

$$T = \frac{1000}{0.61065 \log R + 0.64860}$$

where R is the thermistor resistance in ohms
T is the temperature in °K

Under such conditions the temperature rise of the block due to the dropped samples, ordinarily two to three degrees, is affected by a maximum error of ±0.001° C. This accuracy is slightly lower than that of classical resistance thermometers[14] or of thermocouples in series.[10, 13] In the present instance, however, the thermosensitive element is cheap and requires only a relatively simple measuring apparatus because of its high sensitivity. Another advantage of it is that it easily permits recording.

As a rule, although thermistors do not seem to be adequate at present to be used as absolute thermometers, they are of great interest in the measurement of small temperature differences. Yet they have just started being used for that purpose.[17, 18] Frequent recalibrations of the thermistor used for measuring temperature have shown no serious evolution during three years of service.

Automatic Controlling System

Of two needle-shaped thermistors,† one is placed

* Type CS of Comp. Ind. des Céramiques Electroniques, Montreuil, France.
† Type 52 A1 of Victory Eng. Corp. Union, New Jersey.

in a hole, 10, in the calorimetric block and the other in a pickup tube in the water bath. Each one of them constitutes an arm of a stabilized d-c Wheatstone bridge whose unbalanced emf is applied to a proportional-action electronic regulator. That emf is converted into a 50-cycle alternating voltage by means of a synchronous vibrator. The amplitude and the phase of that voltage correspond, respectively, to the magnitude and the sign of the temperature difference between the water bath and the block. After a 3-stage amplification followed by demodulation, a d-c voltage signal is obtained which is proportional to the unbalance and which is used to modify the symmetry of a multivibrator, one anode of which controls the heating of the bath through a thyratron and a relay. The energy is supplied to the water in the form of impulses whose duration is a function of the temperature difference. The same impulses are used to operate a small motor, mechanically connected to the knob of the adjustable autotransformer which supplies current to the heating elements. It is thus possible to change the power fed to the bath from about 40 to 4000 watts within a few seconds. The low figure corresponds to the mere compensation of heat exchanges with the outside and the high figure to the compensation of the considerable heat supplied to the block by the sample. The sensitivity of the controlling device is of the order of a few ten-thousandths of a degree C, and its response time is of the order of one second.

Preparation of the Samples

Proper operation of the calorimeter under adiabatic conditions is warranted for a total temperature rise of the order of 2 to 3 degrees. For a given furnace temperature, the amount of matter to be used is determined so as to fulfill this requirement.

The substance under investigation is put into a capsule whose material depends upon its content and the temperature range.

The capsule is evacuated ($p < 10^{-5}$ mm Hg), filled with helium under such a pressure as to balance roughly the furnace pressure at high temperature, and sealed.

It is suspended in the furnace by means of a built-in hook to a 20% Rh platinum-rhodium or molybdenum wire which is clamped in the furnace head and is released to drop the sample. The clamp is operated through the electromagnet, 3.

CALIBRATION OF THE CALORIMETER

The heat capacity of the calorimeter was determined in the standard manner by measuring the temperature rise caused by the supply of a known quantity of heat. For doing so, a 61.086-ohm manganin resistance, 14, located in the copper block (see Fig. 1) was made part of a circuit containing also a 0.1-ohm standard resistance connected to a 30-v stabilized d-c generator. The current intensity (about 0.5 A) was calculated from the potential drop across the standard resistance which was measured by means of an SKM type MECI potentiometer. Time (about 10 min) was measured with a Jacquet chronometer with an accuracy of 10^{-4}. The switching on and off of the chronometer was controlled by means of an electromechanical device so as to be simultaneous with the closing and opening of the calibration circuit. Relative current fluctuations during a calibration run lay between 5 and 8×10^{-5} in all instances. The water equivalent thus found for the calorimetric block was 1225.3 cal with a standard deviation of 1.28 cal, i.e., about 0.1%. In order to test the performances of the calorimeter, preliminary enthalpy measurements were made at 850 and 1227° C on alumina taken as a calorimetric standard. A sample of crystallized alumina (corundum), containing over 99.9% Al_2O_3, was supplied by the Société d'Electrochimie et d'Electrométallurgie d'Ugine. When all corrections were made, the mean of 10 measurements at 850° C was $H_{850} - H_{25} = 22,352$ cal/mole with a standard deviation of 0.4%. The value recommended by Kelley[1] at that temperature is 22,407 cal/mole. From the data of Ginnings and Coruccini[15] and from the more recent data of Ginnings and Furukawa,[19] the values of 22,357 and 22,345* cal/mole, respectively, are found. At 1227° C, the mean of five measurements obtained was $H_{1227} - H_{25} = 33,934$ cal/mole with a standard deviation of 0.46%. Kelley's recommended value[1] is 33,920 cal/mole. Agreement with data from the literature is therefore excellent.

MEASUREMENT OF THE ENTHALPY OF SILICON BETWEEN 1200 AND 1550° C

Thermodynamic data on silicon are scarce and unreliable. A few calorimetric measurements between 0 and 100° C were made before 1920 by various authors. But Magnus[20] in 1923 was the first to measure enthalpy with a good accuracy between 20 and 900° C for 99.2% pure silicon.

Serebrennikov and Gel'd[21] have recently published a series of enthalpy measurements on 99.3% pure silicon between −185 and 1283.1° C. Their work is in good agreement with that of Magnus if the same reference temperature is taken and if temperatures are corrected according to the 1948 International Scale. However, to the best of our knowledge, no precise calorimetric measurements have been made above 1300° C and, in particular, near the melting point of silicon.

The determination of its heat of fusion seems also to be of great interest since, after rough estimations, silicon is considered to be the element having the highest heat of fusion.

Material. All the measurements were made on a single batch of silicon containing over 99.99%, prepared by zone melting at the Compagnie Pechiney (Saint-Jean de Maurienne). It contained only traces of iron, and low traces of boron, magnesium, and copper. The original pieces were crushed and sieved to sort out particles around 0.5-mm diam which were used exclusively for the measurements.

Capsule. It is difficult to think of a metal capable of holding silicon at high temperature and especially in the liquid state. Platinum and molybdenum are ruled out, of course. Furthermore, Economos[22] has shown the great reactivity of silicon between 1500 and 1600° C with pure refractory oxides: Al_2O_3, BeO, MgO, ThO_2, TiO_2, ZrO_2.

In spite of predictable drawbacks due to allotropic transformations, silica seems to be the only appropriate material. Transparent silica glass tube, free of defects and bubbles, was selected, and the capsule (2 to 5 grams) containing a known weight of silicon (2 to 4 grams) was filled with helium and sealed in the way indicated previously. After experiments above 1300° C, the silica capsule showed signs of devitrification the more markedly the higher the furnace temperature. However, the transformation of vitreous silica into cristobalite affected only surfaces, the outer one alone below 1500° C and both the outer and the inner one above 1500° C, the thicknesses involved being of the order of 0.1 to 0.2 mm, which corresponds to much less than a tenth of the total weight of the capsule. The resulting correction cannot be of importance since the enthalpy of vitreous silica at 1500° C is 24.38 kcal/mole while that of β cristobalite is 24.05 kcal/mole. During cooling in the block, the β to α cristobalite transformation is also negligible (200 cal/mole).

In any event, it was preferred to eliminate these difficulties by calculating the enthalpy of silicon at a given temperature as the difference between the heat content of a full capsule and that of an empty one having the same weight and shape. It was therefore necessary to run two successive experiments at the same temperature to obtain each one of the results reported below. An X-ray test on two used capsules showed that their percentage of α cristobalite was the same and, consequently, that silicon was not a catalyst of the transformation.

This technique had the further advantage of suppressing any correction for the platinum suspension wire.

Experimental Results. The data obtained are listed in Table 1. At each temperature, the enthalpy $H_\theta - H_{25}$ of silicon is expressed in cal/gram and in cal/gram atom. In the calculations the atomic weight of silicon was taken to be 28.06 grams because this value is most commonly used.[21] Certain tables nowadays quote the figure 28.09 grams, although recent calculations by Batuecas[23] would lead rather to adopt 28.08 grams.

The experimental data expressed in cal/gram atom of silicon are plotted in Fig. 3 together with

* With 1 abs j = 0.239006 cal.

Table 1. Enthalpy of Silicon between 1200 and 1550 °C

Temperature, °C	Physical State of Si	$H_\theta - H_{25}$, cal/gram	$H_\theta - H_{25}$, cal/gram atom*
1194.0	Solid	255.2	7,160
1259.5	Solid	271.1	7,605
1278.9	Solid	274.4	7,700
1299.3	Solid	277.5	7,785
1318.6	Solid	283.4	7,950
1344.0	Solid	288.5	8,095
1365.0	Solid	293.8	8,240
1384.7	Solid	299.8	8,410
1405.0	Solid	305.1	8,560
1412.0	Solid + liquid	484.4	13,590
1412.6	Solid + liquid	657.6	18,455
1414.8	Liquid	737.6	20,700
1419.6	Liquid	737.9	20,710
1419.6	Liquid	737.6	20,700
1435.0	Liquid	737.2	20,685
1449.8	Liquid	744.3	20,885
1456.5	Liquid	750.7	21,065
1469.4	Liquid	746.7	20,950
1478.2	Liquid	755.5	21,200
1490.5	Liquid	753.5	21,140
1499.0	Liquid	756.0	21,210
1499.8	Liquid	756.0	21,210
1519.6	Liquid	757.9	21,270
1530.0	Liquid	762.5	21,395
1544.0	Liquid	762.5	21,395
1552.0	Liquid	768.6	21,570

* At. wt of silicon = 28.06 grams.

the data of Serebrennikov and Gel'd [21] above 1100° C and two measurements of the enthalpy of liquid silicon (97–98% Si) at 1600° C obtained by Körber and Oelsen [24] and Oelsen and Middel [25] (21,582 and 21,890 cal/gram atom, respectively) in the course of a study of the heat of formation of iron-silicon and iron-cobalt alloys.

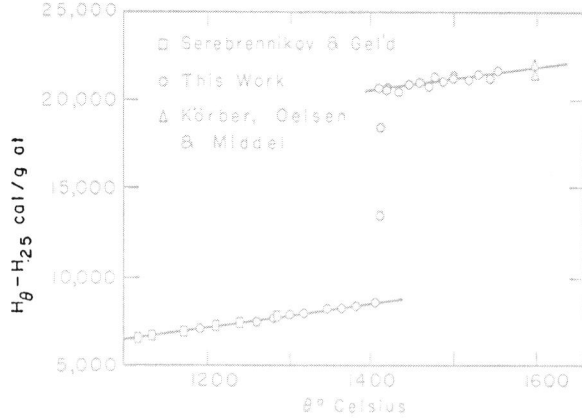

Fig. 3. Enthalpy of silicon between 1200 and 1500° C.

It may be seen from Fig. 3 that the agreement is satisfactory between Serebrennikov and Gel'd data and ours in the range where they overlap. It is also interesting to note that the average of the determi-

nations by Körber and co-workers, 21,736 cal/gram atom, is only slightly lower than the value obtained by extrapolation of our enthalpy temperature curve—21,816 cal/gram atom.

Judging by Fig. 3, our values of the enthalpies of solid and liquid silicon may be represented by two linear functions in the temperature range investigated. The calculation of the two regression lines leads to the following relationships:

Solid silicon (above 1200° C):

$$H_\theta - H_{25} = 6.53\theta - 652 \qquad (1)$$

Liquid silicon (up to 1550° C):

$$H_\theta - H_{25} = 6.12\theta + 12,024 \qquad (2)$$

where $H_\theta - H_{25}$ is in cal/gram atom

θ is in °C

The standard deviations are, respectively, 31 and 70 cal/gram atom, and illustrate the increase in scatter of the measurements when going from solid to liquid silicon.

Melting Point. Calorimetric methods in principle do not permit the accurate determination of melting points. However, Fig. 3 may give a fair estimation of the melting point of silicon. It is certainly located between 1405 and 1415° C. Besides, the two points found between solid and liquid most likely correspond to measurements either on mixtures of solid and liquid or on supercooled liquid (the control of the furnace may cause temperature to exceed slightly the chosen value).

We propose 1412 ± 2° C as the most probable figure. Estimations found in the literature range between 1408 and 1460° C. Our value is in agreement with the most recent determinations of Hoffmann and Schulze:[26] 1410 ± 2° C, and of Gayler:[27] 1415 ± 2° C.

Heat of Fusion. The vertical distance at 1412° C between the two straight lines on Fig. 3 represents the enthalpy of fusion of silicon. By means of equations 1 and 2 at that temperature, a value of *12,095 ± 100 cal/gram atom* is found. It is higher than that of Körber and Oelsen:[24] 11,100 ± 400 cal/gram atom, which was obtained on technical silicon containing 98% Si.

Kelley[28] found figures between 8100 and 14,200 when calculating the heat of fusion of silicon from binary equilibrium diagrams of the systems: Si—Al, Si—Ce, Si—Cu, Si—Au, Si—Be, Si—Mn, Si—Ni. He recommends 9280 cal/gram atom as the most probable value, which is much too low in our opinion.

In studying the equilibrium diagram of the iron-silicon system, Guertler and Tammann[29] have found, by thermal analysis, that the ratio between the heats of fusion of silicon and iron was 3.3. The heat of fusion of iron is not known with great precision. Kelley[1] recommends 3700 cal/gram atom of iron, a value accurate to 2 or 3%. The most recent measurement by Umino[30] yielded 3667 cal/gram atom. It is interesting to note that by using the above ratio the heat of fusion of silicon is found to be 12,210 and 12,100 cal/gram atom, respectively.

Discussion. Although the scatter of the enthalpy measurements on solid silicon appears to be acceptable, that of measurements on liquid silicon is markedly larger. One possible cause of this situation could be the formation of silicon monoxide according to the reaction:

$$SiO_2 + Si \rightleftharpoons$$

$$2SiO \text{ under nonreproducible conditions} \quad (3)$$

It is therefore important to know the order of magnitude of the weight and the physical state of the SiO formed. Careful examination of the used capsules revealed the existence of a brown deposit on the inner wall of the capsule or in a cavity of the silicon ingot. The X-ray examination of the deposit showed that it was a mixture of silicon and α cristobalite; no unidentified line was found. The quantity of substance deposited was roughly estimated to be 0.2 mg in a capsule heated up to 1552° C. With the assumption that the deposit originates from the decomposition of SiO, the data of Tombs and Welch[31] on the equilibrium vapor pressure of SiO over silicon and silica (about 30 mm at 1552° C) would yield approximately 0.1 mg of SiO. There is agreement on the order of magnitude; but no support was found for the opinion of Hoch and Johnston,[32] according to which a mixture of silicon and silica is transformed into a crystalline silicon monoxide above 1250° C, the faster, the higher the temperature. Two experiments in the vicinity of 1500° C were run, respectively, 1 and 2 hr. The corresponding values of the enthalpy of silicon showed no appreciable difference (see Table 1). Geller and Thurmond[33] have suggested that the lines attributed to SiO by Hoch and Johnston with their high-temperature X-ray technique could be due to a mixture of β cristobalite and β silicon carbide.

The effect of SiO formation is therefore considered to be negligible and the largest uncertainty producing the scatter observed above 1450° C rests upon temperature measurements.

SUMMARY

With the purpose of measuring heat contents of substances of interest in steelmaking, a copper-block dropping calorimeter has been built. The experiments were performed under adiabatic conditions, and the determination of the calorimeter temperature was made by means of a thermistor.

Among the elements for which calorimetric data are scarce at high temperatures, silicon was selected first. The enthalpy data presented here were obtained at temperatures as high as 1550° C on a very pure silicon containing over 99.99% silicon.

The results can be summarized in the two following relations where enthalpy is in cal/gram atom and temperature in ° C:

Solid state (from 1200° C):

$$H_\theta - H_{25} = 6.53\theta - 652 \quad (1)$$

Liquid state (up to 1550° C):

$$H_\theta - H_{25} = 6.12\theta + 12,024 \quad (2)$$

The enthalpy vs. temperature curve yielded the melting point of silicon, $1412 \pm 2°$ C, and its heat of fusion, $12,095 \pm 100$ cal.

References

1. K. K. Kelley, Contributions to the Data on Theoretical Metallurgy. High Temperature Heat Capacity and Entropy Data for Inorganic Compounds. *U. S. Bur. Mines, Bull.*, **476** (1949).
2. A. Winkelmann and A. Schott, *Ann. Physik*, **49**, 401 (1893).
3. J. R. Pattison, *J. Iron Steel Inst.*, **180**, 359 (1955).
4. W. Eitel, *Thermochemical Methods in Silicate Investigation.* Rutgers University Press, New Brunswick, N. J., 1952.
5. O. M. Corbino, *Physik. Z.*, **11**, 413 (1910); **12**, 292 (1911); **13**, 375 (1912).
6. M. von Pirani, *Verhandl. deut. physik. Ges.*, **14**, 1037 (1912).
7. A. Perrier and H. Roux, *Arch. sci. phys. et nat.*, **46**, 42 (1918); *Mém. soc. vaudoise sci. nat.*, 109 (1923).
8. P. R. Pallister, *J. Iron Steel Inst.*, **154**, 90 (1946); **161**, 87 (1949).
9. C. J. R. Winckler, *J. Am. Ceram. Soc.*, **26**, 339–349 (1943).
10. W. P. White, *Am. J. Sci.*, **28**, 314 (1909); **47**, 44 (1919). *Phys. Rev.*, **31**, 547 (1910). *Chem. Met. Eng.*, **25**, 17 (1921). *The Modern Calorimeter*, The Chemical Catalog Company, New York, 1928. *J. Am. Chem. Soc.*, **55**, 1047 (1933).
11. W. Nernst, F. Koref, and F. A. Lindemann, *Sitz. ber. preuss. Akad. Wiss. Physik-math. Kl. 1910*, p. 247.
12. A. Magnus, *Ann. Physik*, **48**, 983 (1915); **70**, 303 (1923); **81**, 407 (1926); **3**, 585 (1929).
13. F. M. Jaeger and E. Rosenbohm, *Proc. Acad. Sci. Amsterdam*, **30**, No. 8, 905, 1069–1088 (1927).
14. J. C. Southard, *J. Am. Chem. Soc.*, **63**, 3142 (1941).
15. D. C. Ginnings and R. J. Corruccini, *J. Research Natl. Bur. Standards*, **38**, 583 (1947).
16. W. A. Roth and W. Bertram, *Z. Electrochem.*, **35**, 297 (1929).
17. R. H. Müller and H. J. Stolten, *Anal. Chem.*, **25**, No. 7, 1103 (1953).
18. J. Jach and F. Sebba, *Trans. Faraday Soc.*, **50**, 226 (1954).
19. D. C. Ginnings and G. T. Furukawa, *J. Am. Chem. Soc.*, **75**, No. 3, 522 (1953).
20. A. Magnus, *Ann. Phys.*, **70**, 303 (1923).
21. N. N. Serebrennikov and P. V. Gel'd, *Doklady Akad. Nauk. SSSR*, **87**, 1021 (1952).
22. G. Economos, *Ind. Eng. Chem.*, **45**, 458 (1953).
23. T. Batuecas, *Nature*, **173**, No. 4399, 345 (1954).
24. F. Körber and W. Oelsen, *Mitt. Kaiser-Wilhelm-Inst. Eisenforsch.*, **18**, 110 (1936).
25. W. Oelsen and W. Middel, *Mitt. Kaiser-Wilhelm-Inst. Eisenforsch.*, **19**, 5 (1937).
26. F. Hoffmann and A. Schulze, *Metallwirtschaft*, **17**, 3 (1938).
27. M. L. Gayler, *Nature*, **142**, 478 (1938).

28. K. K. Kelley, Contributions to the Data on Theoretical Metallurgy, *U. S. Bur. Mines, Bull.*, **393**, 97 (1936).

29. W. Guertler and G. Tammann, *Z. anorg. Chem.*, **47**, 163 (1905).

30. S. Umino, *Science Repts. Tôhoku Univ.*, **18**, 91 (1929).

31. N. C. Tombs and A. J. E. Welch, *J. Iron Steel Inst.*, **172**, 69 (1952).

32. M. H. Hoch and H. L. Johnston, *J. Am. Chem. Soc.*, **75**, 5224 (1953).

33. S. Geller and C. D. Thurmond, *J. Am. Chem. Soc.*, **77**, 5285 (1955).

Discussion

RICHARDSON congratulated the author on having a highly precise calorimeter in Europe similar to those which are now in America. He asked about its precision as compared to that of the one Oelsen has built. It was concluded in the discussion that Oelsen's calorimeter was not highly precise when used to obtain heat capacity and heat of transformation data.

ELLIOTT asked the response time of the system, that is, the time for dropping of the sample to the point when the final reading was taken. OLETTE said that this ran from one to one and one-half hours in the usual case. ELLIOTT also asked if there was a possibility of a reaction between the silica containing tube and the silicon.

KOZAKEVITCH stated that this had been checked by holding the capsules at 1500° C for different periods. In Fig. 3 there are two points obtained at 1500° C, one was held for one hour, and the other for two hours before dropping. The fact that they are almost identical would indicate that effect of this reaction, if it occurred, was very slight.

DARKEN proposed that an SiO layer might form and by its presence inhibit further reaction, as in the case with the aluminum-carbide layer which forms between aluminum and a graphite crucible. It was concluded that the presence of SiO caused no significant error in the results obtained.

MC CABE asked whether this equipment or a modification thereof could be used for establishing the heat of fusion of silica, this being of particular interest in the question of silica structure. Richardson's interpretation of the structure indicates that the heat of fusion should be approximately 3600 cal/gram mole, whereas Forland's interpretation indicated that the heat of fusion should be about half that value, or 1800 cal/gram mole. OLETTE indicated that the calorimeter might be modified for the measurement, but operating at an additional 250° C above present range could pose some serious problems. In this matter, RICHARDSON pointed out that there would be an additional complication resulting from a glass forming which would have to be dissolved calorimetrically in hydrogen fluoride.

by P. P. Kozakevitch
and G. Urbain

Surface Tension of Pure Liquid Iron, Cobalt, and Nickel at 1550°C

INTRODUCTION

It has recently been shown that the considerable lowering of the surface tension of liquid iron and iron-carbon alloys by sulfur[1] and oxygen[2] may be of vital importance in such varied phenomena as iron emulsification (iron shot) in slags[3] and formation of nodular graphite.[2, 4] An exact knowledge of the surface tension of pure liquid iron is of course necessary if the lowering action of very small quantities of surface-active substances is to be measured. Unfortunately, even the most recent data vary between 1370 dyne cm^{-1} in 1949 by Becker, Harders, and Kornfeld,[5] 1560 dyne cm^{-1} in 1953 by Kingery and Humenik,[6] and 1720 dyne cm^{-1} by Halden and Kingery[2] in 1955. It is clear that further measurements are desirable. Only one series of measurements on nickel[6] is known to the present authors, and they are not aware of any measurement on cobalt. These metals were therefore included in the present work.

The main features of the present work are: preparation of metal samples in various ways, control of the oxygen pressure in the furnace (we were able to go as low as $P_{O_2} < 10^{-15}$ atm), and use of refractories made of those pure oxides which would not

react significantly with iron, cobalt and nickel in equilibrium with the furnace atmosphere. All the results of experimentally successful runs have been reported, rather than selected or mean values, so as to permit a better evaluation of the dispersion.

EXPERIMENTAL WORK

The sessile-drop method was used, but the apparatus previously described[1] has been improved in many ways. The drop (approx. 1.9 grams) and its support (pure Al_2O_3, CaO, BeO, ZrO_2 or ThO_2) were placed in a gastight vertical alumina tube, heated in a resistance furnace. The atmosphere was either best quality argon dried with magnesium perchlorate and purified with titanium at 850° C, or hydrogen purified in a Deoxo-apparatus (catalysis) and thoroughly dried with perchlorate. To remove all the oxygen adsorbed on the refractories and to reduce the oxide which might be present on the surface of the cold metal, preheating (up to 1450° C) was always operated in a flow of pure dry hydrogen. The image of the liquid drop was then formed on a photographic film by a horizontal beam of X-rays passing through the furnace. Four negatives were taken, the drop being turned 90° each time with a special device. A modification of Dorsey's method previously described[1] was used for measurements and calculation.

The beam of X-rays was generated by a medical-type tube with rotating tungsten target working at 90 to 100 kv. At 8 ma, the exposure time necessary to get a good contrast was approximately 15 sec. Photographic enlargements on high-contrast plates were used for measurements, the absolute dimen-

Dr. Kozakevitch is Head, Dr. Urbain a Member of the Staff, Physical Chemistry Department, Institut de Recherches de la Sidérurgie, Saint-Germain-en-Laye, France.

The authors are especially indebted to Mr. C. Lionnet for the experimental work. They also wish to thank Mr. A. Chaillou and Mr. J. M. Poilleux, respectively, for preparing the pure iron samples and the supports under the direction of Dr. M. Olette, Miss M. Hanin for numerous oxygen determinations, Mr. M. Rouannet for calibration of the pyrometer, and Dr. G. Pomey for advice on X-ray problems.

sions of the drop being determined by comparison with the image of a steel ball whose diameter was exactly known.

The carbon-resistance furnace is represented diagrammatically in Fig. 1. Special elastic joints en-

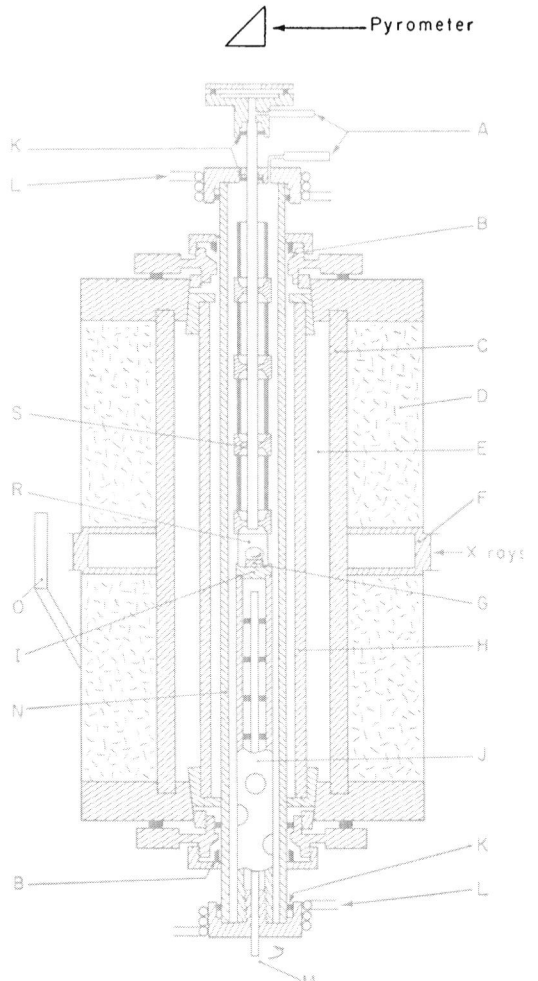

Fig. 1. Schematic cross section of the apparatus. A, Gas inlet; B, tight joint; C, graphite tube; D, coke; E, carbon black; F, graphite; G, drop support; H, carbon resistor; I, refractory stand; K, special elastic joint; L, water-cooling; M, gas outlet to vacuum pump; N, alumina tube; O, photographic film; R, alumina crucible; S, shields.

sured gastightness of the whole system and allowed the drop to be turned together with its support. The temperature was measured with an optical Ribaud pyrometer, the prism correction being taken into account. The contour of the drop could not be seen, an indication that black-body conditions were closely satisfied. All measurements were made under 1 atm of argon or hydrogen, and the evacuation with a rotary oil pump served chiefly to remove rapidly the bulk of the air and to check the gastightness before and after each run.

Hydrogen, when thoroughly dried with magnesium perchlorate, contains about 2 μg H_2O/liter.[7] At 1550° C this gives 10^{-18} atm of oxygen. A direct test indicated that probably the actual vapor pressure was close enough to the anticipated

value. In any case, it is considered that the oxygen pressure at 1550° C was less than 10^{-15} atm. The test consisted in a selective oxidation at 1200° C of a Cr—Ni alloy which was placed in the furnace and submitted to the usual cycle of operations. A microscopic examination of the surface by the method of Moreau and Benard[8] then permitted a rough estimate of the water vapor content of the gas. The test, while reliable in hydrogen with which Moreau and Benard worked, is probably much less so in argon, since oxidation by traces of free oxygen may be kinetically different from oxidation by H_2/H_2O mixtures. Under normal conditions, but at 1200° C, the test indicated 10^{-15} atm O_2 in the argon. The oxygen pressure was rather high when the furnace was evacuated at high temperature. It was even higher when preheating was done under argon. Maintaining the hydrogen flow was therefore essential during the period when the refractories were giving off adsorbed oxygen.

Good supports for the drops should be made of very pure oxides which have little tendency to react with liquid iron, cobalt, or nickel. In addition the supports should have a rough surface because the apparent contact angle (if greater than 90°) is much more favorable on a rough surface than on a smooth one,[9] and this is essential for the method of measurement used here. Supports of Al_2O_3, CaO, ZrO_2 stabilized with lime, BeO, and ThO_2 were used. The upper face of the cylindrical support bearing the drop was slightly concave, chiefly to make sure that the drop always occupied the same position in the tube and to avoid the necessity of keeping this face perfectly horizontal during the run. The X-rays easily penetrate through the refractories except in the case of ZrO_2 and ThO_2, so these had their upper faces almost plane so as to permit measurements with ϕ angles up to 135° (see below). The contact angle on rough surfaces was usually greater than 140°.

Fine-grain alumina (99 to 99.5% Al_2O_3, less than 0.05% silica) was used, while lime was prepared from the purest calcium metal of the Commissariat à l'Energie Atomique. The general manufacturing technique was that of the Department of Caillat at the C.E.A., somewhat modified in this Laboratory by Olette and Poilleux. From a purely thermodynamic point of view, ThO_2 is possibly the best material. Unfortunately, the surfaces of the sintered ThO_2 supports were much too smooth; even when ground with coarse alumina, they had a tendency to be "wetted" by iron, so that favorable contact angles could be maintained for only a few minutes after melting. The angle then rapidly diminished and finally remained stationary at 110°. Little use was therefore made of these supports. To avoid slagging at the contact of the supports with the alumina crucible (Fig. 1), a thin molybdenum sheet was interposed.

The starting material for preparation of the iron samples was pure carbonyl iron. To remove carbon, this iron was treated for 16 hr at 1200° C and 2 to 3 hr at 760° C with wet hydrogen. The water vapor was then removed in vacuum and the iron thoroughly deoxidized (and desulfurized) with a very pure dry hydrogen (10 hr at 1200° C and·5 hr at 760° C). The resulting sintered product usually contained less than 0.001% carbon, not more than 0.0008% oxygen, while sulfur was well below the limits of the combustion method. Impurities other than carbon, oxygen, and sulfur, were: $N_2 <$ 0.0005%, Mn 0.00004 to 0.0007%, Ni 0.004 to 0.015%, As 0.0001%, Al 0.001%, Si 0.001%, Cu 0.0003%, Ag 0.0001%, Mg 0.0002%, Cr, Co, Mo, V, Sn, Pb not detectable, and Ti, Ca, Na traces below the limit of a quantitative determination. The methods employed in the analysis were: semimicrochemical, activation followed by detection of radioisotopes, and spectrographic on solutions. The only noticeable impurity was Ni, which reached 0.015% in the most contaminated samples. But the surface tension of nickel is very high (see below and also[10]), so its presence cannot be harmful. A few samples of less thoroughly deoxidized iron (0.008 and 0.005% oxygen) and of iron with additions of cerium (0.092%) or zirconium (0.21%) were also used. In the latter instances, 400 grams of the purified sintered iron were melted under high vacuum (10^{-5} mm Hg, BeO crucible), the pure alloying element added, and the resulting alloy poured under vacuum into a copper mold. The results for these experiments are listed at the bottom of Table 1.

Cobalt and nickel samples were prepared from spectrographically standardized metal rods (Johnson and Matthey). Cobalt contained $Si \leqslant 0.0002\%$, $Fe \leqslant 0.0005\%$, $Al \leqslant 0.0001\%$, Ca not measurable, $Mg \leqslant 0.0001\%$ and $Cu \leqslant 0.0001\%$. Nickel contained $Fe \leqslant 0.001\%$, Ca not measurable, $Cu \leqslant 0.0002\%$, $Mg \leqslant 0.0001\%$ and $Si \leqslant 0.0001\%$. The two metals exhibited no lines of the following elements: Ag, As, Au, Ba, Bi, Cr, Cd, K, Li, Mn, Mo, Na, Pb, Rb, Sb, Sn, Ti, V, W, Zn, and Zr. Cobalt contained no nickel, and nickel was free from cobalt. The sulfur content of the cobalt and nickel rods was well below the limit of the combustion method, but the total oxygen content was rather high: 0.006% in cobalt and 0.041% in nickel. Each sample was therefore treated for 4 to 5 hr with hydrogen before taking any measurement.

Runs which appeared to be satisfactory during the operation were discarded only in the following cases: a leak in the tube detected after the run, abnormal appearance of the cold drop (spots on its bright surface), or visible dissymmetry of the drop on the photograph. No measurement was ever made on drops originating from discarded runs, and no measurement on experimentally valid drops was discarded, whatever the resulting value of the surface tension.

METHOD OF CALCULATION

The method of Dorsey[11] was followed since the authors know of no certain method of locating exactly the equatorial plane on the photographic image of a drop. On the one hand, AB, the equatorial diameter (Fig. 2) was measured, and on the

Table 1. Capillary Constant of Pure Liquid Iron at 1550° C

Run No.	Deoxidizer	Initial Total Oxygen (Analysis %)	Gas	Support	$a^2 = \gamma/gd$ mm²				a^2 Average Value
					A	B	C	D	
819	None	0.008	Argon	Al₂O₃	26.00	25.80	26.08	25.80	25.92
818	None	0.008	Argon	Al₂O₃	25.93	25.89	26.05	26.03	25.98
817	None	0.005	Argon	Al₂O₃	25.90	26.26	25.67	25.90	25.93
829	None	0.0008	Argon	Al₂O₃	25.80	25.74	25.26*	25.98	25.70
828	None	0.0008	Argon	Al₂O₃	25.83*	26.30*	25.52*	—	25.88
834	None	0.0008	Argon	CaO	26.28	25.72	25.89	25.96	25.96
835	None	0.0008	Argon	CaO	25.96	26.19	26.25	—	26.13
843	None	0.0008	Hydrogen	CaO	26.04	26.28	25.69	25.75	25.94
837	None	0.0008	Hydrogen	Al₂O₃	25.87	26.01	25.94	26.01	25.96
838	None	0.0008	Hydrogen	Al₂O₃	26.01	26.01	25.87	26.51	26.11
844	None	0.0008	Hydrogen	Al₂O₃	26.16	26.12	25.82	26.09	26.10
851	None	0.0008	Hydrogen	Al₂O₃	26.10	25.98	25.77	25.77	25.91
840	None	0.0008	Hydrogen	BeO	26.34	25.90	—†	—†	26.11
841	None	0.0008	Hydrogen	ThO₂	26.06	25.81	—†	—†	25.94
861	None	0.0008	Hydrogen	ThO₂	25.98‡	25.81‡	25.90‡	25.98‡	25.92
862	None	0.0008	Hydrogen	ThO₂	26.07‡	26.16‡	26.07‡	26.07‡	26.09
839	0.092% Ce	0.003	Hydrogen	Al₂O₃	25.98	26.01	25.49	25.72	25.80
824	0.003% Ce	0.002	Argon	Al₂O₃	—	26.10	25.83	—	25.97
820	0.21% Zr	0.001	Argon	Al₂O₃	25.75	25.36	26.15	25.77	25.76
854	0.21% Zr	0.001	Hydrogen	Al₂O₃	—§	—§	26.19‖	25.98‖	26.09
821	0.21% Zr	0.001	Argon	ZrO₂	25.96	26.11	25.78	25.78	25.91
826	0.21% Zr	0.001	Argon	ZrO₂	25.68	25.96	25.75	—	25.80

* Image of poor symmetry.
† Contact angle did not permit use of $\phi = 120°$ and 135° due to smooth surface of sintered oxide.
‡ Sintered oxide with a layer of ThO₂ powder added: good contact angle.
§ Four unusable images of an elongated drop which gradually becomes less asymmetrical.
‖ Fifth and sixth photograph of the drop No. 854, now sufficiently symmetrical.

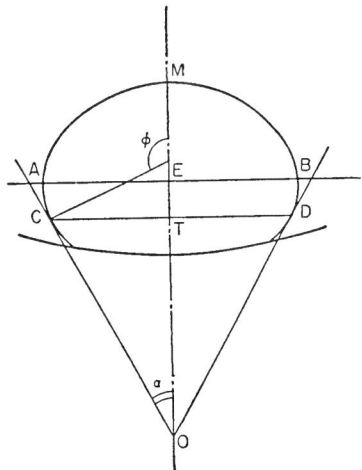

Fig. 2. Geometry of sessile drop by Dorsey's method.[11]

other hand, OM, the distance to the top of the drop from a point O on the axis selected so as to be at the apex of a known angle whose sides are tangential to the meridional contour.

Dorsey used a 90° angle and placed its apex above the top of the drop. But any angle may be used, and, if the shape of the drop permits, it may be placed below the drop just as well as above. Angles of 90° and 60° were used here, and, whenever possible, they were placed in both positions.

The measurements were taken on enlarged images, illuminated through the window of a viewer. The emulsion side of the plate was placed in contact with a grating engraved on a glass plate and consisting of two perpendicular sets of parallel straight lines, the center line of the vertical set bisecting two angles, one of 90° and the other of 60°. The procedure was to make the center line coincide with the axis of symmetry of the meridional section (thus providing a check on the symmetry) and to bring both sides of one of the angles in contact with the contour. The distance from the apex of the angle to the top of the drop and the length of the equatorial diameter were then read off the picture. The coarse grain of the X-ray film did not allow perfectly sharp enlargements to be produced, and we believe this to be the principal drawback of the method. On making several (4 to 10) adjustments and readings, mean values accurate to a tenth of a millimeter could, however, be obtained.

The drop contour changes slightly with 90° rotation, probably because the movement is not smooth enough. So it is believed that the four images of each drop (see Experimental Work) are to be considered as images of separate drops of identical composition. This is supported by the fact, that the dispersion of the results on four images of the same drop (A, B, C, and D, horizontal lines, Table 1) is almost identical with that observed on actually different drops (vertical columns).

Let C and D be the points of tangency of the sides of the angle COD with the contour (Fig. 2), apex below the top of the drop. Let E be the point of intersection of the axis with the normal to the contour in C, and furthermore, let

$$COD = 2\alpha, \quad MEC = \phi, \quad AB = 2r, \quad OM = h$$

Then:

$$\phi = \frac{\pi}{2} + \alpha, \quad OM = OT + TM = -TC\,tg\,\phi + TM$$

and

$$\frac{h}{r} = -\frac{TC}{r}\,tg\,\phi + \frac{TM}{r}$$

If g, d, b, and γ are defined, respectively, as the acceleration due to gravity, the specific mass of the drop, its radius of curvature at the top, and its surface tension, and if all are expressed in cgs units, the shape of the drop depends upon the parameter:

$$\beta = \frac{gdb^2}{\gamma} = \frac{b^2}{a^2}, \quad a^2 = \frac{\gamma}{gd} \quad \text{being the "capillary constant"}$$

The tables of Bashforth and Adams[12] give the ratio TC/b and TM/b as functions of β and ϕ as well as the ratio $p = r/b$ as a function of β. From

these, four tables have been prepared for $\phi = 45°$, 60°, 120°, and 135°, and these give the ratios $u = h/r$ as a function of β for values of β ranging from 0 to 5.0 at intervals of 0.1.[13] By means of interpolation, the tables yield four values of β corresponding to the four measured values of u.

Since the errors vary widely with β and ϕ, weights were assigned to the corresponding values of β by means of the method developed in this Laboratory by Chatel.[13] For most of the data, which correspond to β values in the vicinity of 0.9, the weights used were, respectively, 10, 12, 23, and 35 for $\phi = 45°$, 60°, 120°, and 135°. It becomes clear that, for the results to be reliable, it was essential that there be a possibility of using the angles of 120° and 135°. It should be pointed out that Dorsey's original method, which is limited to the use of $\phi = 45°$, produced very scattered results on our drops.

Finally, a fifth table was prepared giving the values of the ratio $p = r/b$ as a function of β; p was determined by interpolation and b followed. With β and b being known, there remains only to know d to apply the formula $\gamma = gdb^2/\beta$. The value of d, if no reliable data are available, may be determined after the volume of the drop is measured on the enlargement.[1, 13]

The scatter of the average a^2 values obtained on the four negatives of each drop is much smaller than in our previous work on Fe-C-S alloys (compare Table 1 of the present work with Table VII of reference[1]). Better atmospheres, a better X-ray tube, and the use of pure oxides for drop supports are probably responsible for this improvement.

PRESENTATION AND DISCUSSION OF THE RESULTS

The results of all experimentally successful runs on pure iron and on some samples containing deoxidizers (Zr or Ce) are represented in Table 1. Usually, the results on the four negatives A, B, C, and D (Experimental section) are given, but in a few cases some photographic defect of the negative made the measurement impossible.

Because tables of Bashforth and Adams yield directly the values of $a^2 = \gamma/gd$ (see above for method of calculation), it was preferred to tabulate values of a^2 (in mm^2) rather than of γ since the former are not dependent on the existing doubtful values of the density.

The distribution of the experimental results has been tested and found to be random (77 measurements, average value $a^2 = 25.94$ mm^2 standard deviation 0.22). The graphic adjustment is very good, the points following closely the normal Henry linear relationship. It may however be pointed out, that the standard deviations are somewhat higher for the data of the columns B, C, and D (0.22, 0.25, and 0.20, respectively) than for those of the column A (0.16). This is probably because the gas (argon or hydrogen) which initially fills the spaces between the grains of the rough support surface is gradually expelled by the liquid metal of the drop until the appropriate equilibrium contact angles at the irregular surface are attained. The symmetry of the contour, which gives the values of a^2, is then possibly affected by the slight movement of the drop. The mean values remain almost the same (25.99, 25.98, 25.85, and 25.94 for columns A, B, C, and D, respectively), but the deviations become somewhat larger. A few of the highest deviations are certainly due to a nonuniform distribution of grain in the negative near the top of the drop.

No significant deviation of a^2 values depending upon known factors varying from one run to the other could be detected (see for instance the average a^2 values in Table 1). These known factors are:

(a) Total oxygen content of the metal as determined before the run, that is, before preheating in hydrogen (0.0008% in most runs, 0.005% in Run No. 817, 0.008% in Runs 818 and 819, and between 0.001 and 0.003% in the presence of deoxidizers).

(b) Nature of the atmosphere (argon with 10^{-15} atm and hydrogen with less than 10^{-15} atm oxygen).

(c) Nature of the support material: Al_2O_3 with $\Delta G° = -173$ kcal per mole of oxygen at 1550° C, CaO with $\Delta G° = -210$ kcal and boiling point of the resulting metal ~1500° C, BeO with $\Delta G° = -210$ kcal and boiling point of the resulting metal ~1540° C, ZrO_2 with $\Delta G° = -180$ kcal, and ThO_2 with $\Delta G° = -218$ kcal, compared with $\Delta G° = -64$ kcal for FeO.

(d) Deoxidizers: Ce_2O_3 $\Delta G° = -210$ kcal per mole oxygen and Zr with ZrO_2 $\Delta G° = -180$ kcal.

It is not altogether certain, however, that the deoxidizers used here do not slightly lower the surface tension. In any event, the deviations which might be caused by the deoxidizers are less than 1%, and their existence can hardly be proved with our present technique.

Four interpretations could be proposed for this lack of any pronounced dependence of a^2 on factors (a) to (d):

1. The varying concentrations of surface-active substances, most probably [O] and [S] are too small to affect noticeably the results obtained by the present method. In this case, the surface tension of pure liquid iron was actually measured.

2. All results are uniformly affected by the surface-active substances, [S] and/or [O], actually present.

3. In each case, the effects of different impurities compensate each other. For instance, the rise of a^2 due to diminished oxygen content in the presence of H_2 or Zr, as compared with argon, is exactly counteracted by a hypothetical lowering action by H_2 or Zr.

4. The concentrations of known impurities being very small, the values of a^2 are uniformly affected by some unknown but very active substance.

Point 4 would apply to any property depending strongly upon presence of certain impurities. In the case of surface tension, it simply means that the result of any measurement may, in principle, be too low, for traces of impurities can have a noticeable effect on the surface tension only if they lower it. It may be added that no factor is known which would act in the opposite direction, provided measurements are made in a correct manner and on symmetrical drops only.

Point 3 is really highly improbable. So it is necessary to consider only the possible oxygen and sulfur contents of our iron drops under argon or hydrogen and in contact with different support materials. It should, of course, be recognized, that no macro- or microscopic suspension can be surface active. Thus, only "dissolved" oxygen may alter the surface tension of iron, but not the oxygen-bearing inclusions suspended in the liquid metal.

The drops of pure iron with no deoxidizers usually contain, when analyzed after the run, about 0.002% total oxygen, but the limit of the hot-extraction method, when applied to a 1.7-gram sample (drop weight about 1.9 grams) is 0.001% oxygen, corresponding to approximately 0.02 cm³ CO. Thus, the direct drop analysis gives little more than a rough indication: the total oxygen content of the metal after the run lies between 0.001 and 0.003%.* It follows that the iron samples with higher initial oxygen lose a part of it, while the most pure ones possibly pick it up during (or after) the run. Point 2 might therefore apply (uniform contamination with approximately 0.002% oxygen) if this oxygen is really "dissolved" in the liquid iron and not picked up after the run.

The oxygen content in equilibrium with the argon atm should be about 0.0001% and still less with the hydrogen. The preheating was conducted in hydrogen, so that no oxygen could have been picked up from the gas. Aluminum was determined in Drops 837 and 838 (hydrogen, alumina support) and found to be approximately 0.005%. From the results of Gokcen and Chipman,[14] this corresponds to approximately 0.001% oxygen in the melt, so that the alumina support could have been responsible for over a half of the quantity actually found. This is certainly not the case with ZrO_2 and ThO_2.

Melts with Ce and Zr should contain little dissolved oxygen: possibly 0.00002% in the case of Zr.† But a^2 values remain unaffected. Thus, it must be concluded that either the variations in dissolved oxygen contents between approximately 0.001 and

---

* Oxygen determinations before the run were made on 20-gram samples.

† The anticipated activity of oxygen is 0.00002 in the presence of about 0.2% Zr. The measured oxygen content may well be somewhat higher.[15] But we do not think that the difference can be very large, say 20 times or more. Such a lowering of activity would probably be equivalent to formation of a new molecular species. But this would then be ZrO_2, probably insoluble in iron.

0.00002% do not alter the surface tension appreciably or that the bulk of the total oxygen found after the run is due not to dissolved matter but to a contamination by oxygen-bearing impurities suspended in the liquid metal. In either case the measured average value of a^2 may be considered as the minimum "capillary constant" of pure liquid iron at 1550° C, so it does not matter which of these two possibilities is right. The second, however, is supported by the fact that fine particles of support material were always rather numerous in the body of the metal near the bottom of the drop. This layer was, of course, ground away before analysis, but even then fine particles of the abrasive material were always found imbedded in the drop. It may also be pointed out that iron samples with deoxidizers, as analyzed before the run, always gave somewhat higher total oxygen contents than the unmelted pure iron: 0.001% with 0.21% Zr and 0.003% with 0.092% Ce. It is believed that in these cases the bulk of the oxygen was also due to suspended matter, originating from deoxidation of the molten metal with Ce or Zr, or from a reaction between deoxidizers and the crucible during the vacuum fusion.

It has been said (Experimental Work), that the initial sulfur content of our iron samples was well below the limits of the combustion method. A possible source of this contaminant might have been the titanium in the furnace for argon purification, as this metal contained some 0.02% S. If so, the results in hydrogen ought to have been higher than in argon, but this is not the case. Another source of sulfur might have been the alumina tube of the furnace, but this is improbable: the tube was always heated for several hours before use in pure dry hydrogen at 1550° C.

It is thus believed that *all* data of Table 1 can be used to calculate the mean value (low estimate) of the capillary constant of pure liquid iron at 1550° C, so that $a^2 = 25.94$ mm^2, or, rounding off to one-tenth of mm^2,

$$a^2 = \gamma/gd = 25.9 \text{ mm}^2$$

The standard deviation is 0.22 and the total number of measurements 77. As stated above, however, the standard deviations in columns B, C, and D are somewhat higher than in column A, possibly because of the slight movement of the drop which may occur when the liquid metal begins to expel argon or hydrogen from the rough surface of the support. It may be reasonable therefore to take account only of those drops (column A) which have spent least time in contact with the support. The mean value becomes then 25.99 mm^2, or, rounding off to one-tenth of mm^2,

$$a^2 = \gamma/gd = 26.0 \text{ mm}^2$$

The standard deviation is 0.16 and the number of measurements used 20. Although these two values are practically identical, the authors prefer

the second. It is felt (see above) that this value corresponds to an iron with less than 0.001% dissolved oxygen, the actual content varying perhaps between 0.0001 and 0.00002%. The above capillary constant is *half* of the capillary constant used with the capillary rise method: $2\gamma/gd = hr$.

The density of pure liquid iron at 1550° C is 7.21 according to Benedicks, Ericsson, and Ericson[16] and 7.01 according to Stott and Rendall.[17] The authors' estimate (drop volume) is 7.2. It is clear that further density measurements are desirable. From a tentative value of 7.2, the surface tension of pure liquid iron at 1550° C (low estimate) becomes:

$$\gamma = 1835 \text{ dyne cm}^{-1}$$

with standard deviation 15 dyne cm^{-1} and number of measurements 20. This value is higher than the most recent result of Halden and Kingery[2] and our own previous estimate of 1700–1800 dyne cm^{-1} based on an anticipated correction for the effect of 0.005% S.

The results for cobalt and nickel are presented in Table 2. The mean value of a^2 for nickel is 25.8

Table 2. Capillary Constants of Pure Liquid Cobalt and Nickel at 1550° C

COBALT

Run No.	Gas	Support	$a^2 = \gamma/gd$ mm^2			
			A	B	C	D
845	Hydrogen	Al$_2$O$_3$	25.35	25.25	25.09	25.11
846	Hydrogen	Al$_2$O$_3$	25.35	25.37	25.35	25.19
852	Hydrogen	Al$_2$O$_3$	—*	25.50*	25.59*	25.50*
863	Hydrogen	ZrO$_2$	25.40	25.40	25.28	25.40
870	Hydrogen	ThO$_2$†	25.16	25.22	25.40	25.16

NICKEL

Run No.	Gas	Support	$a^2 = \gamma/gd$ mm^2			
			A	B	C	D
847	Hydrogen	Al$_2$O$_3$	25.77	25.77	25.59	25.49
848	Hydrogen	Al$_2$O$_3$	26.07	25.77	26.07	25.77
868	Hydrogen	ThO$_2$†	—‡	25.81	25.40	26.13
869	Argon	ThO$_2$†	25.67	25.63	26.07	25.75

* Image of poor symmetry.
† Sintered oxide with a layer of ThO$_2$ powder added: excellent contact angle.
‡ Double contour, moving drop.

mm^2 (standard deviation 0.22, number of measurements 15). This figure is almost identical with that for iron. On the contrary, for cobalt a^2 is slightly but definitely lower: 25.3 mm^2 with standard deviation 0.14 and number of measurements 19. No density measurements on liquid cobalt are known to the present authors. Their own rough estimate at 1550° C is 7.8 (drop volume). In contrast, two series of measurements on liquid nickel are available;[16] a slight extrapolation of these gives 7.65 and 7.56. Tentative values to be used for the calculation of surface tension are therefore 7.8 for cobalt and 7.6 for nickel. This gives at 1550° C $\gamma = 1936$ dyne cm^{-1} for cobalt and 1924 dyne cm^{-1} for nickel. As far as the authors are aware, these surface tensions (as well as that for iron) are the highest ever measured on any known substance.

In Table 3 a few physical properties of liquid

Table 3. Physical Properties of Liquid Iron, Cobalt, and Nickel at 1550° C Compared to Those of Copper at 1200° C

Metal	Atomic Weight	Density	Atomic Volume	γ dyne cm^{-1}	a^2 mm^2
Iron	55.85	7.2	7.75 cm^3	1835	26.0
Cobalt	58.94	7.8	7.55 cm^3	1936	25.3
Nickel	58.69	7.6	7.72 cm^3	1924	25.8
Copper	63.54	8.1	7.84 cm^3	1300	16.4

iron, cobalt, and nickel at 1550° C are listed to be compared to those of copper at 1200° C.[1] The most characteristic feature of this table is the sudden drop of the surface tension when one passes from the metals of the iron group to copper. It may also be noticed that, for the elements of the iron group, the surface tension increases with increasing density and that the inverse relationship between surface tension and atomic volume found by Atterton and Hoar[18] seems to be valid for all four metals listed. However, in the iron group, the surface-tension values follow more closely the variations of the atomic weights than those of the atomic volume (See Table 3).

SUMMARY

The surface tension of pure liquid iron, cobalt, and nickel has been measured at 1550° C by the sessile-drop method. By control of the atmosphere (partial pressure of oxygen less than 10^{-15} atm) and use of pure oxide refractories, it was possible to obtain less than 0.001% dissolved oxygen in the drop. The "capillary constant" of iron, $a^2 = \gamma/gd$, was found to be 26.0 mm^2 and the surface tension (density 7.2) $\gamma = 1835$ dyne cm^{-1}. For cobalt, $a^2 = 25.3$ mm^2, and $\gamma = 1936$ dyne cm^{-1} (density 7.8). For nickel, $a^2 = 25.8$ mm^2, and $\gamma = 1924$ dyne cm^{-1} (density 7.6).

References

1. P. Kozakevitch, S. Chatel, G. Urbain, and M. Sage, *Rev. mét.*, 52, 139 (1955); *Compt. rend.*, 237, 1690 (1953).
2. F. A. Halden and W. Kingery, *J. Phys. Chem.*, 59, 557 (1955).
   W. vor dem Esche, *"Bestimmungen der Oberflächenspannung an reinem und legiertem Eisen,"* Ph. D. Thesis, Aachen, 1955.
   W. vor dem Esche and O. Peter, *Archiv Eisenhüttenw.*, 27, 355 (1956).

B. V. Stark and S. Philippov, *Bull. acad. sci. URSS, Classe sci. tech.* 413 (1949).
3. P. Kozakevitch, G. Urbain, and M. Sage, *Rev. mét.*, 52, 161 (1955); *Iron & Coal Trades Rev.*, 170, 963 (1955); *Compt. rend.*, 239, 166 (1954).
4. K. Grütter and B. Marincek, *Giesserei, Tech. Wiss. Beih.*, 12, 587 (1953); *Archiv Eisenhüttenw.*, 25, 447 (1954).
   J. Keverian, "The surface tension and microstructure of some iron-carbon alloys," Ph. D. Thesis, M.I.T., Cambridge, Mass., 1954.
5. G. Becker, F. Harders, and H. Kornfeld, *Archiv Eisenhüttenw.*, 20, 363 (1949).
6. W. D. Kingery and M. Humenik, *J. Phys. Chem.*, 57, 359 (1953).
7. J. H. Bower, *U. S. Bur. Standards Journal of Research*, 12, No. 2. *Research Publication*, 649 (1934).
8. J. Moreau and J. Benard, *J. Inst. Metals*, 83, 87 (1954).
9. N. K. Adam, *Physics and Chemistry of Surfaces*, Oxford University Press, London, 1949.
10. P. Kozakevitch and G. Urbain, *Compt. rend.*, 245, 335 (1957).
11. N. E. Dorsey, *J. Wash. Acad. Sci.*, 18, 606 (1928).
12. F. Bashforth and J. C. Adams, *An Attempt To Test the Theories of Capillary Action . . .* , Cambridge University Press, Cambridge, Eng., 1883.
13. S. Chatel, *Représentation analytique approchée d'un ménisque quasi sphérique avec quelques applications. Mesure de tension superficielle des métaux.* Ph. D. Thesis, Paris, to be published in 1957.
14. N. A. Gokcen and J. Chipman, *Trans. A.I.M.E.*, 197, 173 (1953).
15. *Basic Open Hearth Steelmaking*, Second Edition, Physical Chemistry of Steelmaking Committee, A.I.M.E., New York, Chapter XVI (1951).
16. C. Benedicks, N. Ericsson, and G. Ericson, *Archiv Eisenhüttenw.*, 3, 573 (1930).
17. V. N. Stott and J. H. Rendall, *J. Iron Steel Inst.*, 175, 374 (1953).
18. D. V. Atterton and T. P. Hoar, *Nature*, 167, 602 (1951).

Discussion

KING noted that one must be careful in applying statistics in interpreting results. As a result of a systematic error, the average value may be considerably in error, and as a consequence, one must be careful to use the best value rather than the average value. URBAIN pointed out that when a large number of experimentally valid runs is available statistical methods should lead to a more reliable result. KING also recommended the use of an optical technique whereby the light emanating from the drop was used to obtain a picture of it. Experimentally it is much more difficult to collimate an X-ray beam to the precision required for such work.

KINGERY* noted that in his measurements for iron and nickel, results were not quite as high as were found by the authors. He said that it has always been his tendency in analyzing surface-tension data to accept the highest results obtained by a reasonable method as those being most correct. Unless there was some systematic deviation in the experimental method, he would have to apply that criterion here. The best values which he had obtained using the same basic method but with an optical technique are 1730 dynes per cm for iron and 1725

* Prof. W. D. Kingery is with the Department of Metallurgy, Massachusetts Institute of Technology.

dynes per cm for nickel. He was surprised to see that there was such a difference from the author's results because he took considerable pains to be sure that reasonably pure materials were used. In studying the effect of low concentrations of additives, he found that the effect of them on the surface-tension values was not large. Plotting the value of surface tension versus the log of the concentration of the impurity gives a curve as shown in Fig. 3. It is difficult with this type of curve to extrapolate back to 1835 dynes per cm for oxygen, although sulfur may cause this amount of decrease.

KINGERY continued by saying that more recently he and his colleagues had studied the effect of carbon additions on the temperature dependence of surface tension of iron and found the effect to be negligible. By plot-

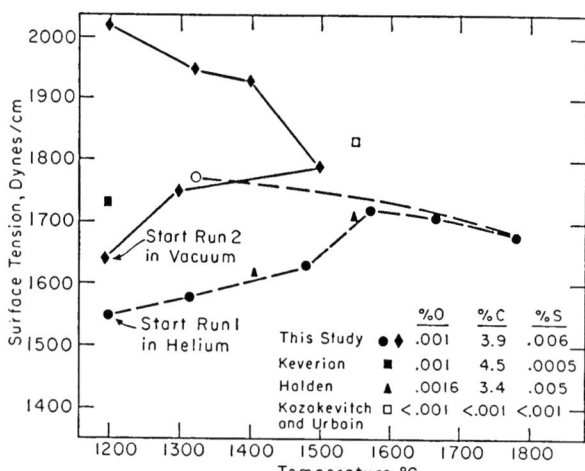

Fig. 4. Surface tension of iron-carbon eutectic composition as influenced by heating and cooling.

ting surface tension versus temperature for the iron-carbon eutectic, results were obtained as shown in Fig. 4. Heating iron-carbon alloys in vacuum from the eutectic temperature gave a series of values of surface tension which increased rapidly from a low value. On cooling, the line came back at a much higher range for the equivalent temperature. Then on reheating again, it retraced the second line. These results appear to be due to very small amounts of volatile materials which are removed by heating in vacuum. On heating two or three times, a value is obtained which is in the neighborhood of 1800 dynes per cm for iron, which is in reasonable agreement with the results of Kozakevitch and Urbain.

URBAIN: "We found that the presence of gases has a very serious effect on the surface tension even at very low concentration, as they tend to lower the surface tension."

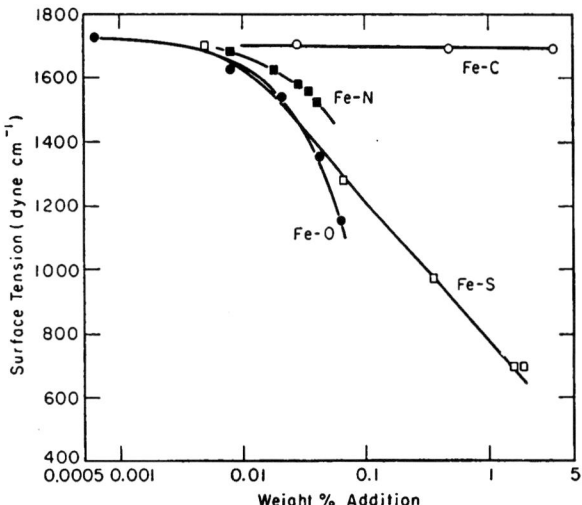

Fig. 3. Effect of C, N, S, and O on the surface tension of liquid iron.

C. R. Taylor

J. Pearson

Presiding

Section 2

Equilibria of Reactions

in Liquid Iron and Steel

by J. F. Elliott

The Carbon-Oxygen Equilibria

in Liquid Iron

The carbon-oxygen equilibria have been selected as the primary topic for this introductory paper to the present session "Equilibria of Reactions in Liquid Iron and Steel" because it appears that the reactions involved will be a major topic of discussion during the Conference. Specifically there seems to be a number of unanswered questions concerning the reasons for the apparent sluggishness of the reaction of carbon with oxygen in iron and steel-making systems. Another reason for taking up the matter is that the products CO and CO_2 can be treated independently of a slag phase. Most other reactions involve a slag phase, and consideration of slag-metal reactions has been reserved for later sessions. These remarks should not be construed to indicate that all is known about the carbon-oxygen equilibria. Some of the areas of uncertainty will be indicated later.

The conventional means of expressing the equilibrium of a reaction is by the equilibrium constant K. For the general reaction

$$A + xB \rightleftarrows AB_x \qquad (1)$$

the constant is

$$K = \frac{a_{ABx}}{a_A \cdot a_B{}^x} = \frac{f_{ABx}C_{ABx}}{f_A C_A \cdot (f_B \cdot C_B)^x} \qquad (1a)$$

As is customary, a is the activity of the component indicated by the subscript, f (or γ) is its activity coefficient, and C its concentration, which can be ex-

* Prof. Elliott is with the Department of Metallurgy, Massachusetts Institute of Technology.

The author is indebted to the authors of many papers reporting experimental results for the basic information used in these calculations.

pressed in any one of a number of ways. The proper evaluation of the equilibrium constant K requires data on the concentrations of the various participating components under well-defined conditions of constant temperature and pressure, and of known chemical potential for each component.

One of the greatest obstacles to our understanding of equilibria in liquid iron solutions in the past has been a deficiency in information on the behavior of components in liquid iron and in slags. Such knowledge is necessary so that one can determine the activity or activity coefficient for a component dissolved in the metal or the slag phase. In fact, many equilibrium studies become, in essence, an evaluation of the activities or activity coefficients of participating components. However, there has been a great deal of interest recently in the behavior of elements in liquid iron, and much has been learned. A recent paper by Chipman,[1] which evaluates atomic interactions in liquid iron, points up the present state of our understanding of the matter.

The first step in this evaluation of the carbon-oxygen equilibria in iron will be to review very briefly the data available on the basic reactions. The second step will be to point out what is known as to the behavior of oxygen and carbon dissolved in iron and to calculate the interaction of the two elements. The third step will be to present data on the equilibria in graphical form.

BASIC EQUILIBRIUM DATA

The thermodynamic values for the various reactions for gases bearing oxygen and carbon which are of interest here are as follows:

$$CO\ (g) + \tfrac{1}{2}O_2\ (g) \rightleftharpoons CO_2\ (g)$$
$$\Delta F° = -66{,}560 + 20.15T \quad (2)$$

$$C\ (gr) + \tfrac{1}{2}O_2\ (g) \rightleftharpoons CO\ (g)$$
$$\Delta F° = -28{,}100 - 20.20T \quad (3)$$

$$C\ (gr) + CO_2\ (g) \rightleftharpoons 2CO\ (g)$$
$$\Delta F° = +38{,}460 - 40.35T \quad (4)$$

These data are taken from Chapter 14 of *Basic Open Hearth Steelmaking*,[2] which in turn summarizes more detailed information from standard tables compiled by the National Bureau of Standards.[3] The use of such data is sufficiently commonplace to need no review here. It is sufficient to note that the equilibrium constant for a reaction can be calculated for any temperature near 1600° C from the appropriate standard free-energy data. In turn, if one knows the activities (partial pressures for gaseous components) of two of the components, the equilibrium activity of the third at the specified temperature can be computed.

BEHAVIOR OF CARBON AND OXYGEN DISSOLVED IN LIQUID IRON

Rist and Chipman have incorporated the results of their study of the reaction*

$$CO_2 + [C] \rightleftharpoons 2CO \quad (5)$$

with a critical review of the data from the other investigations on the same reaction to obtain quantitative information about the behavior of carbon in liquid iron. They recommend the equation

$$\log f_C$$
$$= \frac{4350}{T}[1 + 4 \times 10^{-4}(T - 1770)](1 - N_{Fe})^2 \quad (6)$$

for calculation of the activity coefficient of carbon in liquid iron. The reference state is the infinitely dilute solution of carbon in iron where $a_{[C]}/X_C$, i.e., f_C, approaches one as X_C approaches zero. The activity curves for carbon as shown in Fig. 1 were computed from equation 6. Two scales are used to show the activity curve for the reference where $a_C/X_C \rightarrow 1$ as $X_C \rightarrow 0$ and $a_C/\% [C] \rightarrow 1$ when $\% [C] \rightarrow 0$. The two activity scales are related by a factor of 21.4. It is to be noted that changing the temperature 150° C has a relatively small effect on the position of the activity curve and that the curve increases markedly at higher carbon concentrations.

The behavior of oxygen in carbon-free iron has been evaluated by the use of Dastur and Chipman's [4] data for the reaction

$$\tfrac{1}{2}O_2 \rightleftharpoons [O]\ (\%) \quad (7)$$

as follows:

$$\Delta F° = -27{,}930 - 0.57T \quad (7a)$$

* The component enclosed in brackets is dissolved in liquid iron.

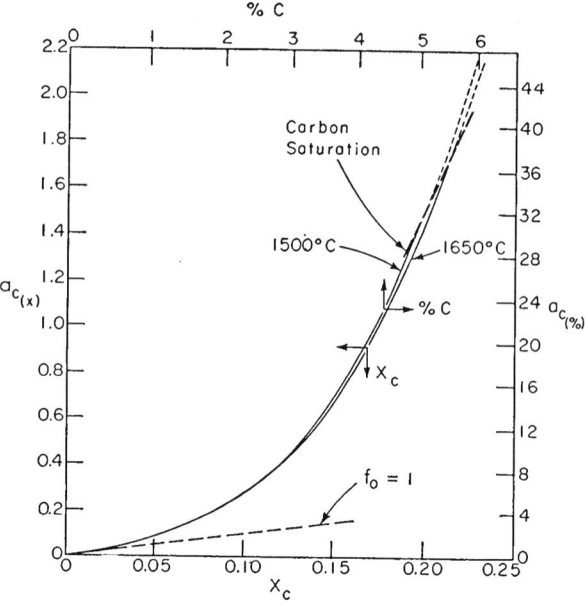

Fig. 1. Activity of carbon in iron-carbon alloys.

From this equation the positions of the full lines in Fig. 2 have been calculated. Equation 7a was evaluated from experiments with an iron melt that was not saturated with respect to liquid iron oxide. The oxygen pressures for liquid iron saturated with liquid iron oxide were obtained from data interpolated [4] from Darken and Gurry's[5] results. The reason for the slight displacement of the saturation value from the line extrapolated from the unsaturated region is not understood at this time. It may be an experimental uncertainty, or it may be the result of oxygen interacting with itself. At any rate, the discrepancy is sufficiently small to give no serious difficulty in later calculations.

Because it is convenient and practical to express

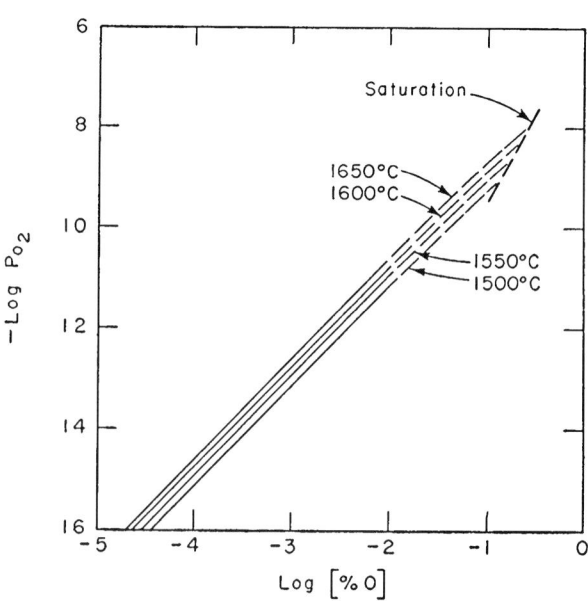

Fig. 2. Solubility of oxygen in liquid iron.

carbon and oxygen concentrations in % by weight, these units will be used in the ensuing treatment. In these units, equation 6 for 1540° C becomes

$$\log f'_C \, (\%) = 0.20[\% \, C] - 0.008[\% \, C]^2 \quad (8)$$

In his recent paper Chipman[1] gives the value of $+0.195\%$ C for equation 8.

By following the same steps shown by Chipman[1] and by using equation 8, the influence of carbon

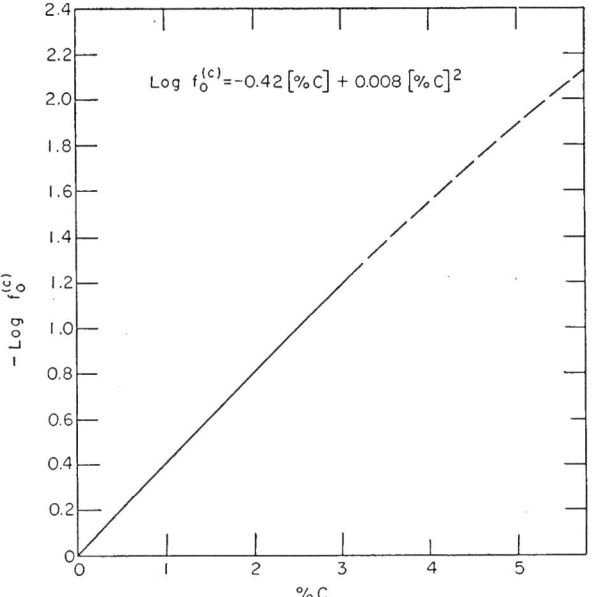

Fig. 3. Influence of carbon on the activity coefficient of oxygen.

on the activity coefficient of oxygen (Fig. 3) is found to be

$$\log f_O^{(C)'} = -0.42[\% \, C] + 0.008[\% \, C]^2 \quad (9)$$

The influence of oxygen on the activity coefficient of carbon as shown in Fig. 4 is

$$\log f_C^{(O)} = -0.315[\% \, O] \quad (10)$$

The individual steps in these calculations are not detailed here, as they are shown in Ref. 1.

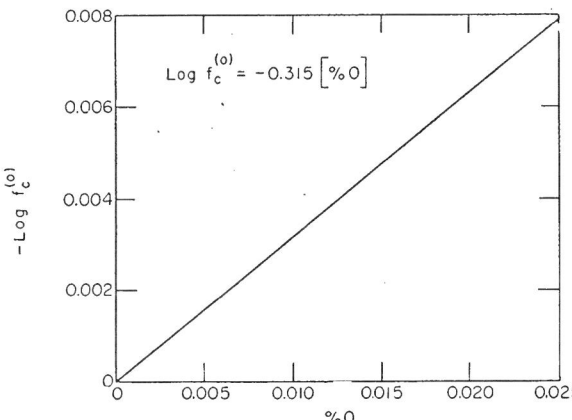

Fig. 4. Influence of oxygen on the activity coefficient of carbon.

Equations 8, 9, and 10 have been computed only for 1540° C, and equation 9 has real validity only out to approximately 2% carbon. These limitations arise because the data of Marshall and Chipman[6] that are used in the calculation extend to 2% carbon at 1540° C. It is probable that the error resulting from using equations 8 and 9 at other temperatures near 1600° C is not large in comparison with the basic uncertainties in the two equations.

The curve in Fig. 3 has been extrapolated to 5.75% carbon in spite of the fact that Wagner's basic derivations of equations[7] for obtaining interaction coefficients are valid only for dilute solutions as all terms but the linear one of the Taylor's series are dropped. With no reliable data to establish the extremity of the line, there is currently no way of estimating the magnitude of the error in the location of the line above approximately 2% carbon. From the plot, it appears that the activity coefficient of oxygen in carbon-saturated iron in the vicinity of 1600° C is approximately 0.01.

EQUILIBRIA

In 1951 Chipman[2] reviewed the then existing data on the carbon-oxygen relations in liquid steel. In the light of recent data it was felt to be of interest to recalculate some of his results and perhaps plot them in various forms.

The Equilibrium Product [% C] [% O]

Equation 3 can be transformed into the following form[2]

$$[C] + [O] \rightleftharpoons CO \quad (11)$$

At low carbon concentrations in the iron melt, the value of the standard free-energy change is:

$$\Delta F° = -8510 - 7.52T \quad (11a)$$

and

$$K_{11} = \frac{p_{CO}}{a_C \cdot a_O} = \frac{p_{CO}}{f_C[\% \, C] \cdot f_O[\% \, O]} \quad (11b)$$

For the ternary system Fe—O—C, equations 8 and 10 can be added to give $\log f_C$, and equation 9 gives $\log f_O$. Therefore, equation 11b can be transformed to

$$\log \frac{[\% \, C][\% \, O]}{p_{CO}}$$
$$= -\log K_{11} + 0.22[\% \, C] + 0.315[\% \, O] \quad (12)$$

This equation can be solved for [% O] in equilibrium with [% C]. The results for 1 atm pressure of the equilibrium gas containing only CO and CO_2 are shown in Fig. 5. The curve extends from oxygen saturation at a very low concentration of carbon to carbon saturation. Calculations indicate that the equilibrium oxygen content of liquid iron-carbon alloys apparently rises from a low of 0.0031% at 2% carbon to approximately 0.0065% at carbon

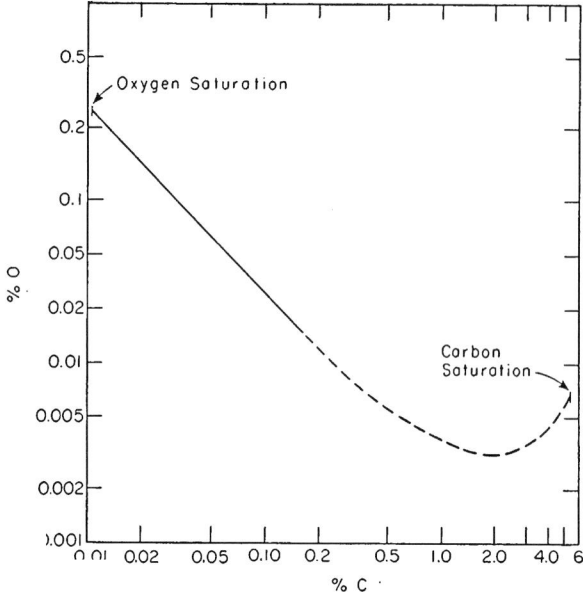

Fig. 5. The carbon-oxygen equilibria at 1600° C [C] + [O] ⇌ CO, 1600° C, $P_{CO} + P_{CO_2} = 1$ atm).

CO—CO$_2$ Ratios

In their analysis of the reaction

$$CO_2 + [C] \ (\%) \rightleftharpoons 2CO \qquad (13)$$

Rist and Chipman[1] concluded that the value of log K for it is best obtained as follows:

$$\log K_{13} = \frac{-7280}{T} + 6.65 \qquad (13a)$$

From this equation and Fig. 1 the CO/CO$_2$ ratios in equilibrium with iron-carbon alloys were determined as shown in Fig. 7. The curves are for total

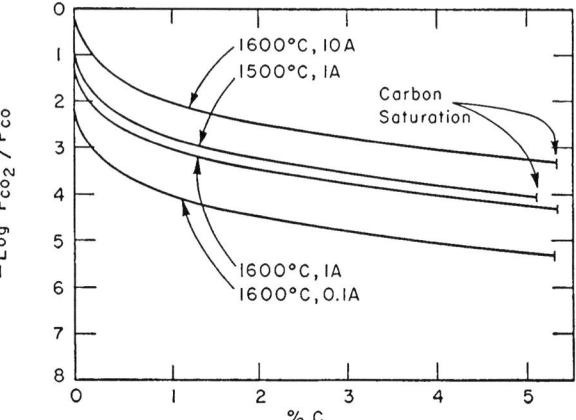

Fig. 7. CO$_2$/CO Ratios in equilibrium with iron-carbon alloys.

saturation. Turkdogan's calculations[8] showed a similar effect. Direct experimental evidence is needed to validate these calculations.

As a matter of interest, curves of the solubility products [% C] [% O] at 1 atm total pressure of CO and CO$_2$ at obtained from equation 12 are plotted in Fig. 6 as a function of temperature and

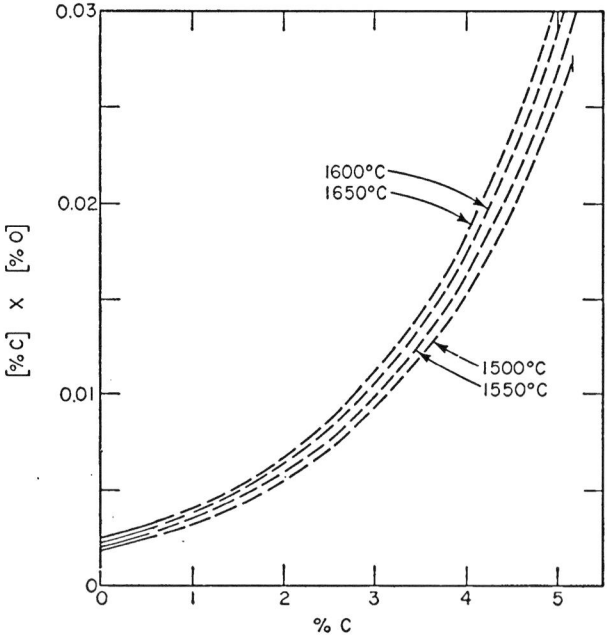

Fig. 6. Estimated carbon-oxygen product at 1 atm pressure of CO.

carbon content. The curves are dotted above 0.75% carbon because of the uncertainties inherent in the calculations.

pressure of $p_{CO} + p_{CO_2}$. Changing the total pressure by a factor of 10 also changes the ratio by 10. Because the pressure of CO$_2$ is small except for the 1600° C–10A curve below 0.15% carbon, the plotted curves represent equally well the total pressures of the equilibrium gas or also the pressure of CO in the system. In other words, the slight difference between a p_{CO} curve and the total pressure curve of the same value is within the precision of the calculations and the plot. The 1600° C, 10$A_{p_{CO}}$ curve is slightly higher than the 1600° C–10A total-pressure curve, being −0.07 log unit at 0.02% carbon instead of −0.25 as shown.

Although the plot is too coarse to show it, the curves in Fig. 7 actually terminate short of zero carbon content in the iron. The termination point is the carbon concentration in equilibrium with the oxygen content of the bath at FeO saturation. At 1600° C and 1 atm total pressure, the curve terminates at approximately 0.001% carbon.

Oxygen Pressures. In Fig. 8 is shown the pressure of oxygen gas in equilibrium with iron-carbon melts at various temperatures and several pressures of CO. The basic equation for the calculation is

$$[C] \ (\%) + \tfrac{1}{2}O_2 \rightleftharpoons CO \qquad (14)$$

for which

$$\log K_{14} = \frac{7320}{T} + 2.22 \qquad (14a)$$

This was obtained from equation 3 and the constant for the transfer of the reference state from

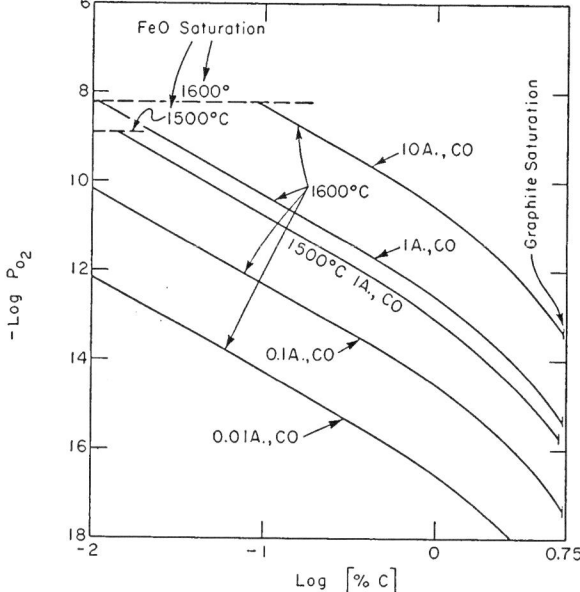

Fig. 8. Pressure of O_2 in equilibrium with carbon in iron.

graphite to the infinitely dilute solution (%) of carbon in liquid iron.

Up to a value of log p_{O_2} of approximately −9, the curves in Fig. 8 can be used equally well either for total pressure (CO plus CO_2) or for the pressure of CO, since p_{CO_2} actually is negligible. Above this oxygen level, the partial pressure of CO_2 in the gas phase becomes appreciable, and, as a consequence, the curves representing total pressure will depart slightly to the left of those shown.

It is interesting to note that the oxygen pressure in equilibrium with carbon in liquid iron at 1600° C changes almost 10,000,000-fold from oxygen saturation to carbon saturation.

By means of the curves in Fig. 8 and equation 7a, the effectiveness of carbon as a deoxidizer under reduced pressure can be illustrated. Reducing the total pressure at 1600° C from 1 atm to 0.1 atm over a bath containing 0.1% carbon (log % C = −1) reduces the equilibrium oxygen pressure from 6.4×10^{-11} atm to 6.4×10^{-13} atm, or one-hundredfold. These figures show the theoretical basis for deoxidation of steel by vacuum and without the use of a metallic deoxidizer. The steel containing 0.1% carbon at 1600° C and at 1 atm pressure would have an equilibrium oxygen content of approximately 0.021% oxygen. If the carbon content were held constant while this steel was thoroughly exposed to a pressure of 0.01 atm and brought to equilibrium with it, the new oxygen level would be 0.00021%, or a one-hundredfold reduction. This is a tenfold lower oxygen level than would actually be realized if, instead, 0.1% aluminum were present as the deoxidizer. It is to be noted that the problem of nucleating a CO bubble

may seriously impair the effectiveness of carbon as a deoxidizer under vacuum.

SUMMARY

The result of this analysis of available data on the Fe—O—C system in the vicinity of 1600° C has been a series of graphical illustrations showing various thermodynamic relationships for the reactions between carbon and oxygen. Consequently, the reader is referred to the figures in the paper as a summary.

References

1. J. Chipman, *J. Iron Steel Inst.*, **180**, 97 (1955).
2. *Basic Open Hearth Steelmaking*, Second Edition, Physical Chemistry of Steelmaking Committee, A.I.M.E., New York, 1951.
3. National Bureau of Standards, *Selected Values of Chemical Thermodynamic Properties*, 1947 and subsequent years.
4. M. Dastur and J. Chipman, *Trans. A.I.M.E.*, **185**, 441 (1949).
5. L. S. Darken and R. W. Gurry, *J. Am. Chem. Soc.*, **68**, 798 (1946).
6. S. Marshall and J. Chipman, *Trans. Am. Soc. Metals*, **30**, 695 (1942).
7. C. Wagner, *Thermodynamics of Alloys*, p. 51, Addison-Wesley Press, Cambridge, Mass., 1952.
8. E. T. Turkdogan, L. S. Davis, and L. E. Leake, *J. Iron Steel Institute*, **181**, 123 (1955).

Discussion

GOKCEN commented on Fig. 5 and observed that the very low oxygen in equilibrium with metal having a carbon content above 1% made the oxygen determination quite difficult, one of the sources of uncertainty being the oxygen on the surface of the sample which was used for analysis. He calculated that a sample of iron one cubic centimeter on an edge could contain approximately ten parts per million of oxygen (0.001%), as a result of the surface oxide. As the carbon and silicon contents of the iron increased this surface, absorbed oxygen would increase by a factor of 4 or 5.

RICHARDSON expressed an interest in what had been developed by connecting all these data together, and he inquired about the primary source of the data for drawing the curve in Fig. 5. In the ensuing discussion, ELLIOTT pointed out that this curve was essentially an extrapolation of Marshall and Chipman's data and that the curve could be used only to indicate the relative effect rather than to show exact numbers. There could be no specific answer to Richardson's question as to what the uncertainty would be in the location of the curve in Fig. 5. RICHARDSON discussed further the extrapolation of the curve for the activity of carbon to high carbon concentrations on a linear basis, as was done by Rist and Chipman. He reviewed the extrapolation that he had used several years ago which involved starting from austenite saturation and calculating the activities at a constant composition as a function of temperature, assuming a constant heat of solution of graphite of about 5 kcal per gram atom of carbon. Rist and Chipman's extrapolation gives similar results.

by S. Matoba
and S. Ban-ya

Equilibrium of Carbon and Oxygen in Molten Iron Saturated with Carbon (II)—The Effect of Silicon in the Range of 0 to 5.20%

INTRODUCTION

The effects of silicon on the equilibrium between carbon and oxygen in molten iron, saturated with carbon and under conditions such that SiO_2 or SiC were not formed, were investigated.

In this case we may express the equilibria between carbon and oxygen by the following three equations for each silicon concentration in molten iron:

$$C\ (s) + CO_2\ (g) = 2CO\ (g) \qquad (1)$$

$$[O] + CO\ (g) = CO_2\ (g) \qquad (2)$$

$$C\ (s) = [C] \qquad (3)$$

EXPERIMENTS AND RESULTS

Details of the method of experiments are almost the same as those described previously.[1] Here we repeat briefly as follows: iron samples of various silicon contents were melted in a carbon crucible under the controlled atmosphere of CO and CO_2 at several temperatures until the equilibrium was attained in each case and then were quenched for analysis. A summary of the experimental results is tabulated in Table 1 and is also shown graphi-

S. Matoba and S. Ban-ya are members of the Faculty of Engineering, Tohoku University, Sendai, Japan.

Table 1. Effect of Silicon Content on the Activity Coefficient of Carbon in Iron

°C	1600		1500		1400		1300	
Si %	[C] %	[O] %	[C] %	[O] %	[C] %	[O] %	[C] %	[O] %
(0.01)*	(5.53)	(0.0032_4)	(5.27)	(0.0029_1)	(5.00)	(0.0025_8)	(4.74)	(0.0022_4)
1.42	4.99	0.0019_0	4.84	0.0018_6	4.54	0.0019_3	—	—
3.27	4.44	0.0014_0	4.21	0.0013_0	3.99	0.0013_7	—	—
4.14	4.19	0.0009_3	3.99	0.0010_6	3.70	0.0011_0	—	—
5.20	3.92	0.0008_4	3.63	0.0008_6	3.39	0.0009_8	—	

* Figures in parentheses were obtained in previous experiments.[1]

cally in Figs. 1 and 2. These values are all mean values of the results of at least six melts.

DISCUSSION

Effect of Silicon Content on the Solubility of Carbon in Molten Iron

Figure 1 is the plot of [C]% versus [Si]% at 1500° C obtained by the present authors. The results obtained are somewhat higher than the data of Chipman and others,[2] but the two curves are nearly parallel.

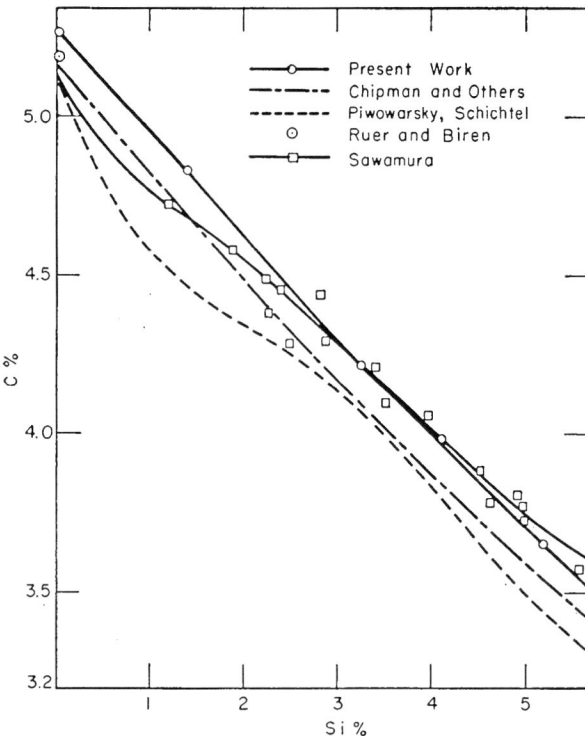

Fig. 1. Effect of silicon content on the solubility of carbon in molten iron at 1500° C.

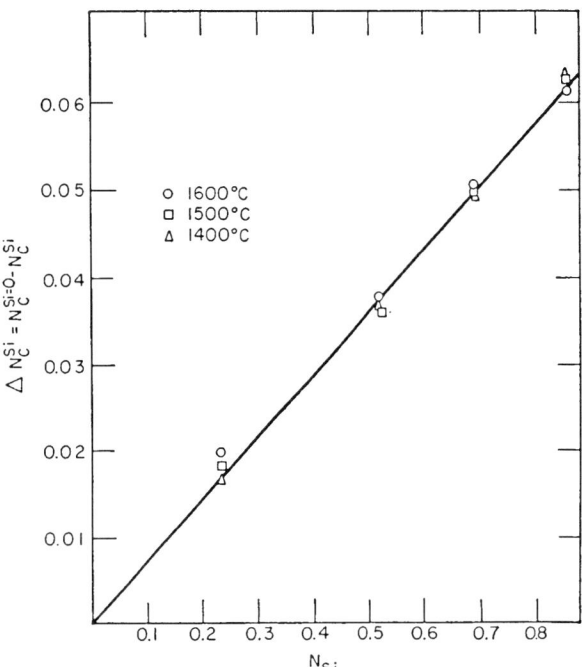

Fig. 3. The change in carbon solubility by the addition of silicon to carbon-saturated iron.

Figure 2 shows the relation between N_C versus N_{Si} for various temperatures. The change of carbon solubilities by the addition of silicon ($\Delta N_C^{(Si)} = N_C^{(Si=0)} - N_C^{(Si)}$) versus N_{Si} is shown in Fig. 3. As was already shown by Turkdogan and others,[3] these relations are not affected by temperatures and are expressed by the empirical formula $\Delta N_C^{(Si)} = -0.72 N_{Si}$.

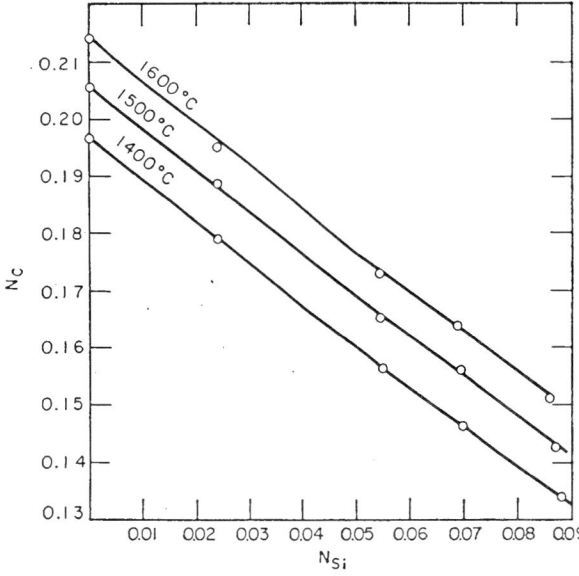

Fig. 2. Relationship between the molar concentrations of silicon and carbon in carbon-saturated iron.

Effect of Silicon Content on the Activity Coefficient of Carbon in Molten Iron

Since the effect of Si on the carbon-saturation value is decreasing, it is expected that the activity coefficient of carbon will be increased by silicon. However, as long as the condition of saturation by carbon is maintained, the activity of carbon in the melt based on pure carbon is always unity. Accordingly we can express approximately the change of activity coefficient of carbon in molten iron by addition of silicon as follows:

$$N_C \cdot \gamma_C = 1 \qquad or \qquad \gamma_C = 1/N_C \qquad (4)$$

where γ_C is the activity coefficient of carbon in carbon-saturated Fe—C—Si alloy, which may be considered as

$$\gamma_C = \gamma_C' \cdot \gamma_C^{(Si)} \qquad or \qquad \gamma_C^{(Si)} = \gamma_C/\gamma_C' \qquad (5)$$

γ_C' is the activity coefficient of carbon in binary Fe—C alloy and $\gamma_C^{(Si)}$ is the change of activity coefficient of carbon by addition of silicon.

When the effect of temperature on the activity coefficient of carbon is neglected, γ_C can be calculated from the present experimental data by equation 4, and γ_C' is reported by several authors.[4] Now, by referring to the data of Ohtani,[4] computing the value of $\gamma_C^{(Si)}$ by equation 5, and then plotting the relations between N_{Si} and log $\gamma_C^{(Si)}$ for 1550° C as shown in Fig. 4, a linear relationship was obtained within the range of our experiments ($N_{Si} < 0.09$). It is noted that the addition of 1 atomic % of silicon yields the value of log $\gamma_C^{(Si)}$ which increases by about 0.045. This is in fairly good agreement with the results reported by Chipman[5] and by Ohtani.[6]

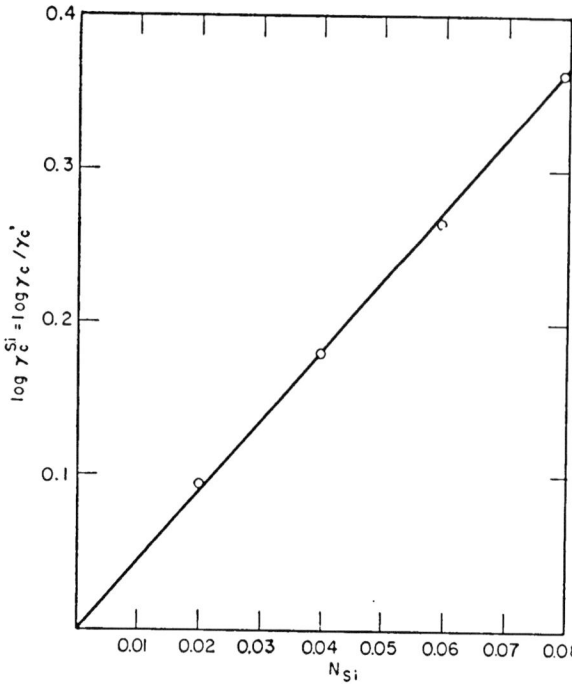

Fig. 4. The influence of the molar concentration of silicon on the activity coefficient of carbon in liquid iron.

Oxygen Content of Silicon-Bearing Molten Iron Saturated with Carbon

The equilibrium relations between carbon and oxygen in carbon-saturated molten iron are shown graphically in Fig. 5 for various silicon concentrations and temperatures. It must be pointed out that the oxygen content is so small, especially in the high silicon range, compared with the limit of reproducibility of our oxygen analysis (which is ±0.0005%), that these results are not to be presented as our final results. However, so far as the data obtained here are concerned, we can assume as follows:

(a) At constant temperature, the equilibrium value of carbon and oxygen both gradually decrease with increasing silicon.

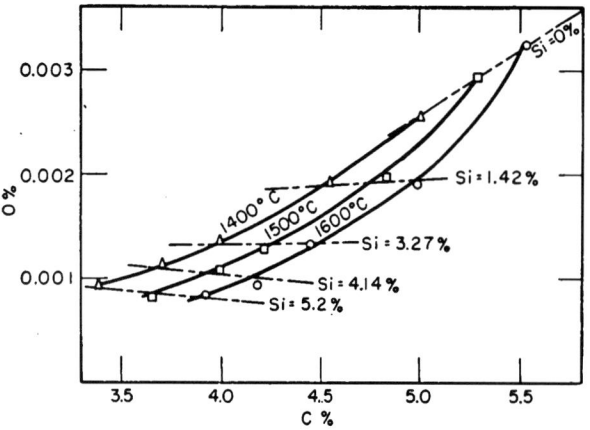

Fig. 5. Oxygen content of silicon bearing carbon-saturated iron.

(b) For equal silicon contents in the lower silicon range, both carbon and oxygen are increased by raising the temperature, but in the higher silicon range, carbon is increased by increasing the temperature while the oxygen contents become smaller.

Activity Coefficient of Oxygen

In the previous experiments,[1] we determined the activity coefficient of oxygen in molten Fe—C—O alloy saturated with carbon by defining

$$a_O = f_O^{(C)} \cdot [\%O] \qquad (6)$$

In the presence of silicon, equation 6 becomes

$$a_O = f_O \cdot [\%O] \qquad (7)$$

in which f_O is the activity coefficient of oxygen including the effect of silicon, and it may be written as follows:

$$f_O = f_O^{(C)} \cdot B, \qquad (8)$$

where B is the factor expressing the change-of-activity coefficient of oxygen in this complicated system.

So long as the condition of carbon saturation is maintained, the equilibrium gaseous composition is always equal for a definite temperature and pressure, and is given by Boudouard's equilibrium. Accordingly, the activity of oxygen in molten iron saturated with carbon at a definite temperature and pressure is also unchanged regardless of the silicon addition. The values for a_O in equation 6 or equation 7 were obtained in the previous experiments[1] in the range of temperature 1300–1600° C and for the carbon contents of 4.73-5.53%.

Assuming the value of $f_O^{(C)}$ is not substantially affected by the temperature but only by the carbon content, we can evaluate the value of B by substituting the values of a_O and $f_O^{(C)}$ previously obtained into the last two equations. Thus we obtain the value for B as 0.9-1.1 for the range of 4.73-5.53% [C]. Now we may deduce that the activity coefficient of oxygen in a high-carbon melt would not be much influenced by the addition of silicon.

From the above-mentioned results we can calculate the value of f_O or $f_O^{(C)}$ for various carbon contents, neglecting the effect of silicon. The relations between [C]% and log $f_O^{(C)}$ thus obtained are shown in Fig. 6. Although there is some scattering, most of the points lie on the straight line through the point $f_O^{(C)} = 1$ with zero carbon and $f_O^{(C)} = 0.017$ with carbon saturation at 1600° C in Fe—C—O alloy, which was determined previously. The slope of the straight line is approximately expressed by the following equation:

$$\partial \log f_O^{(C)} / \partial [C]\% = e_O^{(C)} = -0.32 \qquad (9)$$

This line lies between the extentions of two values (C-I and C-II in Fig. 6) determined for a lower carbon range.

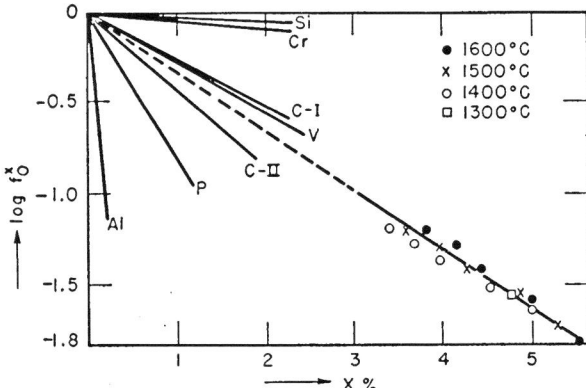

Fig. 6. The influence of carbon and other elements on the activity coefficient of oxygen in liquid iron. (Experimental points for high-carbon Fe—C—O alloys.)

For the sake of comparison, there are also shown in Fig. 6 the effects of various elements[7] on the activity coefficient of oxygen in molten iron.

References

1. S. Matoba and S. Ban-ya, *Tech. Repts. Tohoku University*, **XX**, No. 1, 131 (1955).
2. J. Chipman, R. M. Alfred, L. W. Gott, R. B. Small, D. M. Wilson, C. N. Thomson, D. L. Guernsey, and J. C. Fulton, *Trans. Am. Soc. Metals*, 44, 1215 (1952).
3. E. T. Turkdogan, L. S. Davis, L. E. Leake, and C. G. Stevens, *J. Iron Steel Inst.*, 181, Part I, 123 (1955).
4. S. Marshall and J. Chipman, *Trans. Am. Soc. Metals*, 30, 695 (1942); F. D. Richardson and W. E. Dennis, *Trans. Faraday Soc.*, 49, 171 (1953); M. Ohtani and K. Sanbongi, *Tetsu-to-Hagane*, 39, 483 (1953).
5. J. Chipman, *Discussions Faraday Soc.*, No. 4, 23 (1948).
6. M. Ohtani, *Sci. Repts. Research Insts., Tohoku Univ., Ser. A* 7, No. 5 (1955).
7. J. Chipman, *J. Iron Steel Inst.*, 180, Part 2, 97 (1955).

Discussion

DARKEN asked for a comparison between Fig. 5 of Matoba's paper and Fig. 5 of the review paper presented by Elliott. MATOBA felt that there was general agreement between the two figures. However, ELLIOTT indi-

cated that his figure was based on a calculation and extrapolation and also it was higher than Matoba's experimental results by a factor of approximately 2. Being experimental, Matoba's results were the more significant. DARKEN suggested that atomistically the effect of silicon on oxygen should be strong in view of the fact that silicon is a strong deoxidizer. MATOBA showed in Fig. 6 that the effect of carbon was more significant than the effect of silicon on the activity coefficient of oxygen. However, silicon had a pronounced secondary effect, in view of the fact that it reduced the solubility of carbon in iron.

FUWA* presented some preliminary information which resulted from a study of the carbon-oxygen equilibrium in liquid iron being made at the Massachusetts Institute of Technology. The experimental procedure consisted of equilibrating an atmosphere of CO/CO_2 with carbon-saturated iron in graphite crucibles and in alumina crucibles. The results in both graphite and alumina crucibles showed that 0.0018% oxygen was in equilibrium with a 2.0% carbon solution. In the alumina crucible, a 3% carbon solution had an equilibrium oxygen content of between 0.001 and 0.0022% oxygen. As a result of these experiments, FUWA concluded that the oxygen content of carbon-saturated iron is approximately 0.0017–0.0018%. The time to reach equilibrium was approximately eight hours. Even though the new figures were lower than those given by Matoba, DARKEN's conclusion was that in view of the experimental difficulties, the agreement really was not too bad. FUWA estimated the accuracy in the oxygen determination was approximately 0.0005%. These measurements were made in the range of 1600° C.

PEARSON observed that Matoba's actual measurements agreed with the prediction by Turkdogan to the effect that silicon tends to decrease the solubility of oxygen in carbon-saturated iron and that this effect is temperature sensitive. Both PEARSON and CHIPMAN felt that the location of the phosphorus line in Fig. 6 of the authors' paper was drawn more steeply than they would have expected from other data. DARKEN concluded by observing that Matoba's data tended to confirm results of Marshall's and Chipman's work in that carbon exerts a very strong influence on the activity coefficient of oxygen.

* Graduate Student, Metallurgy Department, M.I.T.

by J. Chipman
and F. C. Langenberg

The Alumina-Graphite Reaction in Liquid Iron and the Aluminum Deoxidation Constant in Steelmaking

INTRODUCTION

There is an increasing knowledge of the interactions between two or more dissolved elements in liquid iron. Besides the usefulness of these data in studying alloy behavior in steel baths, they also make it possible to bring together the data on simple and complex ferrous systems into a coherent body. Knowledge of the interaction between certain elements and dissolved carbon makes possible a comparison of experimental results in carbon-free systems with those obtained under conditions of carbon saturation. Thus data drawn from researches which are primarily aimed at blast-furnace problems may be utilized to supplement experiments on reactions of steelmaking.

One such reaction is the solution of aluminum and oxygen in iron when the iron is in contact with pure solid Al_2O_3. The determination of the de-

oxidation constant, $K = [a_{Al}]^2[a_O]^3$, has been the object of several investigations whose results are in disagreement by several orders of magnitude. Discrepancies have been due in part to experimental difficulties inherent in determining the concentrations of very small amounts of aluminum and oxygen and in assuring the continued presence of pure Al_2O_3 in contact with the iron. In the present investigation the aluminum content of carbon-saturated iron in equilibrium with pure Al_2O_3 was determined; from these experimental results and related data, indirect values of the deoxidation constant at 1600 and at 1700° C have been obtained.

PROCEDURE

The furnace used for the slag-metal equilibrium studies has been fully described by Hatch and Chipman[1] and has been used successfully in subsequent investigations.[2,3] In brief, the experimental method was as follows. Three hundred grams of alumina-saturated slag and 150 grams of Fe—C—Si alloy were stirred together in a graphite crucible under an atmosphere of carbon monoxide. A high-frequency induction furnace was the source of power. Temperatures were read with an optical pyrometer sighted down the stirring tube into a 3/8-in. diam hole drilled in the stirrer. The optical

Dr. Chipman is Professor of Metallurgy, Massachusetts Institute of Technology, and Dr. Langenberg is Supervisor of Pyrometallurgy Research, Crucible Steel Company of America, Pittsburgh, Pennsylvania.

The authors express their thanks to Mr. Donald L. Guernsey and his associates for performing the chemical analyses and to Mr. Herbert Klein for experimental assistance. The investigation received financial support from the Crucible Steel Company of America and the American Iron and Steel Institute.

pyrometer was frequently checked against a National Bureau of Standards certified instrument. Variations in the control of the induction furnace caused a maximum deviation of $\pm 10°$ C during a heat, but the temperature control was usually within $\pm 5°$ C.

When the charge was molten and the desired temperature of 1600 or 1700° C obtained, a 12-in. long, 3/8-in. diam, pure alumina tube was inserted into the melt to maintain alumina saturation. It was important that the tube fitted tightly in the cover hole or else it would slide toward the center of the crucible and interfere with the stirrer. One hour after reaching 1600° C, or one-half hour after reaching 1700° C, the first samples of slag and metal were taken. The slag sample was taken by inserting a cold iron rod on which the slag froze; the metal sample was drawn into a Vycor tube with an aspirator bulb. Approximately eight hours after reaching temperature, another pair of samples was taken. The final slag and metal analyses are presented in Table 1. The three oxide constituents of the slag

Table 1. Analyses of Al₂O₃-Saturated Slags and Graphite-Saturated Fe—C—Si—Al Melts in Equilibrium with the Slags

Temper-ature	Heat No.	Slag Analyses %			Metal Analyses %		
		Al₂O₃	SiO₂	CaO	Si	Al	C*
1600° C	K-52	50.80	24.76	23.90	15.74	0.060	1.17
	K-53	43.50	34.67	20.87	22.38	0.049	0.47
	K-61	46.26	32.58	21.80	20.37	0.042	0.58
	K-62	51.04	24.42	25.42	19.33	0.069	0.95
	L-14	49.23	26.19	24.43	17.06	0.103	0.95
	L-32	45.40	32.44	22.41	21.60	0.079	0.50
	L-35	53.72	19.43	27.24	13.91	0.075	1.56
	L-38	47.71	27.58	24.32	20.46	0.059	0.58
1700° C	L-18	66.83	9.27	23.59	12.91	0.290	1.97
	L-19	67.00	8.96	25.02	13.84	0.296	1.75
	L-20	65.26	14.06	21.55	21.44	0.202	0.58
	L-21	64.50	11.60	23.69	17.19	0.232	1.08
	L-23	62.10	16.14	21.44	23.07	0.164	0.47
	L-25	70.90	2.24	26.79	5.38	0.370	3.97
	L-26	61.20	16.27	22.53	23.32	0.164	0.45
	L-31	57.80	22.62	19.69	26.46	0.149	0.30
	L-36	65.20	10.56	24.65	20.04	0.183	0.72
	L-37	67.37	7.40	25.96	17.19	0.221	1.08
	L-40	68.94	2.24	28.24	7.62	0.355	3.34

* Calculated from the solubility of [C] in Fe—Si melts.[4]

were determined, and the closeness of the total to 100% indicates the absence of lower oxides. The metal was analyzed for silicon and aluminum, and the carbon content was calculated from the data of Chipman and co-workers.[4]

The initial slag compositions were chosen to form slags that were not quite saturated with Al₂O₃ at 1600 or 1700° C. The slag takes Al₂O₃ into solution from the alumina tube and becomes saturated at the liquidus. Concurrently, the transfer of silica between slag and metal changes the slag composition, but the alumina tube causes the slag to follow the liquidus as it gains or loses SiO₂. The method was successful as described by the authors[5] in another paper where the data in Table 1 plus additional data were used to determine the 1600 and 1700° C liquidus lines in the CaO·2Al₂O₃ and Al₂O₃ stability fields of the CaO—Al₂O₃—SiO₂ system. It is also possible to approach the liquidus in the other direction by starting with a slag containing an excess of the primary phase. This procedure, used in heats K-52, K-53, K-61, and K-62, was satisfactory for the present purpose but led to inaccuracies in determination of the liquidus line, and therefore, the approach from the unsaturated side by the tube technique was preferred.

The final slag and metal samples were essentially in equilibrium with respect to the distribution of both silicon and aluminum, as evidenced by the approach to silicon equilibrium from both sides. The results are of interest in determining the activity of silica in the slags. In this respect, Table 1 represents only a part of a large body of data, the discussion of which must be reserved for another occasion.

DISCUSSION

Since all the slags were saturated with Al₂O₃, the reaction and its equilibrium constant are

$$Al_2O_3 \text{ (s)} = 2[Al] + 3[O] \qquad (1)$$

$$K_1 = [a_{Al}]^2 \cdot [a_O]^3$$

The activity of oxygen is fixed by the presence of carbon monoxide and graphite; that of aluminum is obtained from its concentration and its activity coefficient, recently determined by Chipman and Floridis.[6]

In *Basic Open Hearth Steelmaking*[7] the free energy of the following equation was derived from National Bureau of Standards selected values:

$$C \text{ (graphite)} + \tfrac{1}{2}O_2 \text{ (g)} = CO \text{ (g)} \qquad (2)$$

$$\Delta F° = -28,100 - 20.20T$$

Dastur and Chipman[8] have evaluated the free energy of solution of oxygen in liquid iron:

$$\tfrac{1}{2}O_2 \text{ (g)} = [O] \qquad (3)$$

$$\Delta F° = -27,930 - 0.57T$$

By combining equations 2 and 3, the following equation is obtained:

$$C \text{ (graphite)} + [O] = CO \text{ (g)} \qquad (4)$$

$$\Delta F° = -170 - 19.63T \text{ cal}$$

The activities of carbon and carbon monoxide are unity since the iron is saturated with carbon and under 1 atm of carbon monoxide; this leads to an equation for the activity of oxygen:

$$\log a_O = \frac{-37.2}{T} - 4.29$$

from which $a_O = 4.9 \times 10^{-5}$ at either 1600° or 1700° C. It is important to note that the activity of oxygen is referred to the standard state employed by Dastur and Chipman, defined in such a way that $a_O/\%[O]$ approaches unity in the very dilute solution of oxygen in pure liquid iron.

The activity coefficient of aluminum in liquid iron and in Fe—C—Al alloys was determined by means of its distribution between the alloy and liquid silver. In the more complex Fe—C—Si—Al alloys, Si as well as Al was found in both phases, and for this reason the activity data on the quaternary system are of less accuracy than those on the binary and ternary solutions. Chipman and Floridis[6] ignored the effect of silicon on the activity of aluminum in the silver layer and deduced for the iron layer the activity coefficients shown in the lower curve of Fig. 1. It now seems possible to

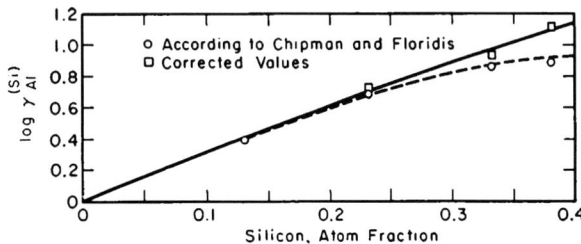

Fig.   Effect of silicon on activity coefficient of aluminum.

make a correction for interaction in the silver layer.

It was found that in the iron layer the activity coefficient of aluminum was about equally affected by increases in concentration of aluminum or of silicon. In dilute solutions, $\epsilon_{Al}^{(Al)} = \partial \ln \gamma_{Al}/\partial N_{Al} = +6.0$ while $\epsilon_{Al}^{(Si)} = \partial \ln \gamma_{Al}/\partial N_{Si} = +6.9$. In the silver layer the corresponding value of $\epsilon_{Al}^{(Al)}$ was found to be +6.7, and it may be concluded that $\epsilon_{Al}^{(Si)}$ is the same or slightly larger. We adopt a round value of $\epsilon_{Al}^{(Si)} = 7.0$ for the silver layer or, in ordinary logarithms, $\partial \log \gamma_{Al}/\partial N_{Si} = 3.0$. The corrections to be applied to the iron layer are set forth in Table 2, and final values for the effect of silicon on the activity coefficient of aluminum in liquid iron at 1600° C are shown in the upper line of Fig. 1.

The activity coefficient of aluminum in carbon-saturated Fe—C—Si—Al solutions is obtained in the following way. The solubility of graphite is taken from the plots of Chipman, Fulton, Gokcen, and Caskey,[9] the concentration of aluminum being too small to have an appreciable effect. The activity coefficient γ_{Al} in the saturated quaternary is obtained by adding the several contributions,

$$\log \gamma_{Al} = \log \gamma_{Al}' + \log \gamma_{Al}^{(Si)} + \log \gamma_{Al}^{(C)} \quad (5)$$

where γ_{Al}' is the activity coefficient of aluminum in binary Fe—Al solutions of the same aluminum concentration, and $\gamma_{Al}^{(Si)}$ and $\gamma_{Al}^{(C)}$ represent the effect of silicon and carbon. The first and last terms on the right are read from the plots of Chipman and Floridis, while the remaining term is taken from Fig. 1. This procedure makes pure liquid aluminum the standard state. It is more convenient to base the standard state on the infinitely dilute solution of aluminum in pure iron. The term f_{Al} will be used for the activity coefficient so defined that $f_{Al} = a_{Al}/N_{Al}$ approaches unity in the infinitely dilute solution. The corresponding f_{Al}' is approximately unity in the alloys used, and hence equation 5 becomes

$$\log f_{Al} = \log \gamma_{Al}^{(Si)} + \log \gamma_{Al}^{(C)} \quad (6)$$

The resulting values of $\log f_{Al}$ for the carbon-saturated solution are shown in Fig. 2. For the

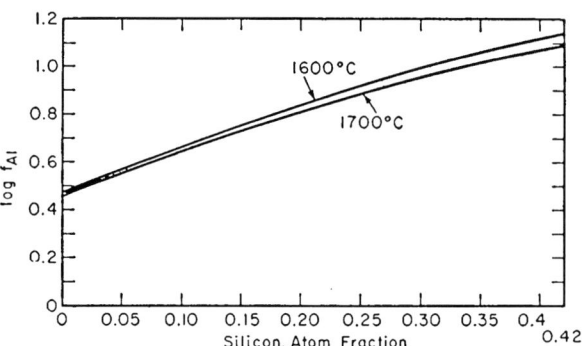

Fig. 2.   Effect of silicon on the activity coefficient of aluminum in graphite-saturated Fe-C-Si-Al solutions.

1700° C line the slight increase in graphite solubility was taken into consideration, and the assumption was made that $\log f_{Al}$ is inversely proportional to T.

Table 2.   Correction of the Data of Chipman and Floridis[6] for the Effect of Silicon on the Activity Coefficient of Aluminum in Liquid Iron

Heat No.	Ag Layer Si, Atom %	Correction $\Delta \log \gamma_{Al}$	Fe Layer Si, Atom %	$\log \gamma_{Al}^{(Si)}$ Reported	$\log \gamma_{Al}^{(Si)}$ Corrected
35	0.14	0.00	13.19	0.39	0.39
36	0.85	0.03	23.08	0.69	0.72
37	2.81	0.08	33.25	0.86	0.94
38	7.82	0.23	38.16	0.89	1.12

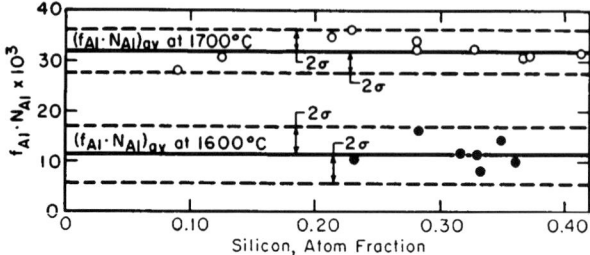

Fig. 3.   Activity of aluminum versus silicon in a graphite-saturated Fe-Si-C alloy in equilibrium with Al₂O₃(s) and under 1 atm of CO.

The activity of aluminum in the alloys, defined as $a_{Al} = f_{Al} \cdot N_{Al}$, is shown in Fig. 3. The 95.5% confidence limits of 2 sigma are shown. The activity of aluminum should be constant, and although the scatter is inherently large, the lack of any trend with silicon content is evidence that the use of the complex alloys has not introduced a systematic error. Average values of a_{Al} are: at 1600° C, 11.4×10^{-3}; and at 1700° C, 31.8×10^{-3}.

THE DEOXIDATION CONSTANT

The equilibrium constant of equation 1 is usually expressed in terms of weight % of the dissolved elements in nearly pure liquid iron. Under such conditions the relation between mole fraction and percentage of aluminum is $N_{Al} = 0.5585 [\% \text{ Al}]/27 = 20.7 \times 10^{-3} [\% \text{ Al}]$. Substitution in the above average values for $a_{Al} = f_{Al} \cdot N_{Al}$ at infinite dilution leads to values of activity defined as $a_{Al} = f_{Al} [\% \text{ Al}]$ at 1600° of 0.55 and at 1700° C, 1.54. The activity of oxygen has been shown to be 4.9×10^{-5} at either temperature. From these results the equilibrium constant is

$$K_1 = [a_{Al}]^2 [a_O]^3 = 3.6 \times 10^{-14} \text{ at } 1600° \text{ C}$$
$$= 2.8 \times 10^{-13} \text{ at } 1700° \text{ C}$$

A comparison of deoxidation constants from various sources, shown as a plot of $\log K_1$ versus $1/T$, is plotted in Fig. 4. Included are the deoxidation products $[\% \text{ Al}]^2 [\% \text{ O}]^3$ determined by Wentrup and Hieber,[10] Hilty and Crafts,[11] Gokcen and Chip-

man.[12] and Geller and Dicke.[13] The calculated deoxidation constant was obtained by the method shown in *Basic Open Hearth Steelmaking*[7]* but slightly modified by the data of Chipman and Floridis. They found the activity coefficient of aluminum in the infinitely dilute solution $\gamma_{Al}°$ to be 0.031 at 1600° C instead of the value of 0.043 used in *Basic Open Hearth Steelmaking*. The calculated equation for the effect of temperature on the deoxidation constant is

$$\log K_1 = \frac{-63,500}{T} + 20.48 \qquad (7)$$

The deoxidation constant obtained in this investigation and the calculated constant agree very well at 1600° C. Although the present data are more precise at 1700° C, the agreement between the calculated and experimental deoxidation constants is not as good. This may be due in part to the extrapolation of the aluminum activity coefficient data determined by Chipman and Floridis[6] at 1600 to 1700° C.

SUMMARY

Al_2O_3-saturated Al_2O_3—SiO_2—CaO slags were equilibrated with graphite-saturated Fe—C—Si—Al melts at 1600° and 1700° C, and the reaction Al_2O_3 (in slag) = 2[Al] + 3[O] studied. Since the slags were saturated with Al_2O_3, $a_{Al_2O_3} = 1$, and the equilibrium constant for the above reaction reduces to $K = [a_{Al}]^2 [a_O]^3$. The activity of oxygen was evaluated by combining equations to obtain $\Delta F°$ for the reaction C (graphite) + [O] (in Fe) = CO (g). The metal was saturated with carbon and under 1 atm of CO, making a_C and a_{CO} unity. Thus, a_O was evaluated without analyzing for oxygen. The activity of Al was determined from the analyses for aluminum and the known effects of carbon and silicon on γ_{Al}. The aluminum concentrations were higher than when pure iron is equilibrated with pure Al_2O_3, and this reduced the uncertainty caused by analyzing for very small quantities of aluminum. Utilizing a_O and a_{Al}, the aluminum deoxidation constant, $[a_{Al}]^2 [a_O]^3$, was found to be 3.6×10^{-14} at 1600° C and 2.8×10^{-13} at 1700° C.

References

1. G. G. Hatch and J. Chipman, *Trans. A.I.M.E.*, 185, 275 (1949).
2. J. C. Fulton, N. J. Grant, and J. Chipman, *Trans. A.I.M.E.*, 197, 185 (1953).
3. J. C. Fulton and J. Chipman, *Trans. A.I.M.E.*, 200, 1136 (1954).
4. J. Chipman, R. M. Alfred, L. W. Gott, R. B. Small, D. M. Wilson, C. N. Thomson, D. L. Guernsey, and J. C. Fulton, *Trans. Am. Soc. Metals*, 44, 1215 (1952).
5. F. C. Langenberg and J. Chipman, "Determination of the 1600° and 1700° C Liquidus Lines in the CaO·2 Al_2O_3 and Al_2O_3 Stability Fields of the CaO—Al_2O_3—SiO_2 System." *J. Am. Ceram. Soc.*, 39, 432 (1956).
6. J. Chipman and T. P. Floridis, *Acta Metallurgica*, 3, No. 5, 456 (1955).

* Page 672.

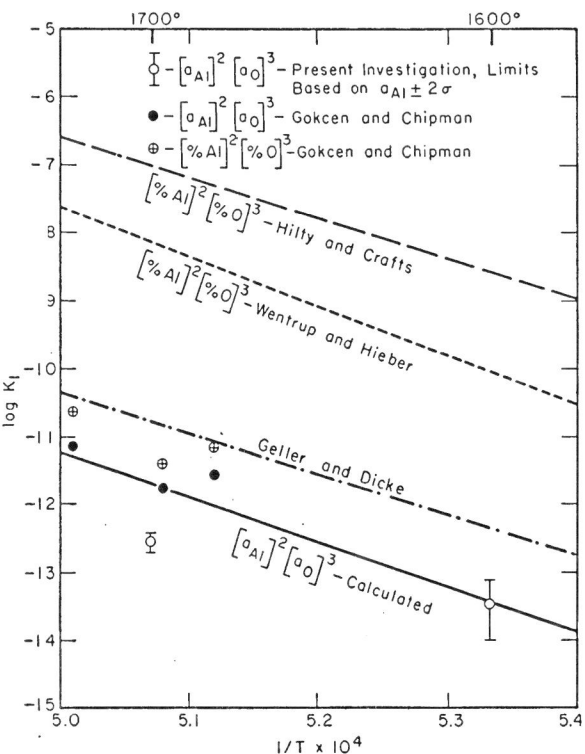

Fig. 4. Comparison of data on aluminum deoxidation constants.

7. *Basic Open Hearth Steelmaking*, Second Edition, p. 571, Physical Chemistry of Steelmaking Committee, A.I.M.E., New York, 1951.

8. M. N. Dastur and J. Chipman, *Trans. A.I.M.E.*, 185, 441 (1949).

9. J. Chipman, J. C. Fulton, N. A. Gokcen, and G. R. Caskey, *Acta Metallurgica*, 2, No. 3, 439 (1954).

10. H. Wentrup and G. Hieber, *Tech. Mitt. Krupp, A. Forschungsber.*, 1, 47 (1939).

11. D. C. Hilty and W. Crafts, *Trans. A.I.M.E.*, 188, 414 (1950).

12. N. A. Gokcen and J. Chipman, *Trans. A.I.M.E.*, 197, 173 (1953).

13. W. Geller, and K. Dicke, *Archiv Eisenhüttenw.*, 16, 431 (1943).

Discussion

PEARSON and HILTY expressed gratification at this paper in that it resolved many of the doubts concerning the activity and activity coefficient of oxygen in iron as it influences the alumina deoxidation constant. HILTY pointed out that the character of the deoxidation product accounts for the discrepancy between actually observed deoxidation by aluminum in steelmaking practice and the line by Hilty and Crafts with that of the calculated line in Fig. 4. In the former cases, the oxide apparently was partially combined with some oxide of iron, whereas in the calculated case the deoxidation product is pure alumina. In answer to DARKEN's question concerning the position of the calculated line in Fig. 4, CHIPMAN stated that he felt that the calculated line was more precise than the points shown on the figure because these points were obtained by several extrapolations, with a resulting inherent lower precision. MC CABE inquired whether the authors were now in a position to attack an evaluation of the activity of alumina in blast-furnace-type slags. To this CHIPMAN replied that a paper was in preparation in which the activity of alumina in the liquid ternary field was derived from the silica-activity data.

by N. A. Gokcen

Discussion of

The Determination of Oxygen and Nitrogen in Metals by Fusion in Vacuum

Dr. Gokcen* presented a somewhat detailed account of the experimental procedures used in the determination of the oxygen and nitrogen contents of steels and iron alloys. His paper is published in the Transactions A.I.M.E. Feb., 1958. *Certain points in the discussion that followed his presentation are included here. Editor.*

An extended discussion of the experimental details of oxygen analysis by a vacuum-fusion technique took place. SHANAHAN asked why chemical methods for measuring CO_2 were used when techniques are available for freezing it out of the system and then determining its volume by gasifying it. The author agreed that liquid nitrogen was a satisfactory material for freezing out CO_2, but he did not have it available in his laboratory.

GOKCEN pointed out that the sources of blank were the analytical train and the fusion system. He estimated his system to have a blank of 0.03 milliliter per hour, which was comparable to that reported by other investigators. However, his blank was measured after the first one or two samples had been dropped, as it is at this time that the highest blank is usually encountered. He felt that others who reported lower blanks determined them at another time. In response to a question, he also stated that the analytical time for his system was between twenty and twenty-five minutes per sample.

* Dr. Gokcen is Associate Professor of Metallurgy, University of Pennsylvania, Philadelphia, Pennsylvania.

CHIPMAN and HILTY reported that in the laboratories with which they are associated, baffling of the fusion crucible was not used. Instead, a long, large-throated funnel led the sample down into the crucible—with the tip of the funnel below the opening of the crucible. This minimized spattering. However, the precaution of dropping the sample when the bath was relatively cold (approximately 1200° C) was followed to help minimize spattering. GOKCEN was emphatic in his stand that the baffle was more certain of avoiding small pellets being entrapped away from the fusion zone.

In answer to another question, GOKCEN stated that he felt that the all-glass pumping system was more satisfactory than a metal system because the glass had a lesser tendency to occlude gases on its surface. To this OLETTE agreed. DERGE asked the value of the use of tin with alloys high in manganese to avoid error. Both GOKCEN and HILTY stated that additions of tin to the fusion bath did help, and HILTY reported the use of tin when running pure manganese for its gas content. However, they both agreed that tin absorbs nitrogen and, as a result, leads to errors where nitrogen is being determined. HILTY also stated that the vacuum-fusion technique was not satisfactory for the analysis of nitrogen in stainless steels. However, GOKCEN felt that the proper technique should overcome difficulties which have been encountered.

In a discussion of whether the cupric oxide oxidizer was a source of blank, GOKCEN stated that the rate at which copper oxide released oxygen was sufficiently low so as not to cause interference. This was corroborated by FUWA. SHANAHAN stated that the equilibrium oxygen pressure over cupric oxide was high as a result of the reaction $2CuO = Cu_2O + \frac{1}{2}O_2$ and could lead to a high blank. At 300° C, $ppO_2 = 3.6 \times 10^{-11}$ mm. At 700° C, $ppO_2 = 0.13$ ml.

D. C. Hilty
D. J. Girardi
Presiding

Section 3

Behavior of Metal Oxides and of Components of Iron and Steelmaking Slags

by F. D. Richardson

Oxide Slags—A Survey of Our Present Knowledge

I propose to review very briefly our present knowledge of slags from both structural and thermodynamic standpoints. Some of the things which I shall mention, particularly about binary silicate melts, do not have any direct bearing on steelmaking; they should, nonetheless, be in our minds and be influencing our thoughts when we come to think of the complex slags which the steelmaker uses.

SILICATE STRUCTURES

Our understanding of the atomic nature of slags is founded essentially on the X-ray studies of the structure of silica, silicates, and silicate glasses by Zachariasen,[1] Bragg,[2] Biscoe and Warren,[3] and others. The fundamental building unit in silica and all silicates is the silicate tetrahedron consisting of four nearly close-packed oxygen atoms or ions, surrounding a small silicon atom (Fig. 1). Each corner oxygen atom of such a tetrahedron has a residual valency, so that an SiO_4^{4-} group on its own carries four negative charges, one associated with each oxygen atom. These tetrahedra can share corners but not edges with each other, so that when every corner (that is, every oxygen atom) is shared, the substance formed has the stoichiometric formula SiO_2. In crystalline silica, the tetrahedral groups are distributed in regular array, sharing all their corners in a manner which may be represented two-dimensionally as in Fig. 1 (left). In molten or vitreous silica, all or nearly all corners remain shared, albeit with some distortion. The long-range order has disappeared, however, and we have the three-dimensional network represented in Fig. 1

Dr. Richardson is Professor of Metallurgy, Imperial College, London, England.

(right). Being tightly linked in all directions, the melt has a very high viscosity. Figure 2 shows what happens when a metal oxide, such as lime or magnesia, is added to molten silica. The oxygen atoms from the added oxide join into the silicate tetrahedra so that a shared corner is freed at every point

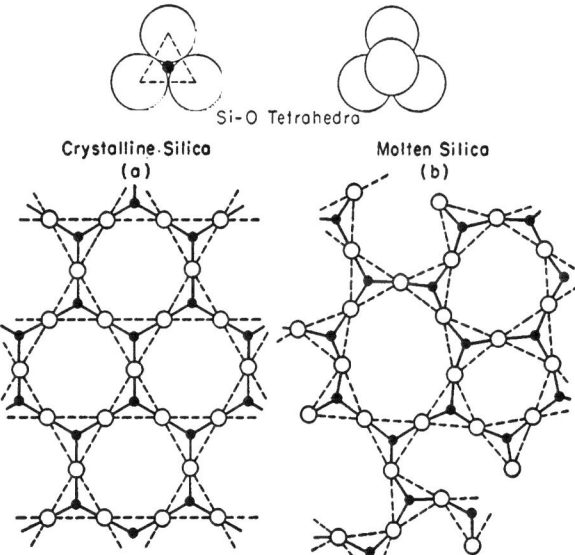

Fig. 1. Schematic representation of the silicate tetrahedron and crystalline and molten silica. The oxygen atoms are shown white and the silicon atoms black. The atoms in the Si-O tetrahedra are drawn with their ionic radii and are viewed from above. In the left-hand tetrahedron the top oxygen has been removed to reveal the silicon atom. The lower diagrams are drawn on the same scale. For the sake of clarity the atom centers only are indicated, and the apical oxygen atoms have been omitted.

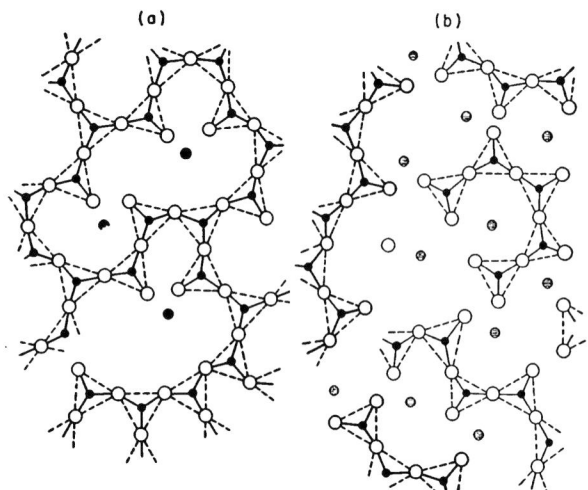

(a)   (b)

Fig. 2. Schematic representation of solutions of a divalent metal oxide in molten silica. Metal ions are represented by the shaded circles. The concentration of metal oxide increases from left to right.

where an oxygen is added. This can be represented by an equation such as:*

$$\text{:Si—O—Si:} + CaO = \text{:Si—O}' + \text{O}'\text{—Si:} + \text{Ca:} \tag{1}$$

Each oxygen atom at such breaks carries a negative charge, and the cations Ca^{2+} or Mg^{2+} are localized near the break, being accommodated in holes in the network. As such oxides are added in increasing concentrations, the silica network is increasingly broken down, and the viscosity falls correspondingly. Finally, the stage is reached at which no corners are shared. The structure of the liquid is then a disordered version of that which has been

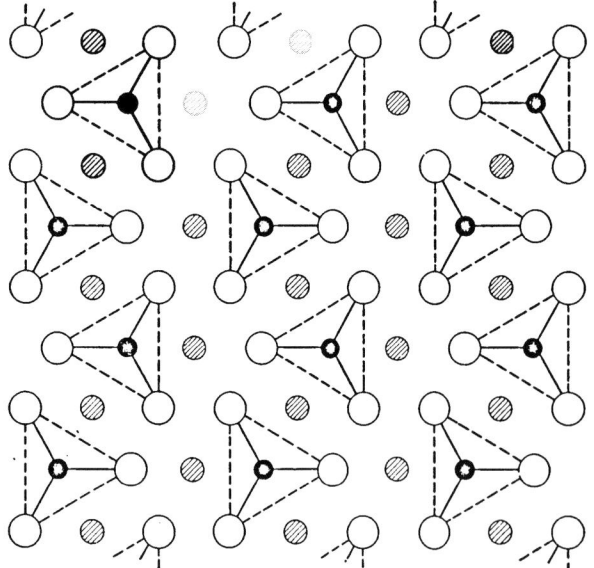

Fig. 3. Schematic representation of crystalline Mg_2SiO_4. The molten silicate would be a disordered version of this.

worked out for the crystalline compound Mg_2SiO_4 and is represented in Fig. 3. Further addition of

*Ed. note: For convenience in writing equations showing bonds and valences, a dot shows positive charge and a prime a negative charge; i.e., Ca: is Ca^{2+}.

oxide results in a melt containing SiO_4^{4+} groups and oxygen ions (O^{2-}), together with the equivalent numbers of cations. It must not be supposed, however, that there are no oxygen ions in slags with more silica than corresponds to the stoichiometric composition $2MO \cdot SiO_2$, e.g., Ca_2SiO_4. At any instant there will be a proportion depending on the thermal dissociation of the silicate groups, as might be represented by the equation

$$2(\text{:Si—O}') = \text{:Si—O—Si:} + O^{2}{}'' \tag{2}$$

and one such oxygen is shown in Fig. 2 (right). Even in highly silicious melts, we may expect the occasional oxygen atom which is unattached to either one or two silicon atoms. Now this leads to the idea that for any melt of a particular stoichiometric composition, we can have a structure which is governed not solely by stoichiometry but also by temperature.

This same idea has been developed for glasses from quite different lines of evidence—viscosities, refractive indices, and heat capacities. I shall mention only the first two here. It has been found that the viscosities and refractive indices of glasses, which are dependent on their thermal history, approach a value characteristic of the temperature at which the glasses have been annealed. This is illustrated in Figs. 4 and 5. The first of these shows

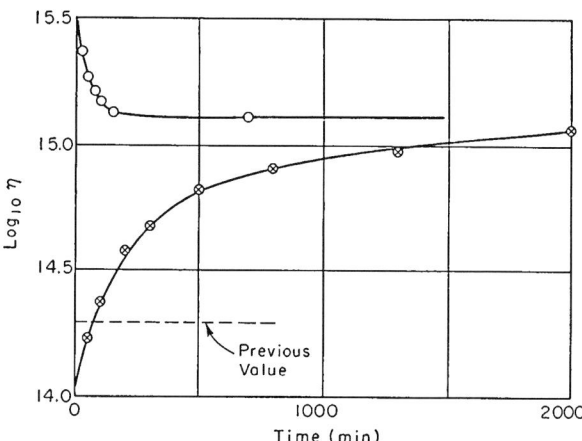

Fig. 4. Typical viscosity-time curves for two glass fibers taken from Lillie.[4]

two typical viscosity-time curves measured by Lillie[4] with glass fibers at 486.7° C. The analysis of the glass was SiO_2 69.73%, Na_2O 20.96%, CaO 9.05%, other oxides 0.26%. The upper curve was obtained with a sample previously heated at 477.8° C for 64 hr; the lower curve was obtained with a newly drawn sample, i.e., one which would have a structure approximating that typical of a higher temperature. The second figure shows typical curves obtained by Winter[5] who studied the changes which took place in the refractive indices (measured at room temperature) of a borosilicate crown glass and of a dense flint glass as they approached their equilibrium configurations at different tempera-

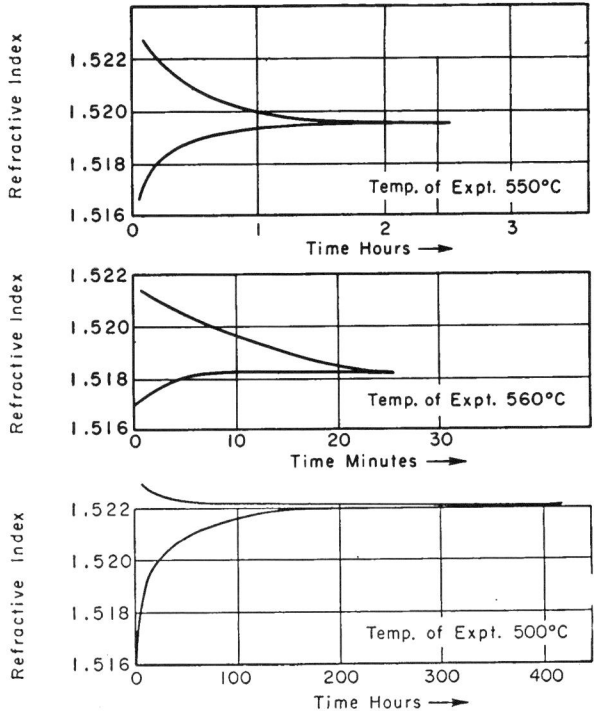

Fig. 5. Typical curves taken from a survey by Douglas[6] showing the changes in refractive index, measured at room temperature, as a glass approaches its equilibrium configuration at different temperatures.

tures. The upper curves are for samples held for many hours below the temperatures of the experiments. The lower curves are for samples quickly cooled from the molten state.

At these much lower temperatures, the conditions are somewhat different from those obtaining for melts at 1600° C. First, the rate of attainment of equilibrium is much slower than at high temperatures. Second, the randomness of atom distribution in the melt will be less than at the high temperatures. In a melt of the orthosilicate composition, for example, the proportion of O^{2-} ions and of silicate ions more complex than SiO_4^{4-} will be much less at low than at high temperatures.

The fact that a change of 10° C in the temperature at which the glass structure has been brought into internal equilibrium can change the viscosity at 500° C by a factor greater than 2 suggests that the equilibrium structures change significantly even with a small rise of temperature. The thermal energy required to alter the proportions of the different silicate groupings significantly is therefore not large, so that a much wider range of groupings may be expected at 1600° C. For an equal change of temperature, however, the configuration will not change nearly so markedly at the higher temperature as at the lower one.

So far I have developed a picture of liquid slags based almost entirely on evidence concerning the nature of solid glasses. How far, if at all, do we need to amend this picture in the light of what has actually been done with liquid slags? Apart from

thermodynamic properties, measurements have been made of electrical conductivities, electrical transport, viscosities, densities, and expansivities. Among the best results are those of Bockris and his co-workers.[7, 8, 9, 10] The electrical measurements confirm that the silicates are ionic melts, and that their conductance is due almost entirely to the metal cations.

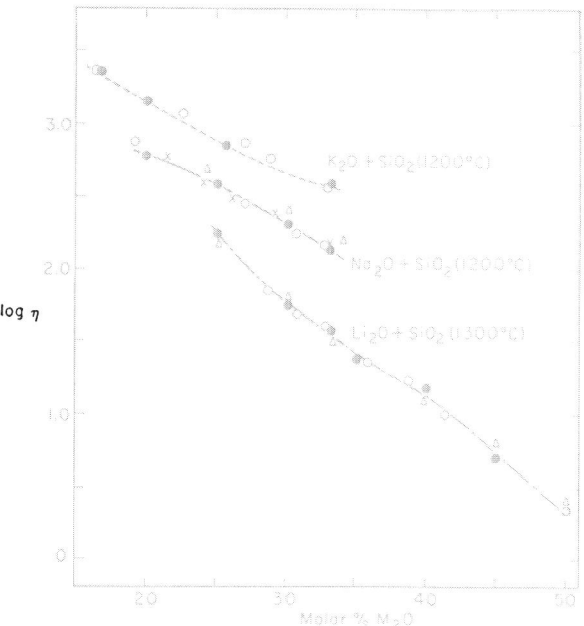

Fig. 6. Viscosity data taken from Bockris, Mackenzie, and Kitchener.[9]

A typical set of viscosity results is given in Fig. 6 which shows the logarithms of the viscosities for three binary mixtures as a function of the concentration of alkali metal oxide. The viscosity falls as the concentration of metal oxide increases, but this is no more than we would expect on the basis of the simple picture I have developed. However, when you come to study the temperature coefficient of the viscosity, and derive from it the so-called heat of activation for viscous flow, it becomes evident that the first 10 mole % of added oxide has a much

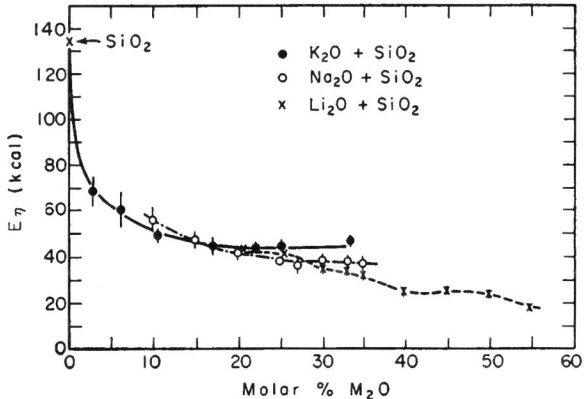

Fig. 7. Heats of activation for viscosity as a function of composition for alkali metal oxide-silica systems taken from Bockris, Mackenzie, and Kitchener.[9]

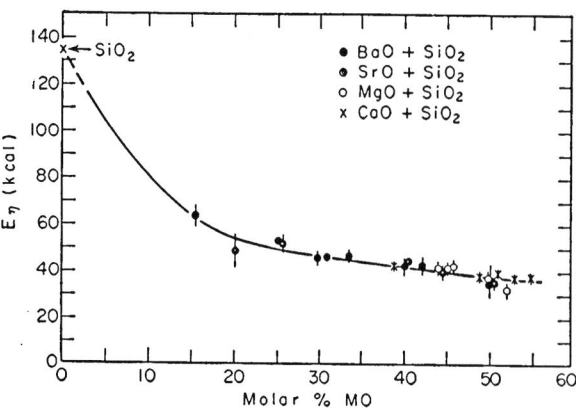

Fig. 8. Heat of activation for viscosity as a function of composition for alkaline earth oxide-silica systems taken from Bockris, Mackenzie, and Kitchener.[9]

greater effect on the heat of activation than any further additions. This is evident from Fig. 7. Similar results are shown in Fig. 8 for binary melts containing alkaline-earth oxides, but with this series

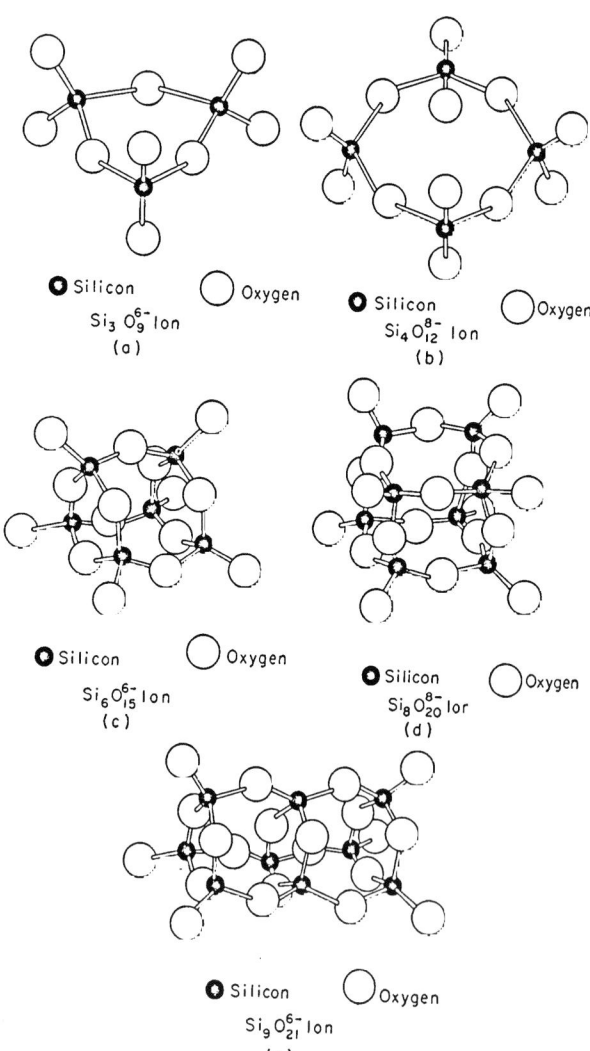

Fig. 9. Suggested discrete ions in liquid silicates taken from Bockris, Mackenzie, and Kitchener.[9]

the mixtures containing 0 to about 20 mole % metal oxide are unattainable due to miscibility gaps. Bockris has taken the results shown in Figs. 7 and 8 as evidence that the structures of binary silicate melts change markedly in the region of 10 to 12 mole % of metal oxide. At lower metal-oxide concentrations, in the range where we have high heats of activations which are falling rapidly, he considers that we have a gradual breakdown of the silica network in the way I have suggested in Fig. 2. At higher metal-oxide concentrations, instead of interlaced silicate chains, he suggests that large globular or ring-type silicate ions are formed and that these become progressively smaller as the metal-oxide content increases. Figure 9 gives possible examples. Once the concentration at which these discrete ions can form has been exceeded, it can be argued that the activation energy should remain more or less constant, each group being associated with the same number of cations. This, however, is debatable. There is, nevertheless, other evidence to suggest that the nature of the melt changes in the vicinity of 12% added oxide.

Bockris, Tomlinson, and White[10] have measured the expansivity of liquid alkali silicates as a function of composition at constant pressure, as shown in Fig. 10. The vertical lines through the

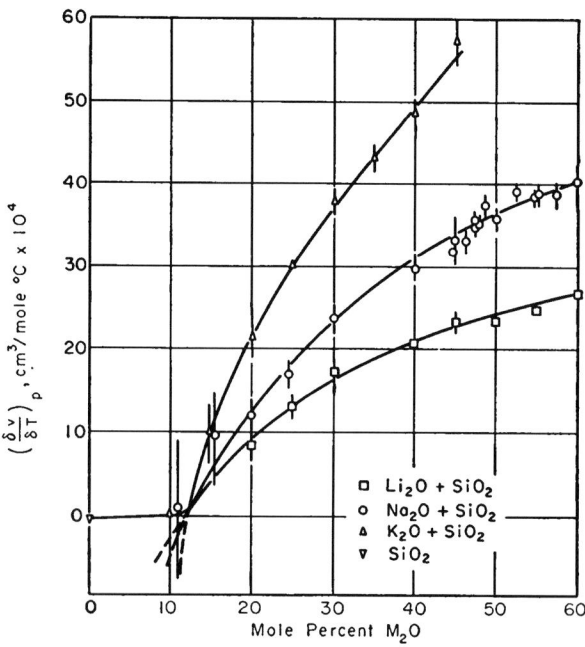

Fig. 10. Expansivity of liquid silicates from Bockris, Tomlinson, and White.[10]

points represent the confidence limits assigned by those authors. It seems doubtful, in view of these confidence limits, that much significance can be attached to the sharp break shown in the curves at 12 mole % added oxide. If, however, the curves were really to run as shown, then it would be reasonable to infer that the structure of the melt changes markedly at the 12 % point. Bockris and others concluded that the high expansivities measured for the alkali-rich silicates arise from expansion of ionic bonds between the large discrete globular or ring-shaped anions. They also concluded that the ionic bonds which must exist in the region 0 to 12 mole % alkali cannot exert their anticipated effect on the expansivity, as they are completely "boxed-in" by Si—O—Si bridges which must have the nearly zero expansivity which was measured for molten silica. An example of "boxed-in" ionic groups is given in Fig. 11. If these conclusions are

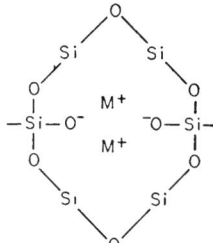

Fig. 11. Ionic bonds in a silicate "boxed-in" by silicon-oxygen bridges, taken from Bockris, Tomlinson, and White.[10]

correct, the self-diffusion coefficient of cations in a molten silicate should increase sharply at the concentration at which the structure changes; it would seem worth attempting measurements to test this.

In view of the high temperatures of silicate melts and the dissociation equilibria which I have already represented in equation 2, I consider any abrupt change in silicate properties with composition to be most unlikely. The curves shown in Fig. 10 more probably have an S shape, merging from near horizontal to an upward slope, in the range 5 to 15 % silica.

THERMODYNAMIC PROPERTIES OF SILICATES

I now want to consider the present state of basic thermodynamic knowledge of simple slags. This is summed up in Fig. 12 which shows the integral free energies of mixing of those melts which have so far been studied. To this we can add a curve which Dr. M. W. Davies in our Group has just measured for MnO—SiO$_2$ melts: at 1600° C it lies about 0.25 kcal below that for MgO—SiO$_2$. The ideal free energies of mixing of two substances, as given by Raoult's law, are represented, for the purpose of comparison, by the uppermost curve. It is a property of these diagrams that the activities of both components at any melt composition can be read from the intercepts made on the free-energy axes by a tangent to the curve at this same composition. The intercept on the ΔG scale is equal to $-RT \ln a$. The activity scale shown on the right-hand side of Fig. 12 has been calculated from this relationship for 1600° C. There is no correlation between the order of stability of these silicate melts and any obvious property of the metal atom—such as the divalent ion radius or the oxygen-ion attrac-

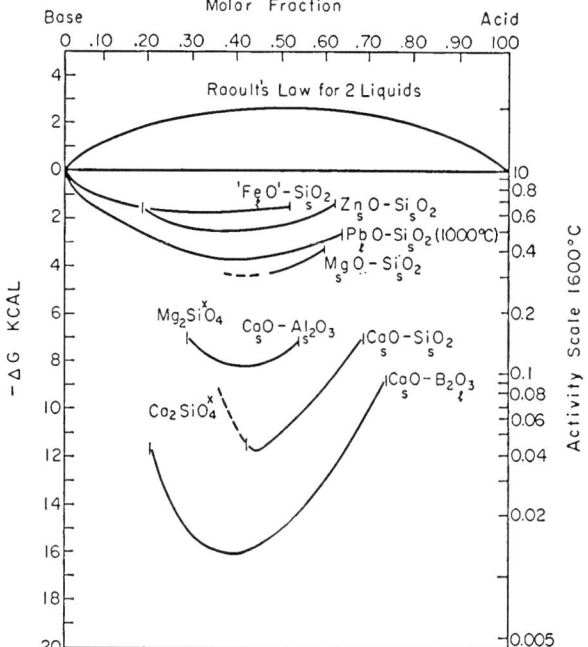

Fig. 12. The Gibbs free energies of formation of binary oxide melts per mole of oxide ($nSiO_2 + nMO = 1$) at 1600° C (except for PbO—SiO$_2$).

tion. The only trend that can be distinguished is the shift of the minimum, which moves toward the higher silica composition as the energy of interaction between oxide and silica increases. A point of general interest, however, is that the curves are not markedly peaked in shape. In theory, therefore, any melt can be divided into a heterogeneous mixture of two melts, one richer and the other poorer in silica, with *only* a small increase in free energy and probably a correspondingly small one in total energy. For instance, it would cost less than 1 kcal ΔG to divide a mole of PbSiO$_3$ melt into molten Pb$_2$SiO$_4$ and PbSi$_2$O$_7$. This is small compared with RT (2.9 kcal at 1200° C), so a wide range of possible silicate groupings in the homogeneous melt is to be expected. The corresponding figures for CaSiO$_3$ are 3 and 3.5 kcal, respectively, so the same argument applies.

Our knowledge of ternary silicates is much more scanty than of the binaries. We have first of all the classic work of Fetters and Taylor and Chipman[11, 12] who among them established FeO activities in FeO—CaO—SiO$_2$ melts, which also contained some magnesia arising from their crucibles. Their results are represented in Fig. 13. The important thing to notice is that the substitution of lime for ferrous oxide at constant (CaO + FeO)/SiO$_2$ ratio markedly raises the activity coefficient of the FeO on the oxide-rich side of the metasilicate line. In the Nuffield Research Group, Dr. M. W. Davies has shown that the activity coefficients of MnO are similarly raised in MnO—CaO—SiO$_2$ melts at 1600° C, and Dr. T. C. M. Pillay has found a similar pattern of behavior for PbO—CaO—SiO$_2$ mixtures at 1200° C. Other measurements made by

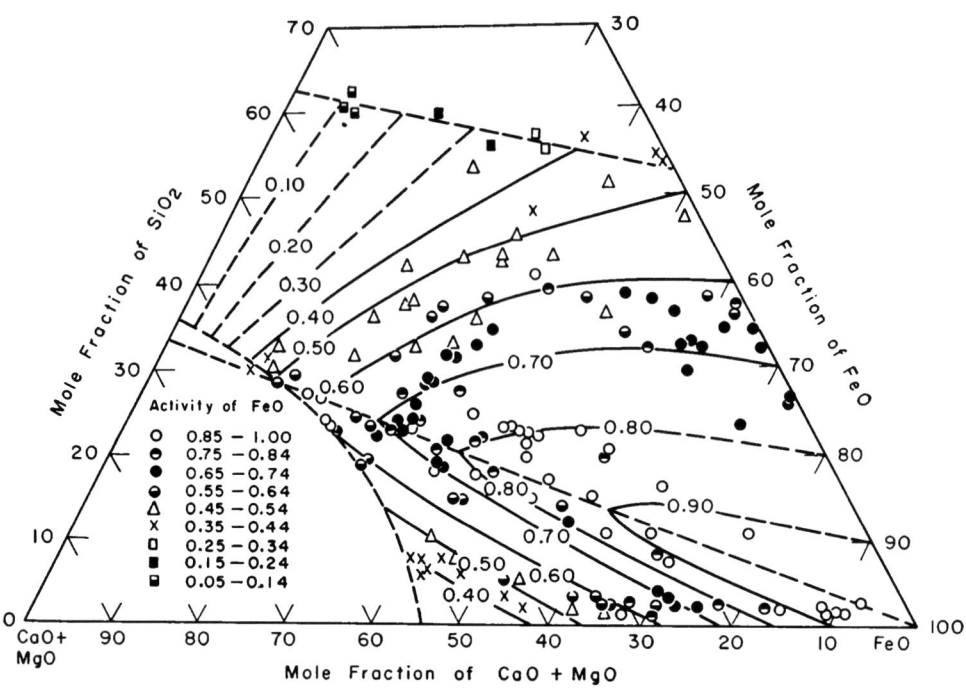

Fig. 13. Activities of FeO in CaO(+MgO) — FeO + SiO₂, slags taken from Taylor and Chipman.[12]

Dr. Pillay on PbO—MgO—SiO₂ and PbO—ZnO—SiO₂ melts at 1200° C substantiate qualitatively the theory of "ideal" mixing for ternary silicate melts which I am putting forward at this Conference in a separate paper. This appears to work quantitatively in the metasilicate region and qualitatively in the orthosilicate region. It leads to the conclusion that (at constant silica mole fraction) the activity coefficient will be *raised* in the ternary for the metal oxide which in the binary gives the *higher* silica activity, and vice versa. The theory provides a basis for making approximate calculations of activities in ternary and more complex slags; furthermore, by providing an "ideal" solution law, it leads one to look intelligently for explanations of gross deviations.

There is another chemical equilibrium which has, I suggest, been of great interest both to those concerned with theory and to those primarily interested in practice. I refer to the studies which have been made in the Nuffield Research Group[13, 14] and subsequently at M.I.T. on the gas-slag equilibrium:

$$\tfrac{1}{2}S_2 + (O^{2-}) = \tfrac{1}{2}O_2 + (S^{2-}) \qquad (3)$$

This holds at constant slag composition at the low oxygen concentrations which primarily interest the steelmaker: pO_2 less than 10^{-5} atm. Our results for slags of different compositions are plotted in Fig. 14, where C_S, the so-called sulfide capacity, is defined by the equation

$$C_S = (\text{Wt \% S}) \sqrt{\frac{pO_2}{pS_2}} \qquad (4)$$

Apart from changes in activity coefficients, the manner in which C_S varies with composition reveals the

manner in which the concentration of replaceable oxygen varies with composition.

The question arises whether the oxygen atoms participating in the equilibrium come from one or all of the three types, O″, :Si—O′, or :Si—O—Si: The sulfide capacities of all the silicate melts investigated decrease as the silica content increases and appear to follow the metal oxide activities. In the case of CaO—SiO₂ mixtures at 1650° C, for example (see Fig. 14), the sulfide capacity decreases by a

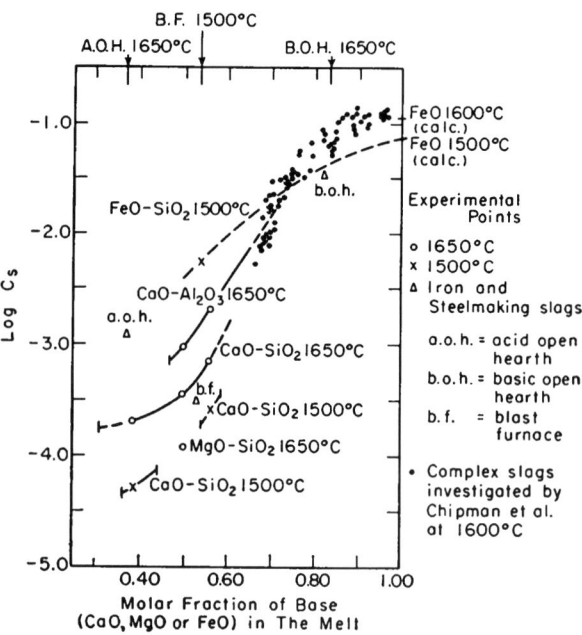

Fig. 14. Sulfide capacities of various silicate and aluminate slags, with curves extended (broken lines) to cover full liquid ranges, taken from Fincham and Richardson.[18]

factor of about 630 as the molar fraction of CaO falls from 0.90 to 0.30. It appears, therefore, that sulfur does not readily replace oxygen atoms bonded to two silicons. Between 0.3 and 0.6 molar fraction of CaO, the number of oxygen atoms bonded to one silicon atom only in a mole of melt ($n_{CaO} + n_{SiO_2} = 1$) is increased by, at most, a factor of 2. Owing to the equilibrium in equation 2, the increase will actually be rather less. An increase in C_S of about twofold would thus be expected if the most important oxygen atoms participating in the sulfide equilibrium were those bonded to one silicon only. The actual increase is about elevenfold at 1500° C and eightfold at 1650° C. If the silicate equilibrium given in equation 2 were to hold in the melt, the increase in (O^{2-}) could well lie in this range, depending on the value of K_2. It thus seems extremely likely that the only oxygen atoms which are readily replaceable by sulfur are those not bonded to silicon at all and that there is an equilibrium among all three types of oxygen.

It seems probable that, in the case of aluminate melts also, the important oxygen atoms are those not bonded to any aluminum. Our structural knowledge of aluminate melts is, however, nowhere nearly as clear as that of silicate melts, so that conclusions in this case cannot be so definite.

SOLUBILITIES OF NEUTRAL METALS

A point which is commonly overlooked is that metals can dissolve in slags as neutral atoms. The solubilities, so far measured, are not very great, but they may be sufficient to upset conclusions drawn from slag-metal equilibrium studies in which certain metal oxides may be present in slag only in small amounts. They are of course of interest to the nonferrous metallurgist to whom small losses of metals in slags can be important. In such cases the dissolved neutral metal might nearly equal the dissolved oxide. Billington[15] found that in a $CaO - SiO_2 - Al_2O_3$ slag (30, 31, 39 wt %) at 1525° C the solubility of neutral copper is about 0.05 wt % and the solubility of neutral silver 0.24%, both increasing with temperature. It looks as though the metal atoms go into holes in the silicate network; it also appears that the solubilities of the metals increase with their vapor pressures and decrease with their atomic radii. The higher the vapor pressure of a metal, the more readily it can pass into any other state; the larger its size, the more difficulty it has getting into the holes available in the silicate melt.

WATER IN SLAGS

Another solubility which is of both technological and fundamental interest is that of water in slags. Tomlinson[16] of the Nuffield Group has recently measured the solubilities of water in the molten sodium silicate $Na_2O \cdot 2SiO_2$ and a more exhaustive study on steelmaking slags has, I understand, just been completed in the Metallurgy Department at M.I.T. Tomlinson found that the solubility of water in the sodium silicate was not much influenced by temperature, and that at 1100° C and 1 atm pressure of vapor it was about 3 cc (STP) per gram. It also looks from Tomlinson's results as though the solubility is proportional to the square root of the pressure of water vapor, so the water may well be incorporated into the slag as hydroxyl groups—

$$\vdots Si-O-Si\vdots + H_2O = 2(\vdots Si-OH) \qquad (5)$$

A solubility of water vapor in slag is probably the means whereby the hydrogen content of the metal in the open hearth rises, after reaching a minimum at the end of the boil. As a result of this solubility, water vapor from the furnace gases can pass through the slag to the metal surface where it can react to give hydrogen and iron oxide, both of which may then dissolve in the metal. The hydrogen pressure over the metal can in theory reach a value equal to about half the water vapor pressure obtaining at the slag-gas interface.

CONCLUSION

It is evident that during the past fifteen years our understanding of the fundamental principles of slag behavior has developed very considerably. We have still a long way to go; as the picture unfolds, so does it become both more complex and more interesting. Those who pursue research on silicates and slags can still expect a stimulating as well as a difficult time.

References

1. W. H. Zachariasen, *J. Am. Chem. Soc.*, 54, 3841 (1932).
2. W. L. Bragg, *The Structure of Silicates*, Akad. Verlag, Leipzig, 1930.
3. J. Biscoe and B. E. Warren, *J. Am. Ceram. Soc.*, 21, 287 (1938).
4. H. R. Lillie, *J. Am. Ceram. Soc.*, 12, 619 (1933).
5. A. Winter, *J. Am. Ceram. Soc.*, 26, 193 (1943).
6. R. W. Douglas, *J. Soc. Glass Technol.*, 141, 51 (1947).
7. J. O'M. Bockris, J. A. Kitchener, and S. Ignatowicz, *Trans. Faraday Soc.*, 48, 75 (1952).
8. J. O'M. Bockris, J. A. Kitchener, and A. E. Davies, *J. Chem. Phys.*, 19, 255 (1950).
9. J. O'M. Bockris, J. D. Mackenzie, and J. A. Kitchener, *Trans. Faraday Soc.*, 51, 1734 (1955).
10. J. O'M. Bockris, J. W. Tomlinson, and J. L. White, *Trans. Faraday Soc.*, 52, 299 (1956).
11. K. L. Fetters and J. Chipman, *Trans. A.I.M.E.* (1941), Tech. Pub. 1316.
12. C. R. Taylor and J. Chipman, *Trans. A.I.M.E.* (1942), Tech. Pub. 1499.
13. C. J. B. Fincham and F. D. Richardson, *Proc. Roy. Soc. (London) A*, 223, 40 (1954).
14. F. D. Richardson and C. J. B. Fincham, *J. Iron Steel Inst. London*, 178, 4 (1954).
15. F. D. Richardson and J. C. Billington, *Bull. Inst. Mining and Metallurgy*, 593, 273 (1956).
16. J. W. Tomlinson, *J. Soc. Glass Technol.*, 40, 25 (1956).

Discussion

SHANAHAN raised the question whether a doubly-charged cation might provide bridging between two oxygen ions which are attached in turn to silicon ions, thereby causing an increase in viscosity instead of a lowering, as has been observed. RICHARDSON agreed that this undoubtedly occurred, but considered that other more dominant effects caused the general lowering of viscosity with the increase of metallic oxide concentration. He felt that this effect mentioned by Shanahan was involved in the observation that with the divalent cation the lowering of viscosity was not as marked as was observed with the monovalent cations. That is, per equivalent of metallic oxide added, CaO does not cause as much lowering of viscosity as does sodium oxide, for example.

SHANAHAN also inquired about the procedure for measuring the neutral solubility of the metals in the slags. RICHARDSON replied that with copper, a bead of copper was surrounded by the slag which did not quite cover it. The whole was contained in an alumina crucible. The atmosphere passing over the bead was a controlled mixture of CO and CO_2. With pure copper, microscopic particles of copper were found dispersed through the slag, and it was concluded that this was due to the cycling of the furnace temperature. At the low point in the cycle, the slag was supersaturated with copper, and a mist of copper would precipitate within the slag. The mist apparently did not completely redissolve on the rising side of the temperature cycle. This fog interfered with the measurements, so palladium was added to the copper to lower the activity of copper to below 0.8, and supersaturation was avoided. As a result of the experimental work, it was proved that palladium was not soluble in the slag. The solubility of copper in the slag was measured as a function of the CO/CO_2 ratio.

By plotting the copper content of the slag versus the CO/CO_2 ratio, a linear plot passing through the origin would indicate that all the copper was in the cuprous form. The fact that the line passes above the origin is interpreted as resulting from the solubility of the neutral copper in the slag. With silver, no ionic solubility was encountered, and the neutral solubility was about 0.25%. An interesting point in the procedure is that if the slag covers the metal bead, a very long time is required for equilibrium when ionic solubility occurs and metal oxides have to be formed. With copper, for example, there is no easy means by which the oxygen can be carried to the metal through the slag cover. KING suggested that the use of iron or vanadium in the slag might provide an oxygen carrier which would avoid the need for adjusting the slag level to just below the surface of the copper pellet, as no doubt this is a very difficult adjustment to make experimentally.

KOZAKEVITCH queried Richardson about the structural characteristics of alumina in slags. RICHARDSON stated that, in theory, the alumina could build into the silicate network with an aluminum ion at the center of a tetrahedron having all four corners shared with the silicate lattice. It would be necessary that the aluminum have nearby a positive monovalent ion such as sodium, or half a calcium ion, to compensate for the net negative charge on the alumina tetrahedron. There is some evidence that alumina builds in this way, and also that phosphates do. With the phosphate it is necessary to find a companion negative charge to go with the tetrahedron containing phosphorus. KOZAKEVITCH said that Menkin has found in the $CaO - Al_2O_3 - SiO_2$ system that the effect of alumina on the viscosity of the highly silicious slags was predicted from the fact that the aluminum ion would fit into the center of the tetrahedron. However, it appears that this is not so with basic slags.

by L. Yang
C. L. McCabe
and R. Miller

A Summary of Some Experimental Work on the Activity of Silica in Liquid Silicate Systems

The activity of silica in the liquidus of the CaO—SiO_2 system at 1637° C has been determined by the use of the Knudsen[1] Cell to measure the pressures of SiO and O_2 above the melt and above pure cristobalite (Figs. 1, 2, 3). Details of the experimental procedure and calculations will be given in a later publication. From a consideration of the equation

$$2SiO_2 \rightleftharpoons 2SiO\ (g) + O_2\ (g) \qquad (1)$$

it can be shown that

$$a_{SiO_2} = \sqrt{p_{SiO}^2 p_{O_2} / p_{SiO}'^2 p_{O_2}'} \qquad (1a)$$

where primes refer to pressures above pure cristobalite and absence of prime is above the melt in question. It should be noted that allowance was made for O atoms in the gas phase at these temperatures and pressures.

The results of the experiments are shown in Fig. 4. In Fig. 5 there is a comparison between the results of this investigation and those of previous investigations, both experimental and theoretical. In general, the trend of the activity versus mole fraction plot follows the expected behavior. The activity of silica obtained in this investigation, at the point where dicalcium silicate forms as a separate phase, lies between the activities of silica calcu-

This work was supported by the National Science Foundation. It was performed at the Metals Research Laboratory, Carnegie Institute of Technology.

The authors are associated with the Carnegie Institute of Technology, Pittsburgh, Pennsylvania.

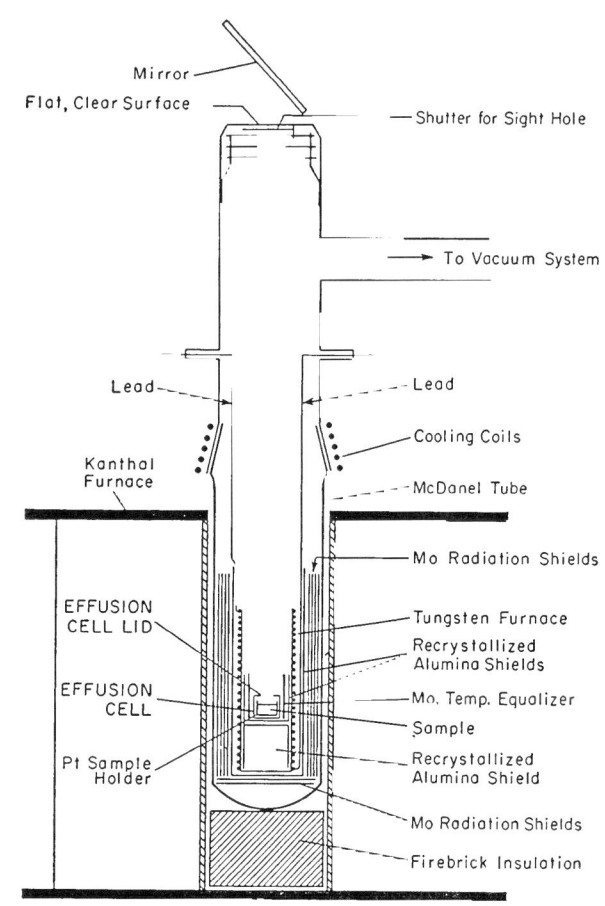

Fig. 1. Experimental furnaces

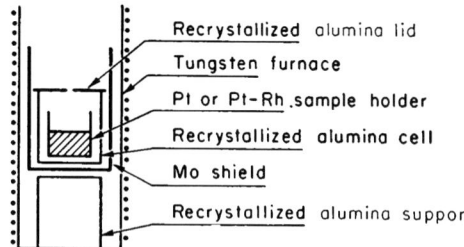

Recrystallized alumina lid
Tungsten furnace
Pt or Pt-Rh sample holder
Recrystallized alumina cell
Mo shield
Recrystallized alumina support

Fig. 2. Effusion cell.

lated by Darken[2] and by Richardson.[3] However, it is considerably higher than the results of Chipman and Fulton.[4] The source of this discrepancy is at present unresolved.

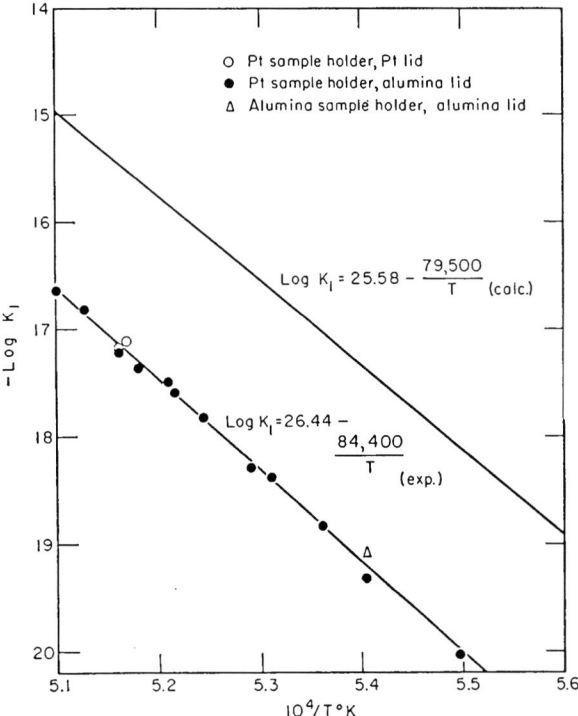

O Pt sample holder, Pt lid
● Pt sample holder, alumina lid
△ Alumina sample holder, alumina lid

$\text{Log } K_1 = 25.58 - \dfrac{79,500}{T}$ (calc.)

$\text{Log } K_1 = 26.44 - \dfrac{84,400}{T}$ (exp.)

Fig. 3. Vapor pressure of silica.

References

1. M. Knudsen, *Ann. phys.*, 28, 75 (1909).
2. L. S. Darken and R. W. Gurry, *Physical Chemistry of Metals*, p. 340, McGraw-Hill Co., New York, 1955.
3. F. D. Richardson, "The Physical Chemistry of Melts," *Bull. Inst. Mining Met.*, 83 (1953).
4. J. C. Fulton and J. Chipman, *Trans. A.I.M.E.*, 200, 1136 (1954).

Discussion

GOKCEN asked for some of the details of the experimental procedure. MC CABE replied that consistent results were obtained with various orifice sizes, various lid materials, and cell geometry. There was an extended discussion among CHIPMAN, DARKEN, GOKCEN, RICHARDSON,

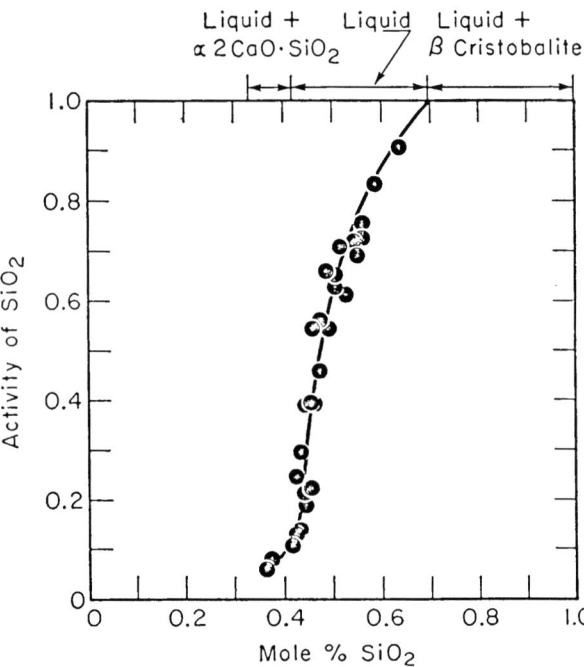

Liquid + α 2CaO·SiO₂   Liquid   Liquid + β Cristobalite

Fig. 4. Activity of silica in liquid region of CaO-SiO₂ system, 1637° C.

and MC CABE with regard to the reasons for the activity for silica reported in the paper being substantially above that of the paper by Chipman and Fulton, especially in the low silica range. CHIPMAN suggested that the error might possibly be in the effusion rate for pure silica rather than in the data for the lower silica samples, and DARKEN felt that the formation of SiO might be one source of error. There was no resolution of the difference in the data. RICHARDSON pointed out that McCabe's results were consistent with some measurements that had been made at the Nuffield Foundation Group, and that there was some possibility that the data by Chipman and Fulton may be in error perhaps because of uncertainties in the thermal data used in the calculations. (See Discussion of next paper.)

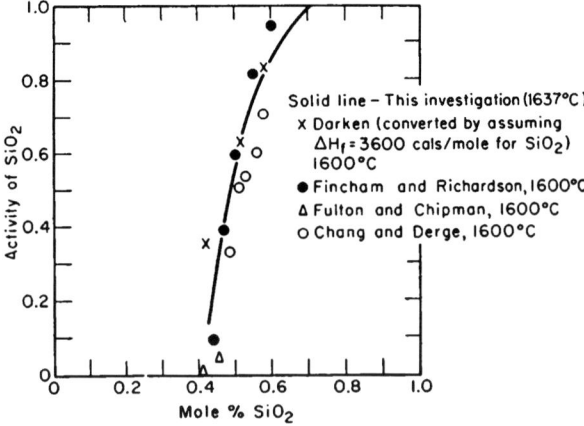

Solid line – This investigation (1637°C)
x Darken (converted by assuming ΔH_f = 3600 cals/mole for SiO₂) 1600°C
● Fincham and Richardson, 1600°C
△ Fulton and Chipman, 1600°C
○ Chang and Derge, 1600°C

Fig. 5. **Comparison of activity measurements of silica in the CaO-SiO₂ system.**

by F. C. Langenberg

H. Kaplan

and J. Chipman

The Activity of Silica in Lime-Alumina-Silica Slags at 1600°C

The distribution of silicon between graphite-saturated Fe—Si—C alloys and blast-furnace-type slags has been described in previous publications.[1,2] The silica-silicon relation was established at temperatures of 1425 to 1700° C for slags containing up to 20% Al_2O_3. This preliminary note presents briefly the results of additional silicon-distribution studies at 1600° C which permit determination of the activity of silica over a wider range than previously reported. The new results are part of a large body of data which extend the silicon-distribution data at 1600 and 1700° C for CaO—Al_2O_3—SiO_2 slags over a range from zero % Al_2O_3 to Al_2O_3 or $CaO \cdot 2Al_2O_3$ saturation. The upper limit of SiO_2 is still set by the occurrence of SiC as a stable phase when the metal contains 23.0 or 23.7% silicon at 1600 or 1700° C, respectively. The total results will be combined in the near future to calculate the activity of CaO and Al_2O_3 in these slags from the activity of SiO_2.

Dr. Langenberg is affiliated with the Crucible Steel Company of America, Pittsburgh, Pennsylvania. Mr. Kaplan with the Allegheny Ludlum Steel Corporation, Watervliet, New York, and Dr. Chipman is Professor of Metallurgy, Massachusetts Institute of Technology.

This work was supported by the American Iron and Steel Institute.

The experimental apparatus and procedure have been fully described in previous investigations.[1,3] In the present investigation the only variation of technique was in those heats where Al_2O_3 or $CaO \cdot 2Al_2O_3$ saturation was desired; for these heats a pure alumina rod was immersed in the slag to ensure saturation.[4] A typical distribution plot is shown in Fig. 1 for slags with a CaO/Al_2O_3 ratio of 0.96. The number of heats necessary to establish a single distribution curve can be seen.

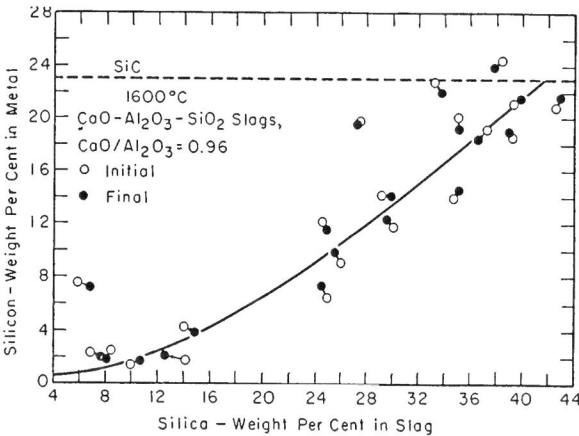

Fig. 1. Silicon distribution at 1600° C for graphite-saturated Fe—Si—C alloys in equilibrium with CaO—Al_2O_3—SiO_2 slags ($CaO/Al_2O_3 = 0.96$) under 1 atm of carbon monoxide.

ACTIVITY OF SILICA

The free-energy change for the following reaction has been recorded by Fulton and Chipman:[2]

$$SiO_2 \text{ (crist.)} + 2C \text{ (graph.)} = Si \text{ (l)} + 2CO \text{ (g)} \quad (1)$$

$$\Delta F^\circ = +161,500 - 87.4T \quad (1a)$$

Since the melts were made under 1 atm pressure of CO and were graphite saturated, the equilibrium constant for equation 1 reduces to $K_1 = a_{Si}/a_{SiO_2} = 1.77$ at 1600° C. Through K_1, the activity of silica is directly related to the activity of silicon in the equilibrium metal. The standard state for silica is taken as pure cristobalite.

The activity of silicon can be obtained from the atom fraction of silicon and the activity coefficient of silicon in graphite-saturated iron as determined by Chipman, Fulton, Gokcen, and Caskey.[5] Because the metal is saturated with carbon, a fixed silicon content represents a definite silicon activity; also, the slags of varying composition in equilibrium with melts of the same silicon content will have the same silica activity.

Distribution curves at 1600° C, similar to Fig. 1, have been obtained at CaO/Al_2O_3 ratios of 0.58, 0.96, and 1.50, and for slags saturated with Al_2O_3 or $CaO \cdot 2Al_2O_3$. Interpolation in these curves and in those published by Fulton and Chipman[2] for slags of 0, 10, and 20% Al_2O_3 yields the points shown in Fig. 2. The lines of constant silicon con-

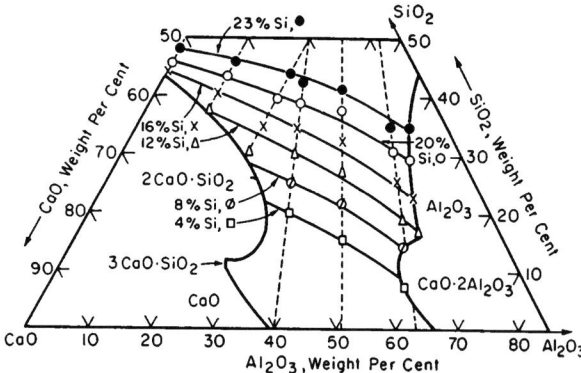

Fig. 2. Composition of $CaO-Al_2O_3-SiO_2$ slags in equilibrium with alloys of constant silicon content. The figure shows location of distribution lines and accuracy of data used to determine iso-activity curves for silica.

tent in the metal represent isoactivity lines for SiO_2 in the slag.

In Fig. 3 curves are drawn for selected values of a_{SiO_2}: The dashed lines for $a_{SiO_2} = 0.00011$ is in the range of silica content where the distribution curves are not as well defined as at higher silica levels. The isoactivity line was determined by calculations based on the free energy of formation of

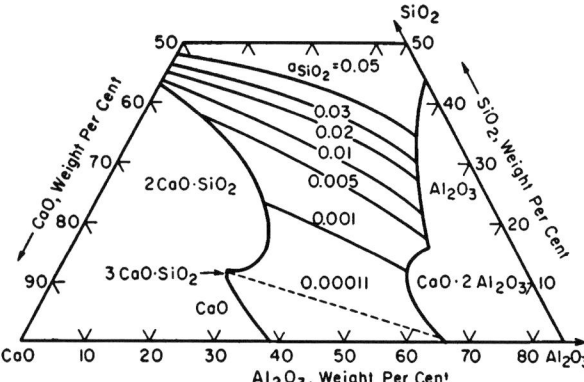

Fig. 3. Isoactivity lines of silica at 1600° C in the $CaO-Al_2O_3-SiO_2$ system.

dicalcium silicate and by agreement with the shape and trends in the family of higher isoactivity lines. From the data of King,[6] Todd,[7] Bureau of Mines Bulletins 476[8] and 542,[9] and recent data provided by Kelley,[10] the following equation was obtained:

$$2CaO \text{ (s)} + SiO_2 \text{ (crist.)} = Ca_2SiO_4 \text{ (s)} \quad (2)$$

$$\Delta F^\circ = -23,540 - 5.6T \quad (2a)$$

A slight extension of the 1600° C isotherms for $2CaO \cdot SiO_2$ and CaO yields a point of intersection where the activity of CaO is unity. By substituting $a_{CaO} = 1$ in the equilibrium constant of equation 2, the value of $a_{SiO_2} = 0.00011$ was calculated. The calculated value of a_{SiO_2} is in good agreement with the trend established by the experimentally determined silica activities at higher silica contents.

The agreement of the experimentally determined silica activities and the free energy of formation of Ca_2SiO_4 can also be verified by calculating a_{CaO} at points where the isoactivity lines intersect the $2CaO \cdot SiO_2$ isotherm. The results are plotted in Fig. 4 where the solid circles are experimentally

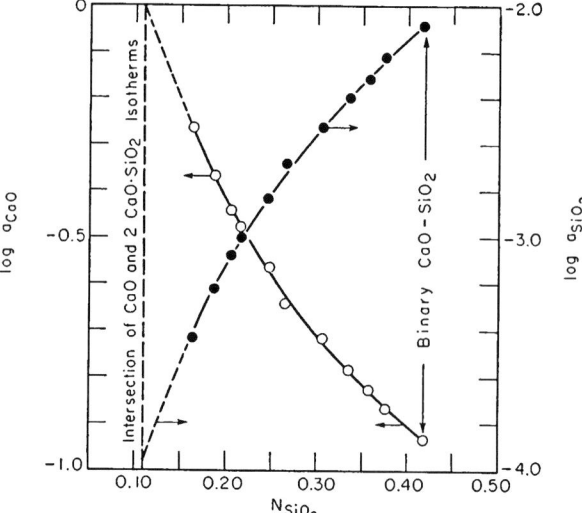

Fig. 4. Activity of SiO_2 and CaO along liquidus line of $2CaO \cdot SiO_2$ at 1600° C.

determined silica activities and the open circles are the lime activities calculated from equation 2. It can be seen that the lime activities along the $2CaO \cdot SiO_2$ liquidus increase in a regular manner and approach unity as the CaO isotherm is approached.

References

1. J. C. Fulton, N. J. Grant, and J. Chipman, *Trans. A.I.M.E.*, **197**, 185 (1953).
2. J. C. Fulton and J. Chipman, *Trans. A.I.M.E.*, **200**, 1136 (1954).
3. G. G. Hatch and J. Chipman, *Trans. A.I.M.E.*, **185**, 274 (1949).
4. F. C. Langenberg and J. Chipman, *J. Am. Ceram. Soc.*, **39**, 432 (1956).
5. J. Chipman, J. C. Fulton, N. Gokcen, and G. R. Caskey, *Acta Metallurgica*, **2**, 439 (1954).
6. E. G. King, *J. Am. Chem. Soc.*, **73**, 656 (1951).
7. S. S. Todd, *J. Am. Chem. Soc.*, **73**, 3277 (1951).
8. *U. S. Bur. Mines Bull.* No. 476 (1949).
9. *U. S. Bur. Mines Bull.* No. 542 (1954).
10. K. K. Kelley, Personal communication.

Discussion

RICHARDSON inquired concerning the accuracy of the data on the activity of silica in this ternary system. LANGENBERG said that it was difficult to estimate such accuracy because of the many data which were used in calculating the results. He felt, however, that the greatest error was in the calculation of the activity of silicon which was dependent upon distribution data. The maximum spread in plotting the activity of silica in Fig. 3 is roughly 3%, indicating the order of the error that is encountered.

HILTY asked about the possible presence of calcium carbide in these slags. In the discussion, LANGENBERG stated that they used a simple water treatment of a crushed sample of the slag to evaluate the presence of calcium carbide by which acetylene would be evolved. Several slags were tested, and only one master slag showed signs of calcium carbide from this test. This particular slag was discarded. FULTON corroborated this viewpoint and said that in the early work in this research, some slags appeared to have some carbon. These slags were analyzed and found to contain no more than 1% carbon which apparently was in a suspended form and not as a carbide. The analysis of the slags containing carbon were recalculated to 100% by excluding the carbon content. No problem of carbon pickup was encountered unless the slags were held in contact with graphite crucibles for an extended time above 1700° C. This might occasionally be the case when master slags were made up, but most slags contained under 0.1% carbon. Master slags which showed a flecking of black particles after they were made up became glass-clear on being equilibrated for 6 or 8 hr in the graphite crucible in contact with carbon-saturated iron. As a consequence, it has been concluded that the carbides do not form below 1700° C in slags of these compositions.

RICHARDSON raised the point that the slags might contain SiO. LANGENBERG said that the slags showed a total analysis close to 100% and it was likely that very little SiO, if any, was present. DARKEN suggested a check of this could be made by dissolving a sample and testing it for reducing action by titrating it with a permanganate solution. CHIPMAN pointed out that one independent check of the precision of these calculations was available from the extrapolation of the calculated activity of lime to pure lime. The fact that it extrapolates to 1 gives a reasonably good check. He noted that a previous source of uncertainty was the free energy of formation of dicalcium silicate. To a large extent this has been eliminated by new data which were obtained from K. K. Kelley's Laboratory and are included in the paper.

by F. D. Richardson

Activities in Ternary

Silicate Melts

An understanding of the manner in which the activities of metal oxides and silica vary with composition in ternary and complex silicate melts is of great importance to both the extraction metallurgist and the chemist interested in molten salts.

Apart from phase-diagram studies, few measurements have been made of the equilibrium properties of ternary silicate mixtures. The first important thermodynamic study was made by Taylor and Chipman,[1] who measured the activities of ferrous oxide in CaO—FeO—SiO_2 mixtures (containing some MgO) at $1600°$ C. Their results have been used by Darken[2] to calculate an approximate free energy of mixing curve for the binary CaO—SiO_2, and by Elliott,[3] to calculate approximate values of the activities of CaO and SiO_2 by application of the Gibbs-Duhem relationship. Activities of SiO_2 in the ternary system CaO—SiO_2—Al_2O_3 at 1500 to $1700°$ C have been measured by Fulton and Chipman.[4]

DEDUCTIONS FROM PHASE DIAGRAMS

Let us consider pseudobinary mixtures of the type $(AO)_ySiO_2$—$(BO)_ySiO_2$. Such solutions can exist in the crystalline state as well as in melts. For example, $CaSiO_3$ forms a very wide range of solid solutions with "$FeSiO_3$" (which is unstable with respect to Fe_2SiO_4 and SiO_2 and so is unobtainable in the pure state). In a like manner, $MgSiO_3$ forms a nearly complete range with "$FeSiO_3$," and Mg_2SiO_4, a complete range with Fe_2SiO_4. The relevant phase diagrams are shown in Figs. 1, 2, and 3. Similarly, $CaSiO_3$ forms a complete series

Dr. Richardson is Professor of Metallurgy, Imperial College, London, England.

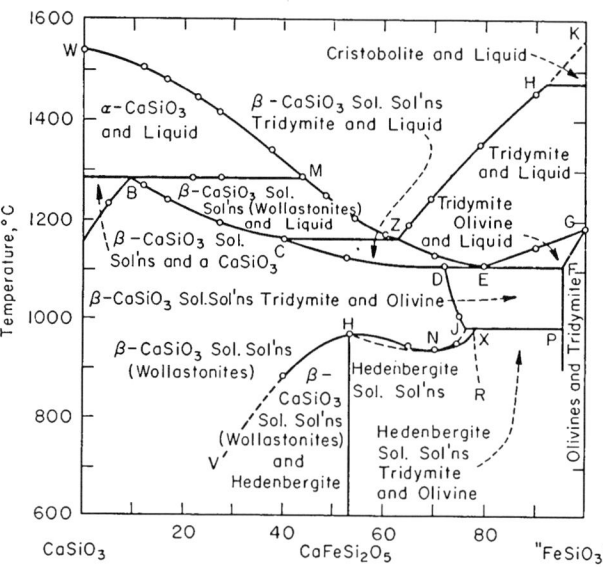

Fig. 1. $CaSiO_3$—"$FeSiO_3$" phase diagram of Bowen, Schairer, and Posnjak.[5]

of solid solutions with $SrSiO_3$ and with $MnSiO_3$, and an incomplete series with $MgSiO_3$. Also, Ca_2SiO_4 forms a wide range with Fe_2SiO_4, and Mn_2SiO_4 forms a complete series with Fe_2SiO_4.

It is known from the calorimetric measurements of Sahama and Torgeson[6] that the heats of formation of the crystalline solid solutions $MgSiO_3$—"$FeSiO_3$" and Mg_2SiO_4—Fe_2SiO_4 are negligibly small at room temperature. It is thus reasonable to assume that the heats of mixing are negligible for the corresponding series of melts. If it is assumed that the melt structure and the anion system in each series of melts remains substantially the same from one limiting composition to the other,

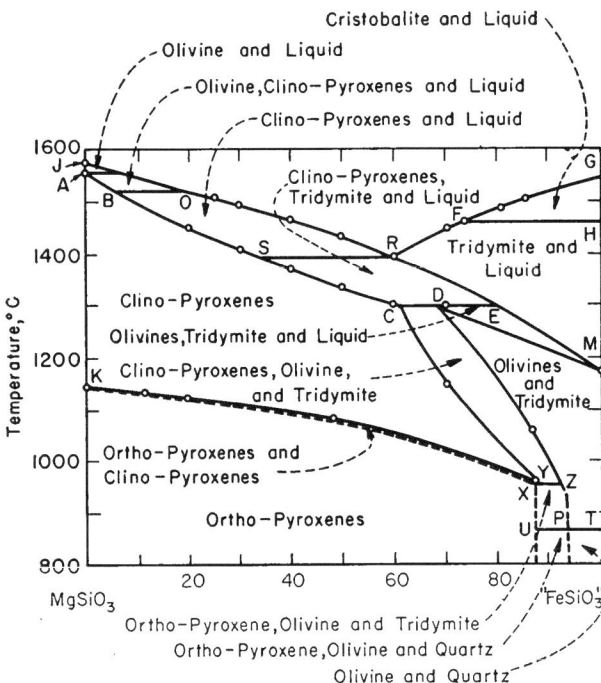

Fig. 2. MgSiO₃—"FeSiO₃" phase diagram of Bowen and Schairer.[7]

and that the entropy of mixing arises solely from the configurational entropy of mixing cations in a random manner, then for reasons explained elsewhere,[8] the relationships between activities (a) and mole fractions (N) are:

$$a(Mg_2SiO_4, Fe_2SiO_4) = N^2(Mg_2SiO_4, Fe_2SiO_4) \quad (1)$$

$$a(MgSiO_3, "FeSiO_3") = N(MgSiO_3, FeSiO_3) \quad (2)$$

A partial check on this conclusion can be made by means of the phase diagram for Mg_2SiO_4—Fe_2SiO_4 shown in Fig. 3. This diagram can be

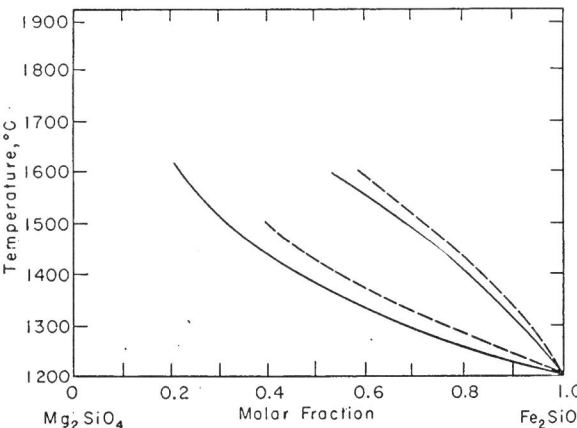

Fig. 3. Mg₂SiO₄—Fe₂SiO₄ phase diagram of Bowen and Schairer.[9] Broken lines indicate calculated phase boundaries.

calculated from equation 1 applied to both liquid and solid solutions, and the melting points and entropies of fusion of the two pure orthosilicates. The entropy change for Fe_2SiO_4 is 14.78 cal per degree[9] and that for Mg_2SiO_4, which has not been

measured, can be assumed to be the same. The calculated liquidus and solidus curves, which are shown by the broken lines in Fig. 3, agree reasonably with those actually measured.* The activities in the liquid, which can be calculated from the measured phase diagram, and the same entropies of fusion, on the assumption that the *solid* solutions are ideal, are shown in Fig. 4. The deviations

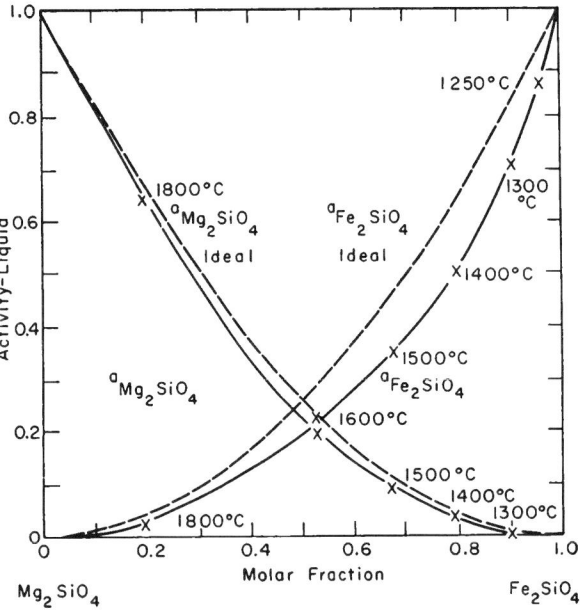

Fig. 4. Activities in the liquid mixtures of Mg₂SiO₄ and Fe₂SiO₄, calculated on the assumption that the solutions are ideal. Broken lines show the ideal values which would result from random mixing of Fe and Mg ions.

from ideality, which are not very great, would be compatible with the free energy of mixing curve for the solid solutions, which is ideal, and one for the liquid solutions, which has a maximum excess free energy of mixing (per mole Fe_2SiO_4 + Mg_2SiO_4) of about 700 cal at about 1500° C. A complication arises when equation 2 is applied to the solid solutions $MgSiO_3$—"FeSiO₃" and an attempt is made to calculate $MgSiO_3$ activities in the liquids. Magnesium metasilicate melts incongruently at 1557° C, so that although its entropy of fusion may be reasonably taken as equal to that for $CaSiO_3$, 8.3 cal per degree,[10] the value of the function $\Delta T \Delta S$ which is required in the calculations is uncertain. From the phase diagram for MgO—SiO_2,[11, 12] it is, however, possible to estimate from the slope of the liquidus line on the silica-rich side of the metasilicate composition that the theoretical congruent melting point would be about 1567° C. From this figure the activity values shown in Table 1 can be calculated, and from these it can be seen that over the range 0 to 54% "FeSiO₃," the $MgSiO_3$

* If equation 2 is applied, that is, cations are not assumed to be fully random, there is no agreement; for example, at 1600° C, solid phase Fe₂SiO₄ = 0.15 molar, liquid 0.74; 1400° C 0.39 and 0.93.

Table 1. Activities in Metal–Oxide—Silicate Systems

System	Molar Fraction of First Component	Temperature, °C	Calculated Activity of First Component		Remarks
1. $MgSiO_3$—"$FeSiO_3$"	0.75	1400	0.79		Solid solutions assumed ideal
	0.46	1500	0.47		
2. $CaSiO_3$—"$FeSiO_3$"	0.88	1500	0.91		
	0.80	1450	0.81		No solid solutions α-$CaSiO_3$
	0.69	1368	0.62		
	0.60	1284	0.51		
	0.60	1284	0.465		
	0.54	1245	0.38		Solid solutions β-$CaSiO_3$—$FeSiO_3$
	0.48	1200	0.29		
	0.41	1167	0.22		
3. $CaSiO_3$—$SrSiO_3$			$a(CaSiO_3)$	$a(SrSiO_3)$	
	0.85	1502	0.80	0.10	
	0.525	1477	0.45	0.37	Solid solutions assumed ideal
	0.26	1520	0.21	0.68	
4. $CaSiO_3$—$MgSiO_3$	0.905	1500	0.91		No solid solutions between
	0.80	1450	0.81		α-$CaSiO_3$ and $MgSiO_3$
	0.70	1368	0.62		
5. Mg_2SiO_4—Fe_2SiO_4			$a(MgSiO_4)$	$a(Fe_2SiO_4)$	
	0.81	1800	0.65	0.025	Solid solutions assumed ideal. For
	0.47	1600	0.20	0.20	ideality in these mixtures $a = N^2$
	0.20	1400	0.09	0.34	
	0.09	1300	0.007	0.70	
6. $CaSiO_3$—$MnSiO_3$	0.875	1500	0.89		α-$CaSiO_3$ solid solution contain-
	0.71	1450	0.76		ing up to 10% $MnSiO_3$ as-
	0.57	1400	0.65		sumed ideal with respect to
	0.52	1378	0.60		$a(CaSiO_3)$

activities are almost ideal. The calculation is not very sensitive to the value of the imaginary congruent melting point.

Activities of $CaSiO_3$ in liquid mixtures with "$FeSiO_3$" and $MgSiO_3$ can be calculated unequivocally from the phase diagrams[5, 13] and the entropy of fusion[10] and the melting point of $CaSiO_3$, over the region in which the liquid is in equilibrium with α-$CaSiO_3$ in which no solid solutions are formed, i.e., from 100 to 70% $CaSiO_3$. Beyond this range, β-$CaSiO_3$ forms extensive solid solutions with "$FeSiO_3$" but hardly any with $MgSiO_3$ because of the formation of diopside, $CaO \cdot MgO \cdot 2SiO_2$. If these "$FeSiO_3$" solid solutions are assumed to be ideal, and the thermodynamic properties of β-$CaSiO_3$ are taken as approximately the same as those of α-$CaSiO_3$, liquid $CaSiO_3$ activities down to about 40 mole % can be calculated for $CaSiO_3$—"$FeSiO_3$" mixtures, as shown in Fig. 5. The error involved in this assumption is clearly not great, because at 1284° C and 59.5 mole % $CaSiO_3$ (where pure α-$CaSiO_3$, β-$CaSiO_3$—"$FeSiO_3$" solid solution, and liquid $CaSiO_3$—"$FeSiO_3$" solution are in equilibrium) the value of $a(CaSiO_3)$ is equal to 0.51 if calculated from the α-$CaSiO_3$ equilibrium and 0.46 if calculated from the β-$CaSiO_3$ solid solution.

Activities in the liquid solutions $CaSiO_3$—$SrSiO_3$

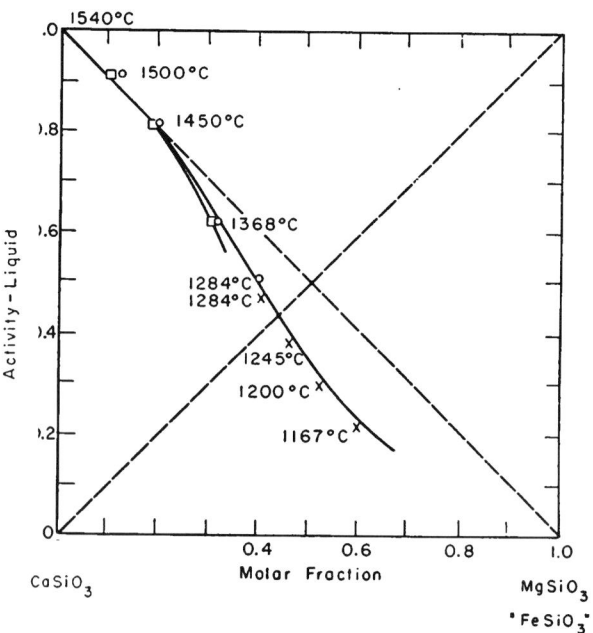

Fig. 5. Activities of $CaSiO_3$ in liquid mixtures.
□ $a(CaSiO_3)$ in $CaSiO_3$—$MgSiO_3$ from equilibrium with pure α-$CaSiO_3$;
○ $a(CaSiO_3)$ in $CaSiO_3$—"$FeSiO_3$" from equilibrium with α-$CaSiO_3$;
✕ $a(CaSiO_3)$ in $CaSiO_3$—"$FeSiO_3$" from equilibrium with β-$CaSiO_3$—"$FeSiO_3$" solid solutions.

can be calculated from the phase diagram,[14] if the entropy of fusion of $SrSiO_3$ is taken as equal to that for $CaSiO_3$ and the solid solutions are assumed ideal. The calculated activities do not depart far from the ideal values, but the minimum in the melting curves indicates that mixing in both liquid and solid phases cannot be ideal. If the solid mixing is taken as ideal, the integral free-energy curve for liquid mixing (per mole $CaSiO_3 + SrSiO_3$) must lie at its mid-point about 675 cal below the ideal curve at 1477° C. All the foregoing results are summarized in Table 1.

Table 1 also gives $CaSiO_3$ activities calculated from the phase diagram for $CaSiO_3$—$MnSiO_3$,[15] with the assumption of ideal solid solutions. This phase diagram is not, however, considered as reliable as the diagrams for the other systems discussed, and better evidence of ideal behavior in $CaSiO_3$—$MnSiO_3$ melts comes from the activity measurements discussed later in this paper.

Activities which can be calculated in the system Ca_2SiO_4—Fe_2SiO_4, for which the phase diagram* is given in Fig. 6, are shown in Fig. 7. Again the

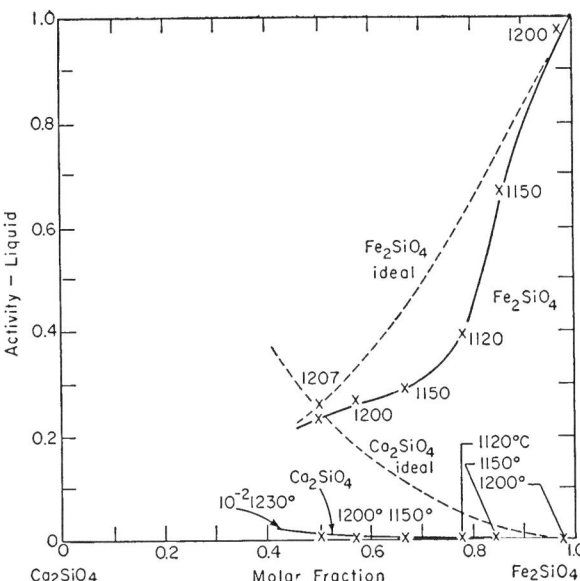

Fig. 7. Activities in the liquid system Ca_2SiO_4—Fe_2SiO_4 calculated on the assumption that the solid solutions are ideal.

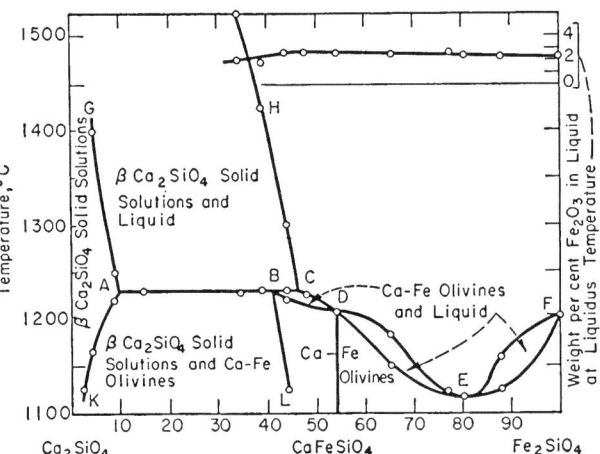

Fig. 6. Ca_2SiO_4—Fe_2SiO_4 phase diagram of Bowen, Schairer, and Posnjak.[16]

solid solutions have been taken as ideal, and it has also been assumed that the entropy of fusion of Ca_2SiO_4 is equal to the established value for Fe_2SiO_4.[9] The mixtures are clearly far from ideal. The extent of the interactions between the two components can be better visualized from Fig. 8, which shows (broken lines) the ideal integral free energies of mixing for this system at 1150° C. The true curve for the free energy of mixing of the solids may lie below but not much above the ideal curve A. This is because it must lie below the ΔG value for the all solid reaction

$$\tfrac{1}{2}Fe_2SiO_4 + \tfrac{1}{2}Ca_2SiO_4 = FeO + CaSiO_3 \quad (3)$$

* There is a slight error in the phase diagram because the solidus and liquidus curves cannot meet at D if the range of Ca—Fe olivines is continuous; instead the curves probably run smoothly from both B and C to E.

as the orthosilicate solid solution is stable with respect to wüstite and calcium metasilicate. The ΔG value shown in Fig. 8 for this reaction was calculated from the heats and entropies of formation of the respective crystalline compounds.[4, 17] For the purpose of the present discussion, the true curve for the solids may be taken as having the form of the full line with a break between Z and Z' because at 1150° C, two phases Ca—Fe olivine, Z, and β-Ca_2SiO_4 solid solutions, Z', are in equilibrium.

Because at this temperature the liquid phases X

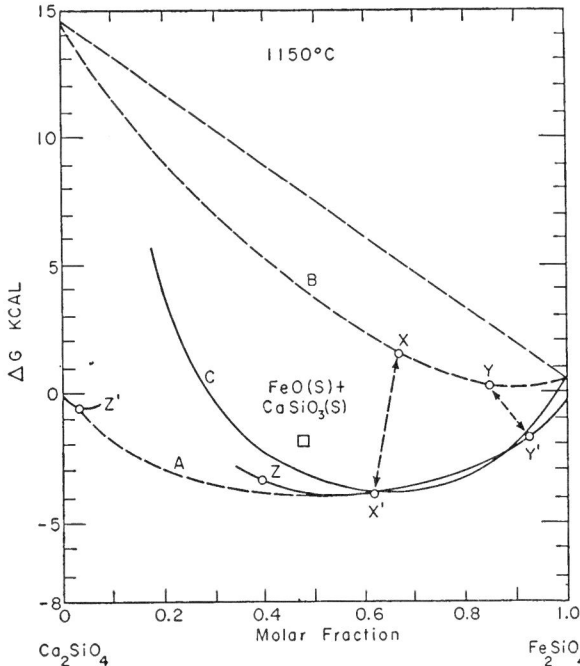

Fig. 8. Integral free energy of mixing curves for Ca_2SiO_4—Fe_2SiO_4 at 1150° C. Broken lines A and B ideal for solids and liquids; full lines probable curves for liquids (C) and solids X, Y', F'.

and Y (curve *B*) are in equilibrium with the solid phases X′ and Y′ (curve *A*), respectively, it follows that (provided the solid curve is correct) the true liquid curve must have some such form as the curve *C*, lying at the mid-composition some 7 kcal below the ideal one. If the true solid curve were lower than the proposed one, the deviation of the liquid curve from the ideal form would be even greater. The value of ΔG (1150° C) assumed for the fusion of Ca_2SiO_4 might be substantially in error as 1150° C is 980° C below the melting point of the orthosilicate, but the heat-capacity data on liquid and solid Fe_2SiO_4 suggest that the error in ΔG is not likely to be more than 30%. Such an error would only reduce the 7-kcal deviation mentioned above to about 4.5 kcal. It would thus seem that there must in any event be strong interactions between these two silicates in the molten state. Similar calculations have been made for the system Fe_2SiO_4—Mn_2SiO_4, but they lead to very low Fe_2SiO_4 activities in the melts if the activities in the continuous series of solid solutions are assumed ideal. No significance has, however, attached to these values, as the phase diagram[18] is admittedly only tentative.

OXIDE ACTIVITIES IN TERNARY MELTS

Since there is evidence from phase diagrams to suggest that the mixing of some silicates, at least, is ideal, it is worth seeing how far such ideal behavior can account for the metal-oxide activities which have been measured in the melts FeO—CaO—SiO_2 by Taylor and Chipman,[1] and in MnO—CaO—SiO_2, PbO—CaO—SiO_2, PbO—MgO—SiO_2, PbO—ZnO—SiO_2 in this laboratory by Drs. M. W. Davies and T. C. M. Pillay. These new results are being published in full detail elsewhere.

McCabe[19] has recently measured silica activities at 1637° C in the molten system CaO—SiO_2 over the range from saturation with solid silica to saturation with solid Ca_2SiO_4. It is thus possible to

derive the corresponding lime activities by application of the Gibbs-Duhem relationship and use of the free energy of formation of crystalline Ca_2SiO_4, which can be derived from the heat of formation of β-Ca_2SiO_4 and heat-capacity data as shown by Fulton and Chipman.[4]* These activities can be assigned to 1600° C without much error, as the heats of mixing are probably not great,[20] and from them a free energy of mixing curve can be constructed. This curve, which is in fair agreement with that derived earlier by Richardson[8, 20] from phase-diagram data (maximum deviation +700 cal), is shown in Fig. 9, together with the corresponding curves for FeO—SiO_2, taken from the results of Taylor and Chipman[1] and Schuhmann and Ensio,[21] and for MnO—SiO_2, derived from the measurements of Davies referred to immediately above.

METHOD OF APPROXIMATION

If it is assumed that the free energy of mixing a binary silicate melt (such as CaO—SiO_2) with another (such as FeO—SiO_2, MnO—SiO_2, and so forth) of similar metal oxide/silica ratio is ideal, i.e., that the heat of mixing is zero and that the entropy of

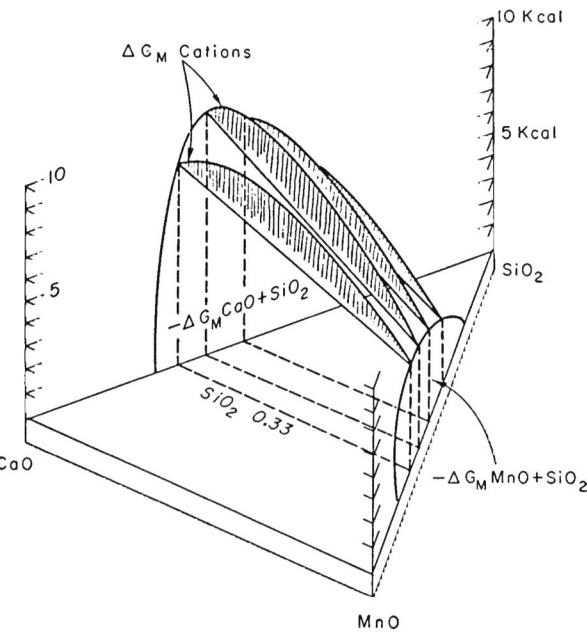

Fig. 10. Model illustrating the build-up of the integral free-energy surface for the formation of CaO—MnO—SiO_2 melts from CaO, MnO, and SiO_2. The surface lies along the upper edges of the sections marked "ΔG_M cations." ΔG_M cations $= RT\,N(MnO)\,\ln \dfrac{N(MnO)}{N(MnO)+N(CaO)} + RT\,N(CaO)\,\ln \dfrac{N(CaO)}{N(MnO)+N(CaO)}$

mixing is only the configurational one of mixing the cations in a constant anionic matrix, it is then obviously possible to calculate the free energies of formation of mixtures (ΔG_M) CaO—MnO—SiO_2, CaO—FeO—SiO_2, and so on, from the three pure oxide components. Figure 10 shows a model illustrating

* Page 1145.

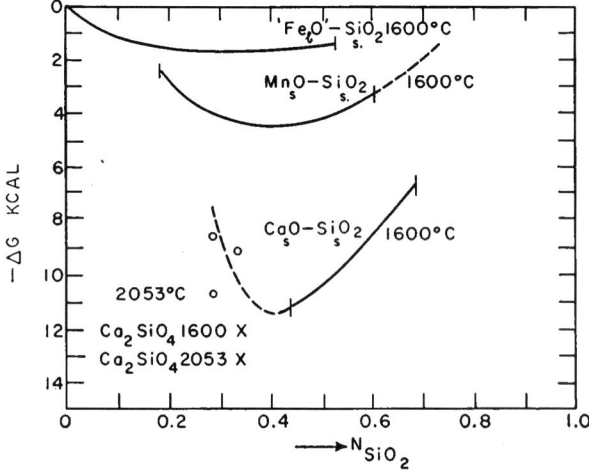

Fig. 9. Integral free energies of mixing at 1600° C per mole of solid (s) or liquid (l) oxide.

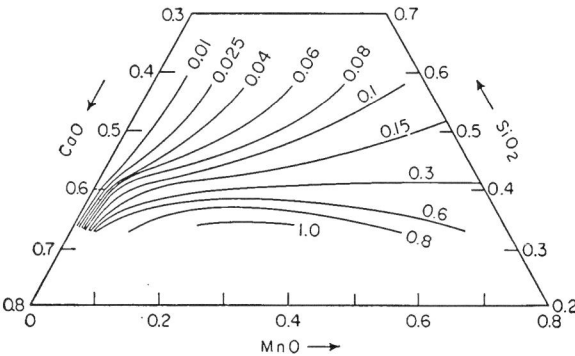

Fig. 11. Calculated isoactivity contours for MnO in CaO—MnO—SiO₂ melts at 1600° C as a function of molar composition.

the build-up of these ΔG_M values for the CaO—MnO—SiO₂ system. The partial molar free energies and hence activities of the separate oxide components, such as FeO, MnO, and so forth, can then be derived by plotting the ΔG_M values at constant CaO/SiO₂ ratio as a function of the mole fractions, $N(\text{FeO})$, $N(\text{MnO})$, and the like, drawing tangents at the desired compositions, and measuring the intercepts on the ΔG_M axis.

As it is difficult to draw the tangents accurately,

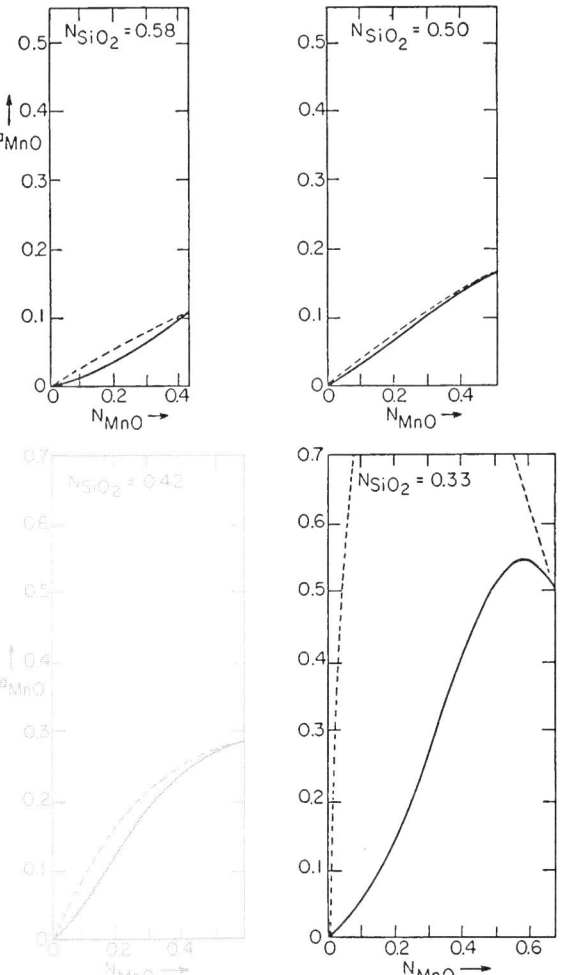

Fig. 12. MnO activities in CaO—MnO—SiO₂ melts. Full lines experimental values; dotted lines calculated.

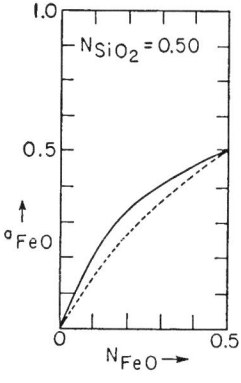

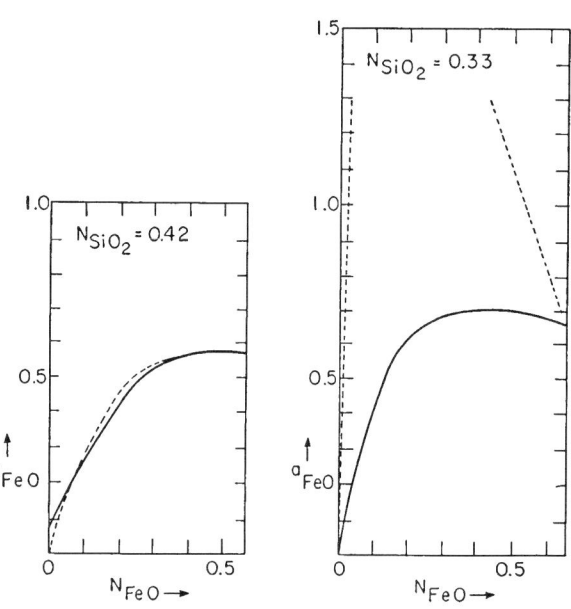

Fig. 13. FeO activities in CaO—FeO—SiO₂ melts. Full lines experimental; dotted lines calculated.

it is better and fully equivalent to calculate the MnO (or FeO) activities as follows. Calculate $a(\text{SiO}_2)$, along constant $N(\text{SiO}_2)$ lines of the ternary, by *linear* interpolations of log $a(\text{SiO}_2)$ between the two binary limits, such as $x\text{CaO}—\text{SiO}_2$ and $x\text{MnO}—\text{SiO}_2$. Then apply the restriction that the activities of both binary silicates in the ternary are ideal, i.e.,

$$a(x\text{MnO}\cdot\text{SiO}_2) = N(x\text{MnO}\cdot\text{SiO}_2)^x \qquad (4)$$

Then,

$$a(x\text{MnO}\cdot\text{SiO}_2) = \frac{a^x(\text{MnO})\cdot a(\text{SiO}_2) \text{ ternary}}{a^x(\text{MnO})\cdot a(\text{SiO}_2) \text{ binary}}$$

$$= N^x(x\text{MnO}\cdot\text{SiO}_2) \qquad (5)$$

The only unknown in equation 5 being $a(\text{MnO})$ ternary, this quantity can be readily calculated. Isoactivity contours calculated in this way for MnO at 1600° C are shown in Fig. 11. The general pattern is similar to that obtained experimentally.

Comparisons between the calculated and measured FeO and MnO activities can best be made by plotting them as a function of composition at constant molar fraction of silica. These plots are shown in Figs. 12 and 13. In agreement with the conclusions reached from the phase diagrams, it can be seen that in the metasilicate region mixing is approximately ideal, whereas in the orthosilicate region there are marked negative deviations.* These conclusions are the reverse of those commonly held hitherto, as it has been customary to consider the high activity coefficients of FeO in the CaO—FeO—SiO_2 melts as due to positive deviations from ideal mixing.

EXPLANATION OF OBSERVED BEHAVIORS

The reasons for the marked negative deviations from ideality in the case of Ca_2SiO_4—Fe_2SiO_{24} melts (much greater than for Mg_2SiO_4—Fe_2SiO_4 melts) are uncertain. There are two general possibilities. First, the anionic matrix may maintain its same general structure from one composition limit to the other but permit cations of different sizes† to pack into it much better than uniform cations; this could lead to correspondingly closer cation to $-O'$ distances and greater coulombic binding energies. Alternatively, there may be a marked change of the anionic structure across the composition range. One would expect the first possibility to exercise some influence on the solid solutions and to lead to ordered structures at particular compositions. This appears to be the case with diopside, $CaMg(SiO_3)_2$, and monticellite, $CaMgSiO_4$,[22] but X-ray evidence on other mixed anhydrous silicates and silicate solid solutions is unfortunately lacking. Evidence for or against closer packing in the melt could be provided by density measurements, and it is interesting to note that the melt $CaSr(SiO_3)_2$ has at 1700° C a density $3.1 \pm 0.2\%$ less than the mean of $CaSiO_3$ and $SrSiO_3$.[23] This might well account for the small negative deviations from ideal mixing inferred for these mixtures from their phase diagram. Similar changes might also account for the relatively small deviations from ideality such as occur in the mixtures approaching the metasilicate composition which have been considered previously.

Marked changes in the anionic matrix which might account for the larger discrepancies observed in the Ca_2SiO_4—Mn_2SiO_4 and Ca_2SiO_4—Fe_2SiO_4 melts are harder to visualize, but evidence might be sought for them also in density measurements. The orthosilicate groups might join together in rings to give metasilicate groups associated mainly with calcium ions, leaving residual oxygen ions associated with manganous or ferrous ions. Such a change could be represented by the equation

$$Ca_2SiO_4 + Fe_2SiO_4 = 2CaSiO_3 + FeO \quad (6)$$

For the pure liquids in equation 6, Fig. 9 shows that $\Delta G°$ (1600° C) is about -3.1 kcal, so such a reaction might well occur to a substantial degree in the melt. In spite of the fact that the change would actually give more recognizably "free FeO" in the melt, it could lead to a lowering and *not* to a raising of the FeO activity coefficients, as it would *lower* the values of ΔG_M from the pure oxides. This, however, is not an attractive explanation, as in the case of Mn_2SiO_4 the corresponding free-energy change is slightly positive ($+0.7$ kcal). It seems more probable that in these cases also it is markedly better packing of mixed cations in the anionic matrix which gives rise to the lower activities. It is suggestive that the compounds $CaMgSiO_4$ (monticellite) and $Ca_3Mg(SiO_4)_2$ (merwinite) have heats of formation from the pure orthosilicates (forsterite and β-Ca_2SiO_4) at 25° C of -4.9 ± 0.3 and -4.2 ± 0.6 kcal.[24] These compounds consist of discrete $SiO_4{}^{4-}$ ions surrounded by cations which are ordered at low temperatures,[22] their structures being similar to but not identical with those of the pure orthosilicates. This is probably the reason why no solid solutions exist between Ca_2SiO_4 and $MgCaSiO_4$, and only an incomplete range between $CaMgSiO_4$, and Mg_2SiO_4, for even at 1400° C the free energy of formation of an ideal random solid solution would only be -4.6 kcal. If such heat changes can arise in the formation of mixed crystalline silicates, it is quite likely that similar or even greater heat changes can arise in the formation of liquid silicate mixtures.

It is interesting to note that the activity calculations based on ideal cation mixing indicate that miscibility gaps should exist when the differences between the binary free energy of mixing curves are great. They also indicate that gaps should have the same form as those observed by Oelsen and Maetz in the system CaO—"FeO"—P_2O_5[25] where one liquid phase in each conjugate pair is rich in "FeO," i.e., in the less basic oxide, and the other is rich in CaO and P_2O_5, i.e., in the more basic and the acid oxide. Indeed, such immiscibility would occur in metal-oxide-rich parts of the CaO—FeO—SiO_2 and CaO—MnO—SiO_2 systems were it not for the interactions between the orthosilicates, and the high melting temperature of MnO.

For ideal mixing it has been stated that if two silicates of equal silica mole fraction are mixed, the

* In the system CaO—FeO—SiO_2, the accuracy of the comparison is somewhat impaired by the 10 mole % MgO present in the ternary slags for which measurements were made by Tayor and Chipman.[1] In the orthosilicate region, the accuracy of the calculated values in both ternary systems is in greater doubt than in the metasilicate region because the ΔG_M curve for CaO—SiO_2 has to be estimated at lime concentrations greater than 0.58 molar—the saturation limit for the melt in equilibrium with Ca_2SiO_4 at 1600° C. This estimation depends on the two points shown in Fig. 9. The value of ΔG (formation) for liquid Ca_2SiO_4 at 1600° C was calculated from ΔG (fusion) for Ca_2SiO_4 at 1600° C and ΔG (formation) of the crystalline orthosilicate at this temperature. This term for the liquid containing 28 mole % SiO_2 was calculated first for 2053° C, at which temperature it is the eutectic in equilibrium with crystalline Ca_2SiO_4 and lime. This value was then corrected to 1600° C, on the assumption that the heat of formation of this mixture is approximately zero.[20]

† The cation radii for Ca^{2+}, Mn^{2+}, Fe^{2+}, and Mg^{2+} for sixfold coordination with oxygen as occurring in silicates are 1.06, 0.91, 0.83, and 0.78, respectively.

log of the activity of the silica is a linear function of the proportions of the two metal oxides. It follows that the activity coefficient will be *raised* for the metal oxide which in the binary gives the *higher* silica activity, and *lowered* for the metal oxide which gives the *lower* silica activity. These changes will be greater, the greater the differences between the silica activities in the binaries.

The recent work already mentioned on the activities of PbO in the ternary silicate melts at 1000 to 1200° C shows that, whereas CaO ond MgO both increase the PbO activity coefficients over the range 0.2 to 0.6 mole fraction SiO_2, zinc oxide raises it at silica mole fractions less than 0.35 and lowers it at higher values. The actual results at 1200° C are shown in Fig. 14. This is quite reasonable, for

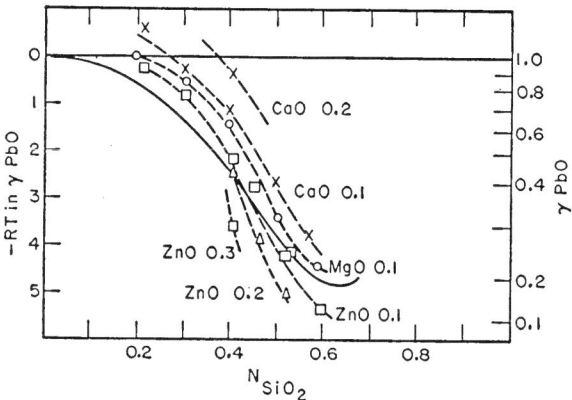

Fig. 14. The influence of CaO, MgO, and ZnO on the activity coefficients of PbO in ternary melts with SiO_2 at 1200° C.

although the ΔG_M curve for ZnO—SiO_2 [8, 20] almost certainly lies above that for PbO—SiO_2 [26] even at 1200° C, it is quite possible that at low silica concentrations the values of $\Delta \bar{G}(SiO_2)$ are more negative in ZnO—SiO_2 mixtures than in $PbO \cdot SiO_2$ mixtures, and vice versa. The direction of the effects observed for MgO and CaO are as expected.

SUMMARY

Reasoning from silicate phase diagrams and thermal data suggests that mixtures of two molten metasilicates behave nearly ideally (relative to the separate silicates) even when there are large differences between the sizes of the cations and the heats of formation of the separate silicates. In the orthosilicates, mixing is nearly ideal in the case Fe_2SiO_4—Mg_2SiO_4, where the cation sizes are much the same, but far from ideal when they are significantly different, such as Fe_2SiO_4—Ca_2SiO_4, Mn_2SiO_4—Ca_2SiO_4.

Measured activities of FeO and MnO in the ternary melts FeO—CaO—SiO_2 and MnO—CaO—SiO_2 support this conclusion, and mixing appears to be approximately ideal over the composition range $N(SiO_2) = 0.58$ to 0.42. It is suggested that the excess free energies of mixing near the ortho-

silicate compositions arise from better packing of mixed than of uniform cations in the anionic silicate matrix.

In a mixture of two melts of equal silica molar fraction, the logarithm of the activity of silica is a linear function of the proportions of the two metal oxides, in the region of "near-to-ideal" mixing. The activity coefficient is raised for that metal oxide which in its binary melt (with silica) has the higher silica activity, and vice versa. The changes will be greater, the greater the differences between the silica activities in the binary melts.

References

1. C. R. Taylor and J. Chipman, *Trans. A.I.M.E.*, Tech. Pub. 1499 (1942).
2. L. S. Darken, *Thermodynamics in Physical Metallurgy*, p. 28, American Society for Metals, Cleveland, 1950.
3. J. F. Elliott, *J. Metals*, 7, 485 (1955).
4. J. C. Fulton and J. Chipman, *J. Metals*, 6, 1136 (1954).
5. N. L. Bowen, J. F. Schairer, and E. Posnjak, *Am. J. Sci.*, 5th Ser., 26, 213 (1933).
6. T. G. Sahama and D. R. Torgeson, *U. S. Bur. Mines Rept. Invest.*, No. 4408 (1949).
7. N. L. Bowen and J. F. Schairer, *Am. J. Sci.*, 5th Ser., 29, 163 (1935).
8. F. D. Richardson, *The Physical Chemistry of Melts*, p. 91, Institution of Mining and Metallurgy, London, 1953.
9. R. L. Orr, *J. Am. Chem. Soc.*, 75, 528 (1953).
10. F. C. Kracek and K. J. Neuvenon, Private communication.
11. N. L. Bowen and O. Anderson, *Am. J. Sci.*, 4th Ser., 37, 488 (1914).
12. J. W. Greig, *Am. J. Sci.*, 5th Ser., 13, 133 (1927).
13. J. F. Schairer and N. L. Bowen, *Am. J. Sci.*, 240, 725 (1942).
14. P. Eskola, *Am. J. Sci.*, 5th Ser., 4, 353 (1922).
15. E. Voos, *Z. anorg. u. allgem. Chem.*, 222, 213 (1935).
16. N. L. Bowen, J. F. Schairer and E. Posnjak, *Am. J. Sci.*, 5th Ser., 25, 281 (1933).
17. F. D. Richardson, J. H. E. Jeffes, and G. Withers, *J. Iron Steel Inst. London*, 166, 213 (1950).
18. J. White, *J. Iron Steel Inst. London*, 148, II, 586 (1943).
19. C. L. McCabe, Private communication.
20. C. J. B. Fincham and F. D. Richardson, *Proc. Roy. Soc. London*, A, 223 56 (1954).
21. R. Schuhmann and P. J. Ensio, *J. Metals*, 3, 401 (1951).
22. W. L. Bragg, *Atomic Structure of Minerals*, Cornell University Press, Ithaca, New York, 1937.
23. J. W. Tomlinson, Private communication.
24. K. J. Neuvonen, *Am. J. Sci.*, 250, Bowen Volume, 373 (1952).
25. W. Oelsen and H. Maetz, *Mitt. Kaiser-Wilhelm-Inst. Eisenforsch. Düsseldorf*, 23, 195 (1941).
26. F. D. Richardson and L. E. Webb, *Bull. I.M.M.*, 584, 529 (1955).

Discussion

KING inquired whether Richardson had a mental picture of the possible mechanism of interaction between the two dissimilar metal cations in the ternary melts. After some consideration, RICHARDSON pointed out that in the metasilicates the packing in the material is dominated by the silicate structure, which is apparently fairly rigid. In the orthosilicates, however, the structure is made up of much smaller silicate groups. It thus seems reasonable to expect that with these structural components closer packing could more easily result from the presence of mixed cations, and this would lead to nega- tive deviations from ideal mixing. Values of the order of 2 kcal for ΔG_M (excess) per mole of oxide are sufficient to account for the seemingly large effects observed on the partial free energies (i.e., activities).

by G. Trömel

Comments on the Equilibrium

Diagram Al₂O₃—SiO₂

More than thirty years ago one of the basic equilibrium diagrams, the system Al_2O_3—SiO_2, was determined by Bowen and Greig.[1] In the last few years the incongruent melting point of the compound "mullite," $3Al_2O_3 \cdot 2SiO_2$, has been questioned by Toropow and Galakhow,[2] and by Budnikow and his collaborators.[3] Figure 1 illustrates

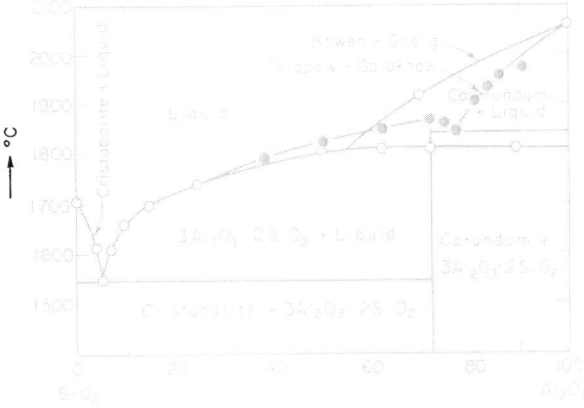

Fig. 1. System SiO₂—Al₂O₃.

the two different interpretations. The fact that it is possible to produce large mullite crystals from the melt with the "Flame Fusion Synthesis" is also most consistent with a congruent melting point.

In co-operation with Konopicky and his co-workers (Forschungs-Institut der Feuerfest-Industrie, Bonn), we have tried to find an explanation for these different results in the system SiO_2—Al_2O_3 because of its great general importance for the knowledge of metallurgic slags, as well as for many

Dr. Trömel is affiliated with the Max-Planck-Institut für Eisenforschung, Düsseldorf, Germany.

ceramic products. In order to avoid errors, purest SiO_2 (from Si-ester) and purest Al_2O_3 (also produced from organic Al-compounds) were used as basic materials. Both products were found to be spectroscopically pure. In particular, they are free from alkalies. We have used the quenching method insofar as it is possible at the necessary high temperatures.

The composition has been checked after the final heating, for example, at 1830° C, since it is known that SiO_2 evaporates easily at such a high temperature. Few SiO_2 losses are noticeable. They become much greater if the sample is completely melted and heated to higher temperatures. Further, it was established that alkalies were not absorbed by the samples by heating up to 28 hr.

LIQUIDUS LINE

Our observations of the mixtures' melting points show a course of the liquid line which, in the main, tallies—within the necessarily large margin of error at these high temperatures—with the statements of Bowen and Greig, particularly in the area of mullite. The line in the SiO_2-rich region is flatter than assumed by Bowen and Greig. As they claimed, the temperature of the eutectic was found at 1545° C. The composition of the eutectic, however, we found at approximately 10% SiO_2.

Samples with Al_2O_3 contents up to 78% showed mullite without corundum, even if they were completely molten and heated for 1 or 2 hr. Thus the transformation of mullite to corundum and a liquid phase described by Bowen and Greig could not be observed, as already stated by Toropow and Galakhow. But after the samples were heated for at least 4 hr at temperatures higher than 1820° C, corundum was observed already at a content of 54% Al_2O_3. If the same experiment is made, not with purest mullite, but with one crystallized under technical conditions and containing especially alkalies, the decomposition appears sooner.

MELTING POINT OF MULLITE

These observations on the influence of the heating time and of the impurities explain the varying interpretations of the melting point of mullite. Toropow and Galakhow as well as Budnikow have heated their samples for only a relatively short time at high temperatures and have melted them in a vacuum.*

Bowen and Greig's diagram shows the real equilibrium which appears between the purest substances as a result of lengthy heating. It may be that the same results can be achieved by a shorter heating period if the substances contain small amounts of alkalies. On the other hand, the findings of Toropow, Galakhow, and Budnikow hold good for higher heating and cooling speeds.

The crystallization of mullite by rapid cooling of the melts with contents of 54 to 78% Al_2O_3 can be expected because of the slow diffusion process which is necessary for the precipitation of corundum. Thus only mullite occurs in these melts by rapid cooling and also by short heating time to a temperature of more than 1820° C.

COMPOSITION RANGE OF MULLITE

Bowen and Greig's diagram has to be improved in other respects. As Shears and Archibald[4] have already suggested, basing their opinions on a survey of publications, an area of homogeneity must be ascribed to mullite. The experiments to determine the boundaries of this area have not yet been completed. Up to now we have not noticed a higher content of Al_2O_3 than 72 to 78% in mullite. However, there may be some with a considerably lower content. It is very difficult to determine the exact boundary of the solid solution area on the SiO_2 side. With a SiO_2 content of more than 60%, the melts no longer crystallize.

From this formation of solid solutions, it is concluded that the composition of mullite cannot be described by a simple chemical formula. The formula $3Al_2O_3 \cdot 2SiO_2$ used until now will probably indicate the boundary of the solid solutions' area on the Al_2O_3 side.

* Under these conditions the formation of SiO can be expected. This effect probably somewhat lowers the melting point.

The reason for the small differences in the position of X-ray interferences between the composition $Al_2O_3 \cdot SiO_2$, sillimanite, and $3Al_2O_3 \cdot 2SiO_2$, mullite, has often been discussed. Scholze[5] recently re-examined this problem. He finds again that sillimanite and mullite have in the main the same structure.

In measuring our X-ray diagrams we established that the described differences in the position of particular X-ray interferences between sillimanite and mullite are dependent on the Al_2O_3 content. This result is still another reason for the assumption that mullite has an area of homogeneity. It possibly extends to the composition $Al_2O_3 \cdot SiO_2$.

CONCLUSIONS

These results are tentatively illustrated in Fig. 2.

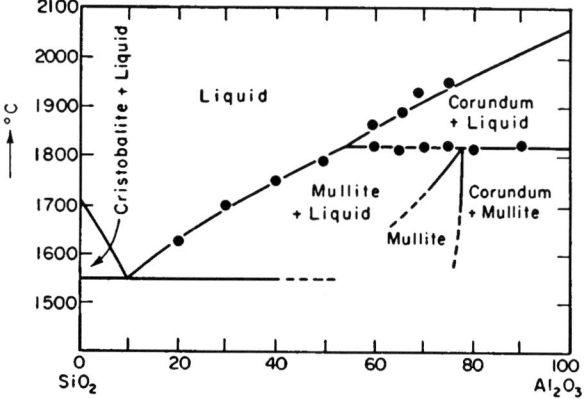

Fig. 2. The results in the system SiO₂—Al₂O₃.

More exact examinations should show where the boundary of the solid solutions' area really lies.

References

1. N. L. Bowen and J. W. Greig, *J. Am. Ceram. Soc.*, **7**, 242 (1924).
2. N. A. Toropow and F. J. Galakhow, *Ber. Akad. d. Wiss. U.S.S.R.*, **78**, 299–302 (1951).
3. P. P. Budnikow, S. G. Treswjatskij, and W. J. Kuschakowskij, *Ber. Akad. d. Wiss. U.S.S.R.*, **93**, S. 281 (1953).
4. E. C. Shears and W. A. Archibald, *J. Iron Steel Inst. London*, **27**, 26–30 and 61–66 (1954).
5. H. Scholze, *Ber. deut. keram. Ges.*, **32**, 381–385 (1955).

Discussion

MUAN reported that some of his recent work substantiated the fact that mullite can dissolve some excess silica, as was reported in the paper. Further, iron oxide can also dissolve in mullite. With iron oxide present, the sluggishness of the reaction encountered in the simple binary system is avoided.

by W. A. Fischer

Electrical Measurements on Oxides and Solid Cells at Temperatures of 1200 to 1800°C

For the study of the behavior of oxides at high temperatures, we used solid cells of the general type $Pt/MeO + Me_2O_3/Me_2O_3/Pt$ and with these carried through measurements in the systems FeO—Al_2O_3 and MgO—Al_2O_3 at 1500° C. A report on these measurements has already been given.[1,2] Therefore this report is restricted to a representation of the essential results. Figure 1 shows the ap-

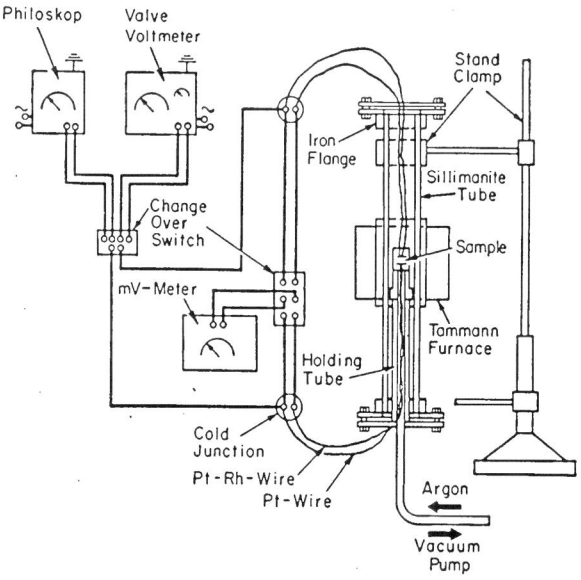

Fig. 1. Apparatus for measuring.

paratus used. It consists of a Tamman furnace with a vacuum-tight sealable ceramic tube of sillimanite

Dr. Fischer is affiliated with the Max-Planck-Institut für Eisenforschung, Düsseldorf, Germany.

in which the specimen is placed. A section of the specimen is shown in Fig. 2. The specimen con-

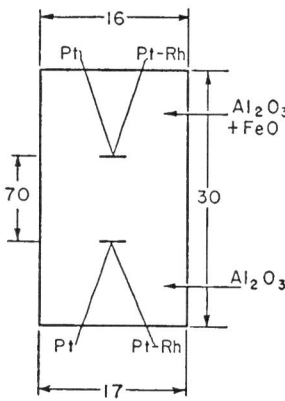

Fig. 2. Section of a specimen, dimensions in mm.

sists half of aluminum oxide and half of mixtures of various composition of ferrous oxide or magnesium oxide and aluminum oxide. At equal distances from the phase boundary (0.5 cm), there are two platinum electrodes, each with Pt—Rh thermocouples welded on. The emf values are measured between two legs of these thermocouples by means of a null-point electronic voltmeter, with both electrodes at 1500° C. All measurements were made in argon at a pressure of 1 atm.

IRON OXIDE—ALUMINA SYSTEM

Figure 3 represents the emf values as dependent on the FeO content measured in the cell $Pt/FeO +$

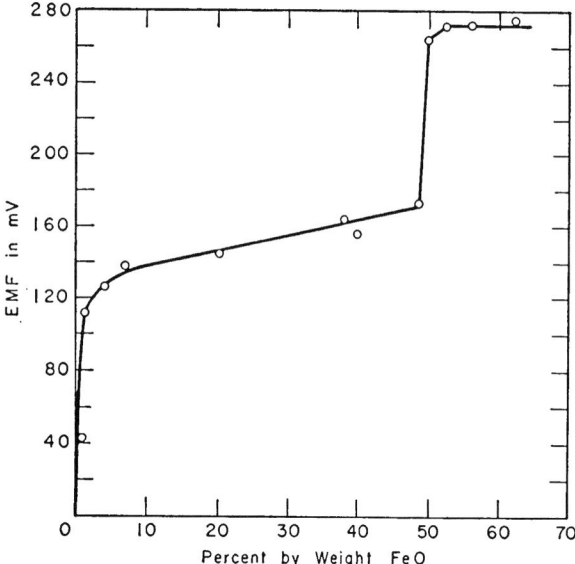

Fig. 3. Potential of cell Pt/FeO + Al₂O₃/Al₂O₃/Pt.

Al₂O₃/Al₂O₃/Pt.[1]. The interpretation of the shape of this curve was as follows: the first steep rise of the emf values corresponds to the formation of an aluminum-oxide solid solution containing FeO. The saturation of this solution will be attained at approximately 1.5% FeO. This solid solution is in equilibrium with the spinel FeO·Al₂O₃. Within the heterogeneous range of these two phases the emf should remain constant, but it does not. The relatively small rise of the emf values might be explained by the fact that the trivalent iron, contained in the FeO, as magnetite forms a solid solution with the spinel FeO·Al₂O₃, and that with increasing FeO contents the portion of magnetite in the spinel solid solution increases. With the first appearance of molten wüstite above 48.8% FeO, a sudden increase of the emf from 170 to 270 mv takes place. In the succeeding heterogeneous phase range between the molten wüstite and the spinel, this

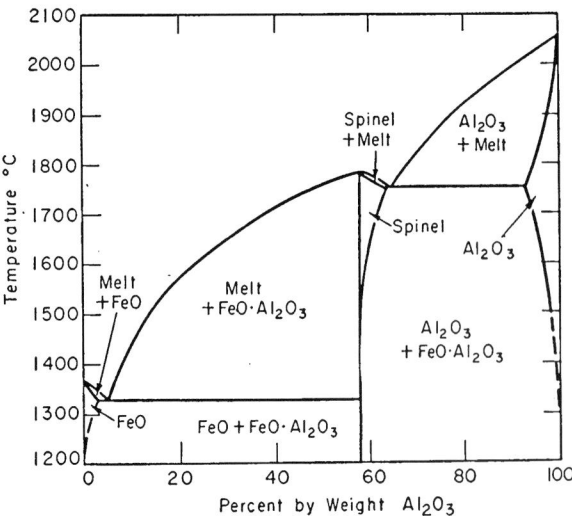

Fig. 4. Diagram of the system FeO—Al₂O₃.

value does not change further until about 63% FeO is reached. Careful thermal analyses as well as metallographic and X-ray examinations have confirmed these conclusions about the system FeO—Al₂O₃ drawn from the emf measurements.

The new diagram FeO—Al₂O₃ is represented in Fig. 4, where, however, the trivalent iron has been ignored.[3]

In addition to the determination of small solubilities, it is possible as well to determine larger ranges of solid solution by means of these emf measurements. Thus, Fig. 5 shows the emf values of the

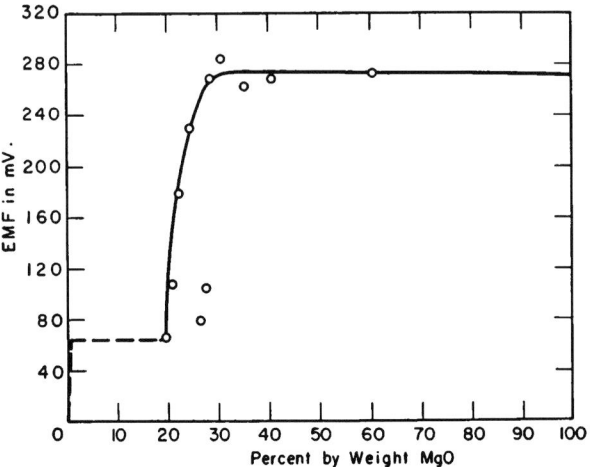

Fig. 5. Potential of cell Pt/MgO + Al₂O₃/Al₂O₃/Pt.

cell Pt/MgO + Al₂O₃/Al₂O₃/Pt as dependent on the MgO-content.[2] This figure enables us to recognize the ample solubility of the spinel MgO·Al₂O₃ for aluminum oxide which, as shown, at 1500° C extends from 71.7% with stoichiometric composition of the spinel to 80.5% Al₂O₃. It is of interest, furthermore, that here—contrary to the situation in the cell Pt/FeO + Al₂O₃/Al₂O₃/Pt—the potential remains constant when the composition of the spinel is exceeded. There exists, evidently, no solubility of the spinel for MgO.

In our opinion the emf's described may be explained by a diffusion of the divalent and trivalent cations. Since the electrode in pure aluminum oxide is always the negative pole of the cell, here more positive charges are leaving than arriving. We concluded therefrom that the diffusion velocity of the aluminum ions is higher than that of the iron ions. This opinion is confirmed by the observation of a Kirkendall effect, which is shown in Fig. 6 by the example of a cell with 40% FeO.[4] On the aluminum-oxide side a constriction and formation of holes adjacent to the phase boundary occurs, and on the side containing FeO there is a slight constriction. As a result of these investigations it can be stated that in the solid cells described cation conduction is predominant. With the transfer numbers being known, the free energy of the spinel

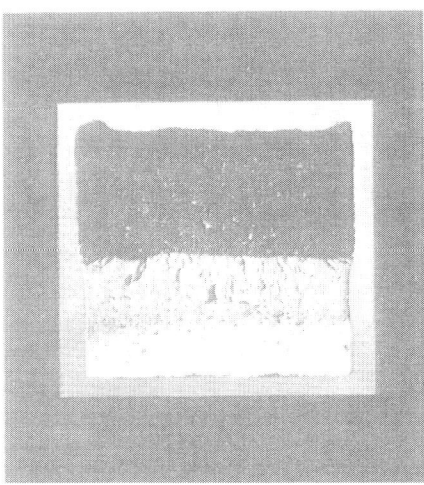

Fig. 6. Kirkendall effect specimen of Al_2O_3 and $FeO \cdot Al_2O_3$ after heating at 1500° C during 50 hr.

$FeO \cdot Al_2O_3$ can be determined from the emf values according to the equation

$$\Delta G = -(n_{Al} - n_{Fe})z \cdot F \cdot E \qquad (1)$$

The change in free energy of the process is equal to the negative difference of the transfer numbers multiplied by the valency, Faraday's constant, and the emf. We have estimated $(n_{Al} - n_{Fe})$ at 0.2 to 0.4. Then ΔG can be calculated at -7500 to $-15,000$ cal/mole.[1] A calorimetric determination[5] resulted in the value of $14,600 \pm 25\%$ cal/mole for ΔH, which is in good agreement with the value of ΔG.

OTHER TRIVALENT OXIDES

The next question of interest was the behavior of other trivalent oxides, where special attention was devoted to Cr_2O_3. Contrary to the results described above, no measurable potentials were obtained with the cells $Pt/FeO + Cr_2O_3/Cr_2O_3/Pt$ having FeO contents of 10% to 50% in the tempera-

ture range of 1200 to 1800° C. This implies that these cells probably are predominantly electron-conducting, a supposition which was confirmed also by diffusion measurements of Lindner.[6]

The following investigations[7] were planned to determine the type of conduction and hence the model of disorder of the Cr_2O_3. Figure 7 gives a survey of the methods which may be applied for this purpose. All were used by us except the Hall-effect measurements.

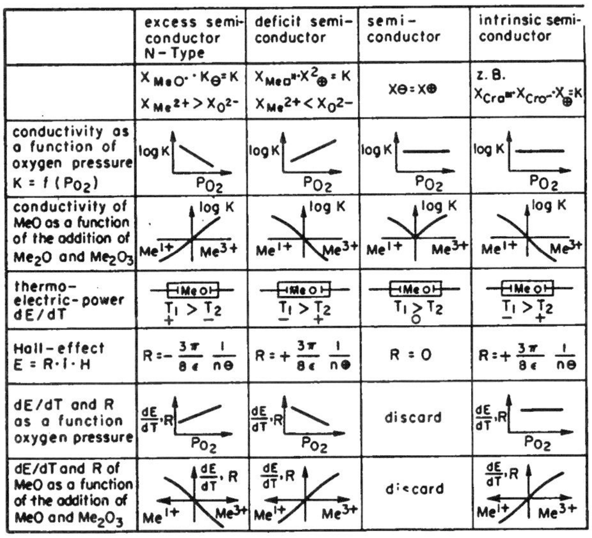

According to the investigations on the conductivity by Hauffe and Block[8] which, however, were carried up to 800° C, the Cr_2O_3 generally behaves like a deficit semiconductor, but, unlike the latter, it shows no dependency of the electric conductivity on the oxygen partial pressure of the atmosphere. This type of semiconductor is called "Eigenstörstellenhalbleiter." In extending these investigations to 1500° C, we found an extremely small dependency of the electric resistance on the oxygen partial pressure. In addition to these measurements of the conductivity, we carried through measurements of the thermoelectric power in the temperature range of 800 to 1500° C. The negative pole of the solid cell $Pt/Cr_2O_3/Pt$ was always the hotter electrode, indicating an electron defect conduction. The thermoelectric power also did not show any dependency on the oxygen partial pressure at all temperatures. The results of these measurements are represented in Fig. 8. The open circles refer to

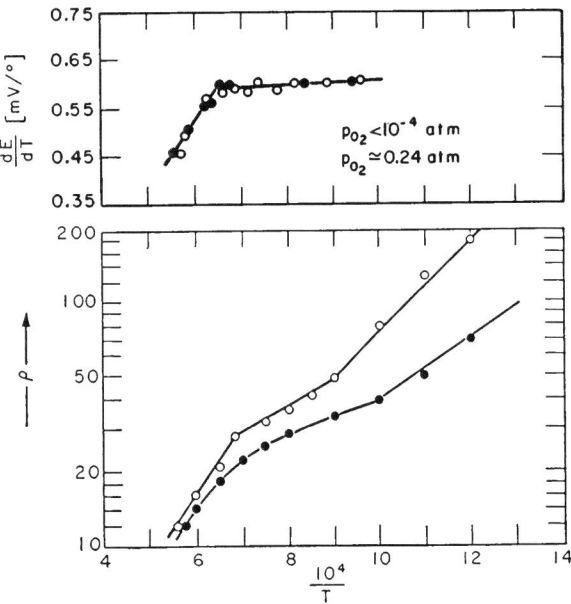

Fig. 8. Thermoelectric power and log of electrical resistance as a function of reciprocal of absolute temperature for various partial pressures of oxygen.

the measurements carried out in argon ($p_{O_2} < 10^{-4}$ atm), the closed circles to those carried out in air at 1 atm. At 1250° C a sharp change in the direction of the thermoelectric power as well as of the resistance curve will be observed for the first time, a

Fig. 7. Experimental methods for identifying semi-conductor types.

result which doubtless indicates a change of the conduction process and hence of the disorder. From these measurements of the thermoelectric power as a function of the reciprocal of the absolute temperature, we calculated the activation energy up to 1200° C as 0.1 ev and in the temperature range of 1200° to 1500° as 2.5 ev. The addition of small amounts of oxides of higher valence shifts this point of the curve at 1250° C to a lower temperature, whereas the addition of oxides of lower valence shifts it to a higher temperature. This is shown in Fig. 9 by an example of 1 mole% Nb_2O_5 and 1

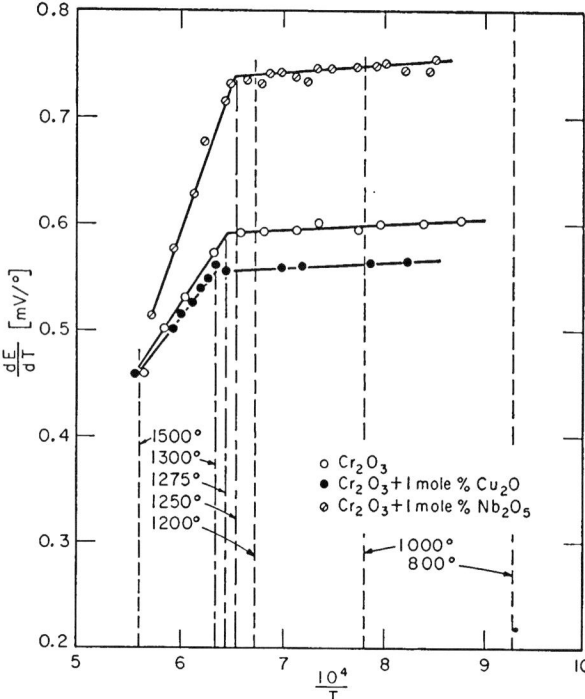

Fig. 9. Thermoelectric power as a function of the reciprocal of absolute temperature for various specimens.

mole% Cu_2O. Furthermore, these additions also change the amounts of the thermoelectric power and of the resistance. In accordance with the references of Fig. 7, Nb_2O_5 increases and Cu_2O decreases the thermoelectric power of the chromium oxide. These experimental data agreed also with our opinion that the conductivity of Cr_2O_3 is carried out mainly by defect electrons up to 1500° C.

Because of these measurements, the model of disorder, as designed by Hauffe and Block,[8] can be accepted for a temperature as high as approximately 1200° C. It is proposed in the following way:

$$0 \rightleftharpoons Cr\,\square''' + Cr\,O^{\cdot\cdot} + \oplus \qquad (2)$$

(that is, the ideal crystal (0) is in equilibrium with a chromium vacancy with three negative charges and chromium on an interstitial site with two positive charges and one defect electron).

The distinct change of the thermoelectric power and of the electric resistance above 1200° C may be

explained by the assumption that now, apart from the defect electrons, excess electrons are taking part in the electrical conduction. For the development of these excess electrons, a change in the valency of the trivalent chromium into a higher and lower rank of the valency is assumed. Such a change of the valency could be expressed by the following equation of disorder:

$$0 \rightleftharpoons (a + b)Cr\,\square''' + a\,Cr\,O^{\cdot\cdot} + a\,\oplus + b\,Cr\,O^{\cdot\cdot\cdot\cdot}$$
$$+ b\,\ominus \qquad a > b \quad \varsigma))$$

(that is, the ideal crystal (0) is in equilibrium with a plus b chromium-vacancy with three negative charges and a chromium on interstitial sites with two positive charges, a defect electrons, b chromium on interstitial sites with four positive charges, and b excess electrons); a and b are functions of the temperature. This disorder equation will be supported by the observation of the lack of dependence of the thermoelectric power and of the electric resistance up to 1500° C on the oxygen pressure in the system.

References

1. W. A. Fischer and A. Hoffmann, *Arch. Eisenhüttenw.*, **26**, 43 (1955).
2. W. A. Fischer and A. Hoffmann, *Arch. Eisenhüttenw.*, **26**, 63 (1955).
3. W. A. Fischer and A. Hoffmann, *Arch. Eisenhüttenw.*, **27**, 343 (1956); A. Hoffmann and W. A. Fischer, *Z. physik. Chem. Frankfurt*, **7**, 80 (1956).
4. W. A. Fischer and A. Hoffmann, *Naturwissenschaften*, **41**, 162 (1954).
5. W. A. Fischer and G. Lorenz, *Arch. Eisenhüttenw.*, **27**, 375 (1956).
6. R. Lindner and A. Åkerström, *Z. physik. Chem. Frankfurt*, **6**, 162 (1956).
7. W. A. Fischer and G. Lorenz, *Arch. Eisenhüttenw.*, **28**, 497 (1957).
8. K. Hauffe and J. Block, *Z. physik. Chem. Leipzig*, **198**, 232 (1951).

Discussion

RICHARDSON voiced the opinion that it might be that ferric iron could be dissolved in the alumina lattice rather than ferrous iron. WAGNER pointed out that the experimental cell which was used had an atmosphere of argon around the platinum electrode in contact with the alumina and iron oxide. He indicated that it would have been more suitable to maintain a fixed oxygen pressure on this electrode. GOKCEN said that he had measured the saturation value of alumina in wüstite melts under carefully controlled hydrogen-water vapor atmosphere and found at 1600° C that the point was 40% by weight Al_2O_3 which in Fig. 4 would indicate that the liquidus line from the eutectic to the spinel compound would be almost straight instead of as shown. FISCHER said that in their experimental program they found a lowering of this liquidus line when the oxygen pressure was high relative to that obtained in the pure argon atmosphere. The melting point of the spinel also dropped with higher oxygen pressures as a result of the formation of the solid solution of magnetite and alumina. HILTY reported that in some work on the iron-oxide alumina inclusions obtained from steels, it was found that the spinel, as identified by X-ray, had a melting point of roughly 1780° C. This is an agreement with the points shown on Fig. 4. MUAN said that a study of the solubility of alumina in iron oxide in an atmosphere of air showed approximately 13% alumina dissolved at 1350° C.

by T. Baak

The Action of Calcium Fluoride

in Slags

Calcium fluoride has primarily a twofold action in slags:

(1) It lowers the melting point of slags, so that slags of higher basicity can be used.

(a) It decreases the viscosity.

It appears that secondary chemical reactions between calcium fluoride and slags can occur.

The lime-silica diagram is important in metallurgy. The action of calcium fluoride on this system is shown in Fig. 1. The ternary diagram has been outlined earlier by Oelsen and Maetz.[1] The diagram given here shows a much more detailed picture and has corrected many points. The $CaO—SiO_2$ binary is from Rankin and Greig[2] and the $CaO—CaF_2$ binary from Baak.[3] The $SiO_2—CaF_2$ system has never been investigated in detail, but the solubility in the terminal phases has been shown to be very small if present at all.[4] Data from a study of the $Ca_2SiO_4—CaO—CaF_2$ diagram by Eitel[5] have been partially corrected and used. The pseudo-binary system $CaSiO_3—CaF_2$ is from Baak and Ölander.[6] In the diagram there is no indication of any new compounds, but in the solid state a compound, cuspidine, $3CaO \cdot 2SiO_2 \cdot CaF_2$, is known to exist.[7] This compound, however, dissociates before melting. We see here a diagram with two eutectic points with a maximum between them.

In the region of the greatest interest for metallurgists, it is seen that relatively small amounts of CaF_2 lower the melting point of phases such as Ca_2SiO_4 very much. This effect is still more pronounced at the sides of the dystecticum. Also

Ca_2SiO_4 exists in the temperature interval under consideration in two forms, α and β, with a transition point at 1420° C.[2] From the diagram it is seen that the action of CaF_2 on the freezing-point depression is different for the two forms. The freezing-point depression is much greater for the β-form. This favorable depression of the freezing point for calcium orthosilicate is one reason for using calcium fluoride to form a fluid slag of high basicity.

On the other side, if one considers the pseudo-binary, $CaSiO_3—CaF_2$, there appears another phenomenon in connection with the action of calcium fluoride in slags. It has been shown[6] that calcium metasilicate in the liquid form is polymerized from three units, i.e., $Ca_3Si_3O_9$. This result has also been confirmed by interpretations of the viscosity data in the system $CaO—SiO_2$.[8] In Fig. 2 it is shown with the help of this known polymerization how one can understand the great effect of small amounts of calcium fluoride on the viscosity.[9] The lower curve is given in weight percentage, while the upper one is given in mole percentage of $Ca_3Si_3O_9$ and CaF_2. The mole-percentage interpretation results in a staight line.

One can look at the problem molecularly in another way. The fluoride ion has a valence of -1, while the oxygen has a valence of -2. Each oxygen ion which is not bonded between two silicon ions joins with a calcium ion. This calcium ion in turn is attached to another oxygen according to the scheme

$$—Si—O—Ca—O—Si— \qquad (1)$$

With the introduction of calcium fluoride, the bonding can be thought to occur in two ways.

The author, formerly with the Department of Metallurgy, Massachusetts Institute of Technology, is now at Pennsylvania State University, University Park, Pennsylvania.

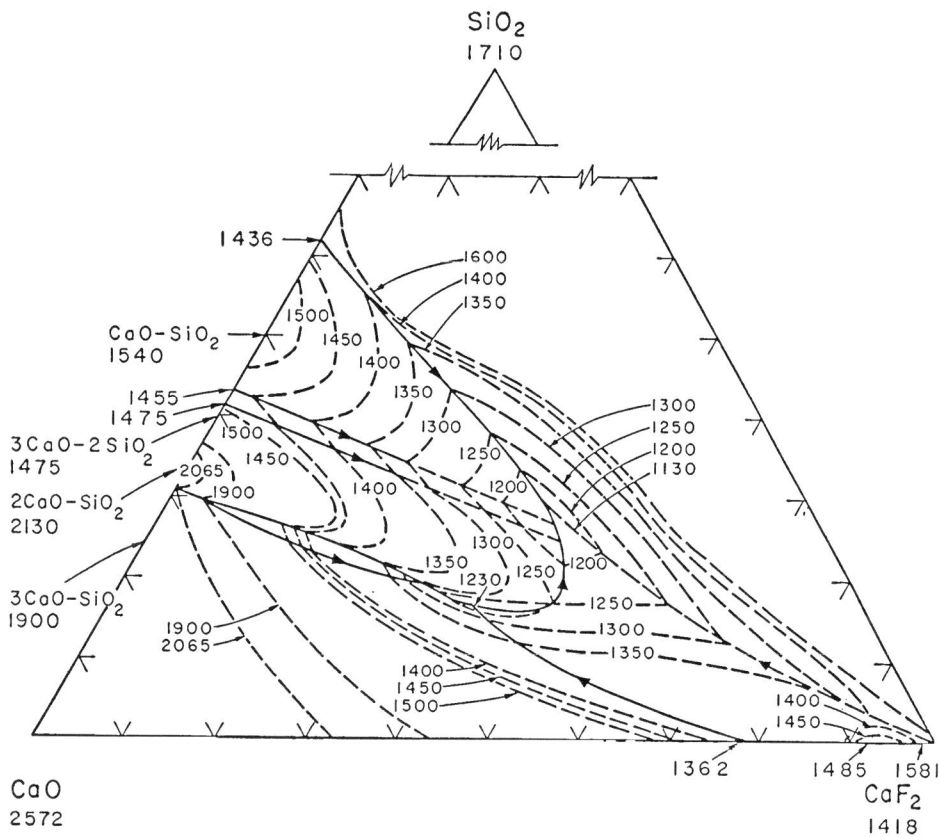

Fig. 1. The liquidus surface of the CaO—CaF₂—SiO₂ System, temperature in ° C.

Either the fluoride ion substitutes for an oxygen ion

$$
\begin{array}{ccc}
& | & & | \\
\text{—Si—F} & & \text{—Ca—O—Si—} & \quad(2)\\
& | & & |
\end{array}
$$

or as follows:

$$
\begin{array}{cc}
| & | \\
\text{—Si—O—Ca—F} \;\bigg|\; \textbf{F—Ca—O—Si—} & \quad(3)\\
| & |
\end{array}
$$

The first case is not very likely, as it would probably show up in the phase diagram as a new compound. In the second case we have solvatation of the silicate by the calcium fluoride. No really new compound is formed, but the different silicate units are shielded from each other by unstable CaF+ ions, which behave like a lubricant between the larger anion units. This argument is supported by the activity-coefficient curves in the system CaSiO₃—CaF₂.[8]

Now we come to the secondary effect, that is, the chemical reaction between CaF₂ and slags. It has long been known that in slags containing fluorspar, fluorine is lost during the process, and it has been demonstrated that SiF₄ is formed.[10] The reaction is supposed to follow the scheme

$$2CaF_2 + SiO_2 = 2CaO + SiF_4 \quad (4)$$

which is equivalent to reaction 2. From a recent study[6] it was found that such a reaction can occur

in a CaSiO₂ slag only in the presence of water. Slags exposed to the moisture of the air will take up some water. If the silicate is first heated to a high temperature to dry it, and then after cooling it is mixed with previously dried CaF₂, the resulting melt of these components will not show loss of fluorine. If this precaution of drying before mixing is avoided, there will be greater or smaller loss of fluorine depending on heating time. A recent investigation[11] of this behavior of fluorine in glass melts has proved that both SiF₄ and HF are vapor-

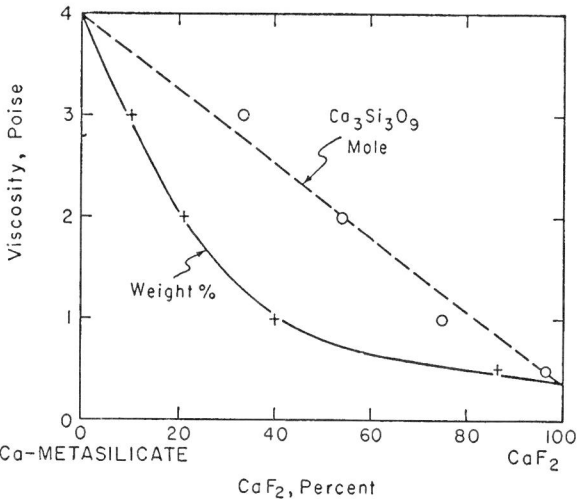

Fig. 2. Effect of CaF₂ additions on viscosity of calcium metasilicate.

ized. A possible scheme for this catalytic reaction is shown:

$$4HF + SiO_2 = SiF_4 + 2H_2O \qquad (5)$$

$$2H_2O + 2CaF_2 = 2CaO + 4HF \qquad (6)$$

which gives the over-all reaction:

$$2CaF_2 + SiO_2 \xrightarrow{H_2O} 2CaO + SiF_4 \qquad (7)$$

The HF can escape if reaction 5 is slow.

Two practical applications are suggested from the foregoing arguments. One is that it may be possible to diminish the water content in a slag by fluorspar additions. Thus a steel of low hydrogen content may be obtained because water is picked up in the slag from the furnace atmosphere and transferred to the steel. The second practical point of interest is that the widely known heavy corrosion of the silica roof in furnaces using slags containing fluorspar can be explained in part by the vaporization of hydrogen fluoride, which attacks silica. Silicon tetrafluoride cannot be responsible for this corrosion of the furnace roof because this compound should build up a layer.

References

1. W. Oelsen and H. Maetz, *Mitt. Kaiser-Wilhelm-Inst. Eisenforsch. Düsseldorf,* 23, 202 (1941).
2. G. A. Rankin and J. W. Greig, *Am. J. Sci.,* 13, 1 (1927).
3. T. Baak, *Acta Chem. Scand.,* 8, 1727 (1954).
4. E. Diepschlag and E. Brennecke, *Feuerungstechnik,* 22, 65 (1934).
5. W. Eitel, *Z. angew. Mineral.,* 1, 269 (1938/39); *Zement,* 27, 469 (1938).
6. T. Baak and A. Ölander, *Acta Chem. Scand.,* 9, 1350 (1955).
7. W. F. McCaughey, K. Kautz, and R. G. Wells, *Am. Mineralogist,* 33, 200 (1948).
8. T. Baak, *Acta Chem. Scand.,* 9, 1540 (1955).
9. C. Herty, F. A. Hartgren, G. L. Frear, and M. B. Royer, *U. S. Bur. Mines Rept. Invest.,* 3232 (1934).
10. J. Görrisen, *Arch. Eisenhüttenw.,* 15, 347 (1941/42).
11. M. A. Tsaritsyn, *Steklo i Keram.,* 12, No. 2, 15 (1955).

Discussion

SHANAHAN described an experiment in which three platinum crucibles were used: one contained silica, another calcium fluoride, and the third a mixture of silica and calcium fluoride. They were all heated together in a muffle at 990° C. Over a period of 3 hr

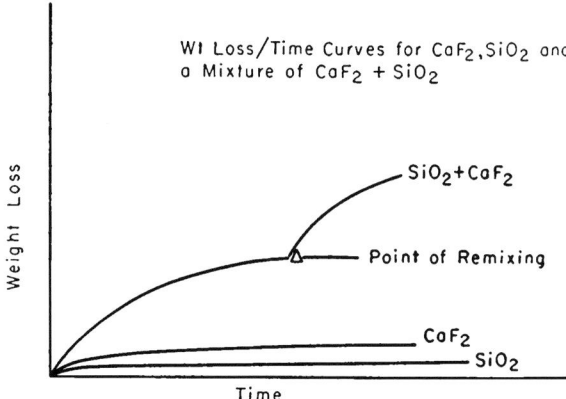

Fig. 3. Loss of weight of slag components when heated to 900° C.

(Fig. 3), the silica showed a very slight loss in weight. Calcium fluoride showed a slightly greater loss. The crucible containing calcium fluoride and silica showed a very rapid increase in the rate of loss of weight which continued over an appreciable period. On cooling the sample, regrinding it, and then repeating the experiment, the high rate of weight loss was reproduced.

K. Fetters

W. A. Fischer

Presiding

Section 4

Slag-Metal Equilibria
in Blast-Furnace and Steelmaking
Furnace Systems

by P. P. Kozakevitch

Formation and Destruction of Foams and Emulsions in Iron and Steelmaking

The presence of foams and emulsions in iron and steelmaking systems has long been a problem with which only the furnace operator has been concerned. Very little attention has been paid to them by the scientist in the interest of understanding how to cause them to form, how to prevent them, and how to destroy them once they form. In general, they are undesirable because a foam, which is a mixture of gas bubbles in a liquid phase, makes heat transfer through the slag layer difficult. An emulsion of metal in liquid slag can cause serious losses in yield. Nevertheless, these conditions provide a tremendous surface area for reaction between the two phases which could be of economic advantage. It is the purpose of this paper to discuss informally some observations which have been made on foams and emulsions.

GENERAL REMARKS

Foam and emulsion *formation* is favored:

(a) By lowering of the interfacial tension (liquid/gas for foams and liquid/liquid for emulsions) with dissolved surface-active substances (formation of an adsorption layer), since pure liquids will never give a rich, stable foam, no matter how low their surface tension may be.

(b) By stirring, which provides the energy necessary to increase the interface.

(c) By gas evolution, which supplies not only the

necessary mechanical work but also, in the case of foaming, the matter to be dispersed in the liquid.

Stabilization of gas and liquid emulsions is favored:

(a) By high viscosity of the liquid medium, which may practically stop the normal destruction of the emulsion through coalescence of droplets or bubbles.

(b) By formation of a viscous or rigid adsorption film at the interface. In most cases, the second factor is the more important, especially in the case of true rigid foams. Thus, the formation of an adsorption film is always necessary to facilitate foaming or emulsifying, but a certain minimum viscosity or rigidity of this film will be required to stabilize the foam or emulsion thus formed.

When the gas evolution is particularly strong and continues without stopping, foaming may become a nuisance even when the nature of the liquid (pure metals, pure water) does not favor it. These cases will not be considered here, the main purpose of the report being a discussion of the physical properties of slags and metals involved in foams and emulsions rather than one of the energy supply through stirring, blowing, spontaneous gas evolution, and so on.

FOAMING SLAGS

Foaming slags will occasionally occur in cupola and even in blast furnaces. They are known to cause trouble in basic open-hearth furnaces with heats high in carbon and silicon.

Dr. Kozakevitch is Head, Physical Chemistry Department, Institut de Recherches de la Sidérurgie, Saint-Germain-en-Laye, France.

Foaming in basic open-hearth furnaces occurs in most cases while the slag is very rich in FeO and relatively rich in SiO_2. Its lime content (dissolved lime) is usually low then. Such a slag is very fluid, so that the foam stabilization can be attributed to an adsorption film only.

If the P_2O_5 content is low (this is usually the case when the boil begins), the only slag constituent known to lower the surface tension of liquid FeO is silica. It is thus probable that the foam is stabilized chiefly by an adsorption film rich in silica.[1]

It is believed that the adsorbed silica has a tendency to constitute a coherent surface network just as it does in the bulk phase when this becomes sufficiently acid. The mechanism of the "thickening" of the film would thus be similar to that of the bulk phase of very acid slags. Since SiO_2 concentration in the surface film is necessarily much higher than in the bulk of the liquid (adsorption), it becomes clear that a coherent, viscous, or even rigid film may be formed on the surface of a FeO—SiO_2 slag whose bulk phase is yet perfectly fluid. It is likely, however, that small quantities of P_2O_5 present in foaming slags also contribute to foam stabilization since P_2O_5 was found to be highly surface active when dissolved in FeO or in FeO—SiO_2 mixtures. It may also be remembered here that the PO_4 groups may, to a certain extent, replace SiO_4^{4-} in the coherent random network. The surface activity of P_2O_5 is lowered by lime.[1]

The surface activity of silica in ternary FeO—CaO—SiO_2 mixtures is lowered by lime additions parallel to the lowering of its thermodynamic activity. This means that the SiO_4 groups bound to Ca^{2+} tend to be adsorbed less than silica in presence of FeO only. This explains why the open-hearth slags do not foam if a sufficient quantity of lime is present.

It seems possible to prevent foaming or to destroy the existing foam by addition of such surface-active oxides as would disintegrate the coherent surface network, thus making the surface film less viscous. Titanium dioxide was found to be surface active in FeO. As TiO_2 is known to fluidify efficiently the acid slags, it was natural to suggest its use to prevent or destroy the foam.[1] An industrial test conducted recently confirmed the soundness of this reasoning.

Slag foam in the open-hearth furnaces may be destroyed by a luminous flame containing carbon particles. The same result may be obtained by pitch briquettes cast into the foaming slag,[2] or by tar injection into the flame.[3] These facts may perhaps be explained by a reduction of FeO by solid carbon: a carbon particle fixing itself on the thin slag film of a bubble will provoke formation of a "secondary" CO bubble. The film will thus become too thin and will burst under the action of the gas pressure inside the bubble.

At what composition of a binary mixture of given components will the foam reach its maximum stability? It appears that this maximum is reached at

surface concentrations of the adsorbed substance somewhat below the adsorption maximum. Since in most cases this last point is reached when all solvent molecules are expelled from the film by the surface-active substance, the presence of a few residual solvent "molecules" in the film ("solvated" film) seems to be essential to the foam stability. The relatively low surface activity of SiO_2 in FeO does not permit a calculation of the maximum adsorption after the Gibbs equation, but one may be practically sure that the maximum difference in composition between the film and the bulk phase is reached at high SiO_2 contents (bulk concentration), probably not less than 60%. It has been found in FeO—SiO_2 mixtures that foam stabilization becomes apparent with 25 or 30% silica in the slag, then goes through a maximum, and drops abruptly to zero near the saturation point. These interesting findings are in good agreement with the existing general knowledge acquired on other foams and may explain why the very acid Bessemer slags will not foam (in the acid open hearth, insufficient gas evolution is a more probable explanation).

In some cases, numerous metal droplets are found in foaming slags. Their stabilization is probably due to an electrical double layer Fe^{2+}/SiO_4^{4-} at the slag-metal interface, as explained in the third section of this paper. The droplets imbedded in the slag will react with the FeO present, gradually lose their carbon, and produce numerous small CO bubbles partially adhering to their surface. Those bubbles will contribute to a rapid swelling of the foaming slag layer, their action being perhaps even more detrimental than that of the regular bubbles originating from the metal bath.

Slag foams will sometimes be stabilized by solid particles in suspension, the most interesting case being perhaps that of Cr_2O_3. That oxide, which is scarcely soluble in FeO-bearing slags, is deprived of any surface activity (although it is very active when dissolved in ordinary glass). Nevertheless, open-hearth furnace slags containing Cr_2O_3 give a particularly stable and troublesome foam which greatly hinders the processing of Cr containing iron. The Cr_2O_3 (or chromite) particles in suspension probably adhere to the bubble surface. The coalescence of adjacent bubbles will thus be practically prevented by the layers of solid particles separating them.

METAL EMULSIONS IN SLAGS

Generally, a metal emulsion can be formed in three different ways:

(a) Direct dispersion of the metal in a slag.

(b) Decomposition of a chemical compound dissolved in the slag resulting in the formation of finely dispersed metal droplets.

(c) Separation, on cooling, of metal initially dissolved in the slag.

It is clear that the second and third processes will usually give finer droplets than the first and that the necessity of some mechanical work to emulsify the metal applies to the first only.

Two kinds of metal emulsions will be considered here:

(a) Iron emulsions in FeO—SiO₂ slags.

(b) Emulsions of sulfur-rich iron in blast-furnace slags.

The first slag formed during the early stages of a basic converter operation has many common features with the open-hearth foaming slags. It is rich in $FeO + MnO$ (35 to 40%) and in silica (35 to 45%), and poor in $CaO + MgO$ (13 to 15%), which are relatively slow to dissolve. The stirring from the blast being very strong, the metal becomes emulsified in the slag to a much higher degree than during a typical open-hearth operation.

Several industrial scale-converter operations have been conducted without lime additions in order to slow down the dissolution of CaO and MgO. The total quantity of emulsified iron amounted to 40% of the weight of the slag when the processed metal was rich in silicon (0.61%). With 0.3% Si in the iron (less silica in slag) the quantity of emulsified metal was smaller (about 28%). The slag becoming more and more basic with time as it dissolved MgO and CaO from the converter lining, the metal emulsion gradually disappeared. At 27% $CaO + MgO$, the slag was completely free from metal droplets although blowing was not stopped.[4]

The most important factor varying during these operations was the $(CaO + MgO)/SiO_2$ ratio. Starting with approximately 0.2, this ratio finally rose to 1.5, and that was enough to destroy the emulsion completely. It is believed that the formation of an electrical double layer Fe^{2+}/SiO_4^{4-} at the surface of the droplets is essential for their stabilization. When the slag becomes rich in $(CaO + MgO)$, the quantity of available Fe_2SiO_4 units diminishes, and the emulsion disappears. It is suggested that the difference in behavior of Fe_2SiO_4 is related to the solubility of FeO and the insolubility of CaO in the liquid metal (Fe^{2+} being capable and Ca^{2+} incapable to "dip" into the liquid metal).

The carbon content of the droplets is always lower than that of the metal bath itself, and small drops are always found to be less rich in carbon than the bigger ones. The velocity of decarburizing by the FeO-rich slag depending on the surface/volume ratio, this phenomenon appears to be quite natural. The carbon content of small drops is often 50% lower than that of the bath. So the CO evolution must be strong in the slag containing metal emulsion and may contribute noticeably to its swelling.

Just as in the case of foams, one may wonder why the iron emulsions cause less trouble in acid converters, since their slags contain no lime at all. It is believed that the adsorbed layer, when it becomes too rich in SiO_2, begins to lose Fe^{2+} because of the natural trend of silica to form a coherent SiO_4^{4-} random network. The Fe^{2+} ions attached to the interface layer may then become less numerous, and this may result in a drastic decrease in the emulsifying power of highly acid slags. Accordingly, there should exist an optimum SiO_2 content for emulsifying, just as seems to be the case for foam stabilization.

It is believed that, when sulfur is low in the slag and relatively high (0.2–0.7%) in the pig iron, the first step of desulfurization of pig with a $CaO—Al_2O_3—SiO_2$ slag is the transfer of stoichiometric quantities of Fe^{2+} and S^{2-} through the slag-metal interface.[5] This transfer of ions or molecules (or "molecules" dissociating as soon as they enter slag) must lower the interfacial tension to a high degree, the action being probably reinforced by the adsorption of $Fe^{2+}S^{2-}$ units, a substance which has been found highly surface active in iron-carbon alloys as well as in pure iron.[6]

Interfacial tension between a synthetic $CaO—Al_2O_3—SiO_2$ slag with no sulfur and a thoroughly desulfurized iron with 3% carbon and 0.005% sulfur is approximately 820 dyne cm⁻¹ at 1450° C. Measurements with the same iron in an industrial blast-furnace slag of similar composition, but containing 1% sulfur, gave approximately 800 dyne cm⁻¹. It seems thus probable that in absence of Fe^{2+}, sulfur (S^{2-}) has little influence on the interfacial tension. These measurements and most observations cited later were made by means of a beam of X-rays passing through an experimental furnace.[7]

When the industrial slag mentioned in the preceding paragraph is poured on a drop of iron containing 3% carbon and 0.7% sulfur, the drop flattens and the (estimated) interfacial tension drops to less than 5 dyne cm⁻¹ (160 times less than with practically sulfur-free iron). As sulfur passes into the slag, the drop surface becomes gradually more and more convex. At sulfur contents less than 0.01%, the drop recovers its normal shape. The measured interfacial tension is then approximately 800 dyne cm⁻¹. In most cases, almost no CO bubbles appear at the slag-metal interface in the early stages of this process. It is believed that:

(a) Little chemical interaction takes place at the interface if the slag is not very basic.

(b) The sudden drop of the interfacial tension is due to the diffusion of $Fe^{2+}S^{2-}$ units (or couples) through the interface.

Normal blast-furnace slags do not wet liquid pig iron: the spreading coefficient is negative (approximately −231 dyne cm⁻¹). On the contrary, the same slags wet an iron with 0.7% sulfur: the spreading coefficient is then positive (approximately +42 dyne cm⁻¹). A positive spreading coefficient is usually observed when diffusion takes place between two liquids.

In exceptional cases, when the pig iron is very rich in sulfur (saturated) and the relative quantities of the metal and slag are such that the transfer of sulfur makes the slag rich enough in $Fe^{2+}S^{2-}$ (solidified slag of dark brown color), a violent boil starts after a while in the mass of the slag and on the graphite walls of the crucible rather than at the slag-metal interface. In some cases the metal layer becomes completely dispersed as droplets by this violent stirring: a coarse emulsion is thus formed. It is likely that the well-known reactions $FeS + CaO$ and $FeO + C$ (SiC or CaC_2) take place in the mass of the slag and on graphite walls. As soon as the transformation is accomplished, everything becomes quiet, the emulsion undergoes self-destruction by coalescence, the slag recovers its normal color, and the metal layer reappears at the bottom. The brown slag, when examined under the microscope before the boil starts contains numerous fine iron droplets visible when magnified at 600×. This finely dispersed emulsion probably results from the decomposition of FeS in the body of the slag, the reducing agents being perhaps SiC, CaC_2, SiO, or graphite particles in suspension.

One might ask what would happen if the free passage of S^{2-} ions into the slag were prevented by saturating the slag beforehand with $Ca^{2+}S^{2-}$ in such a way as to reduce its capacity to absorb sulfur? It has been shown that in this case numerous CO bubbles are formed at the slag-metal interface, the slag "boils" at the interface, and the iron phase disintegrates gradually under the action of these bubbles. An iron emulsion is thus formed by a mechanical dispersion of the metal. It thus seems probable that, when the slag contains much CaS, the chemical interaction starts at the interface and, when the iron is rich in sulfur, it will be easily dispersed in the slag.

The metal drops imbedded in the slag or found on the coke surface at the tuyère level have very irregular compositions: some of them are poor in sulfur (less than 0.1%); in most cases their sulfur content varies between 0.2 and 0.4%, but sometimes contents as high as 0.7% and even more are recorded. The slag-metal interfacial tension, when the metal contains 0.3% sulfur, is very low; at 0.7% it is almost zero. So the amount of work necessary to disperse a given mass of such a metal into droplets should be very small, and the trickling of the iron and slag over the pieces of coke should be capable of providing the work required to form droplets of metal more or less enwrapped in the slag. This would then result in an accelerated transfer of sulfur to the slag, so that in most cases the normal, that is high, interfacial tension might be rapidly established and the coarse emulsion destroyed

by coalescence. But if for some reason the transfer of sulfur becomes too slow, the emulsion will not have time to disappear, and iron shot will be found in the blast-furnace slag. When the principal reason of this slow rate is a high sulfur content of the slag, the chemical interaction will start at the interface, and a finely dispersed, relatively stable emulsion may be formed. In an exceptional case, such a blast-furnace slag contained more than 75% emulsified iron. Large metal drops suspended in this slag were well desulfurized and rich in carbon and silicon, while the small ones contained much more sulfur and were not as rich in carbon and silicon. This difference may be explained by assuming that the large drops were formed by coalescence of those small drops which had already completed their normal reaction cycle and thus acquired a high surface tension favoring coalescence. It is interesting that these relations are just the reverse of what was stated about iron emulsions in the basic converter: those emulsions, as described, were in the process of being formed, and the difference in size of the droplets was due to the hazards of the stirring only, while the emulsion in the blast furnace considered here is undergoing slow self-destruction, so that the size of drops is related rather to the progress of coalescence.

References

1. P. Kozakevitch. *Rev. mét.*, **46,** 505 and 572 (1949).
2. G. Grenier, *Rev. Tech. Luxembourg*, **44,** 65 (1952).
3. M. Mallevialle, *Rev. mét.*, **47,** 465 (1950).
4. P. Kozakevitch, and P. Leroy, *Rev. mét.*, **51,** 203 (1954).
5. G. Derge, W. O. Philbrook, and K. M. Goldman, *Trans. A.I.M.E.*, **188,** 1111 (1950).
6. F. A. Halden and W. D. Kingery, *J. Phys. Chem.*, **59,** 557 (1955); P. Kozakevitch, S. Chatel, G. Urbain, and M. Sage, *Rev. mét.*, **52,** 139 (1955).
7. P. Kozakevitch, G. Urbain, and M. Sage, *Rev. mét.*, **52,** 161 (1955); *Iron & Coal Trades Rev.*, **170,** 963 (1955).

Discussion

FETTERS stated that the presence of calcium fluoride in slags tended to increase the likelihood of foaming and asked if the author had studied its influence. KOZAKEVITCH said that in the laboratory work they had found strong indication that calcium fluoride increased the tendency towards foaming. SHANAHAN reviewed briefly a problem that had been encountered where soda-ash slags were being used for desulfurization and it was desired that a foamy slag remain so that the operation of removing it by rabbling was facilitated. In spite of all attempts, he said that they were unable to form a foam, and asked for a formula for doing so. KOZAKEVITCH said that the addition of chromite to a slag would cause a very stable foam to form which then was almost impossible to collapse.

by C. W. McCoy
and W. O. Philbrook

Slag-Metal Reactions of Chromium

in Carbon-Saturated Melts

The iron-chromium-oxygen system has been studied intensively under conditions typical of the refining, deoxidation, and solidification of steel.[1, 2, 3] As to the chemistry of chromium in metallurgical slags and slag-metal distributions, the literature deals for the most part with the relatively oxidizing conditions of steel refining in open-hearth or electric furnaces. There is no published information of quantitative value on the reactions of chromium under the more strongly reducing conditions prevalent in smelting processes, and statements on the recovery of chromium in the iron blast furnace are meager and conflicting. The present study gives a general picture of the behavior of chromium in distributing itself between slag and metal under the low oxygen pressures that prevail for carbon-saturated iron, and for blast-furnace-type slags. In addition to its theoretical value, this information may be of practical interest in connection with the smelting of chromium-bearing iron ores, such as the laterites, or the manufacture of ferroalloys.

Of the previous studies on the slag-metal distribution of chromium, which are reviewed in greater detail elsewhere,[4] the work most closely related to the present paper was done by Körber and Oelsen.[5]

C. W. McCoy is with the Metallurgical Department, Electro-Metallurgical Co., Division of Union Carbide and Carbon Corporation, Niagara Falls, New York, and W. O. Philbrook is Professor of Metallurgical Engineering, Carnegie Institute of Technology, Pittsburgh, Pennsylvania.

This work was part of a doctoral thesis at Carnegie Institute of Technology. The authors are most appreciative of fellowship support provided one of them by the Union Carbide and Carbon Corporation and the Carnegie Metals Research Laboratory.

They used virtually carbon-free iron under silica-saturated FeO—MnO—SiO_2 slags and varied the effective oxygen pressure by the addition of such deoxidizers as manganese and silicon. Their work showed that the recovery of chromium in the metal increased as the FeO content of the slag decreased, and the distribution ratio (% Cr)/[% Cr] appeared to extrapolate to zero at zero FeO content. Furthermore, as the oxidation level of the slag decreased, the fraction of the total chromium content of the slag that was present in the divalent form, CrO or Cr^{2+}, increased to approach 100%. These trends have been confirmed by the work reported here. In addition, a rational explanation can now be given for the relationship between chromium distribution and the content of manganese or silicon in the metal that Körber and Oelsen observed, as well as for similar distribution "index numbers" that have been reported for titanium, manganese, and sulfur.[6, 7, 8] The results of the chromium study are also entirely consistent with the observed effects of silicon[9] and silica[10] on desulfurization kinetics.

High-temperature thermodynamic properties of the lower oxides of chromium have not been measured, and the estimates of Maier[11] are evidently unreliable because they lead to the unreasonable prediction that Cr_2O_3 is more stable than CrO at elevated temperatures. Thermodynamic data on the chromium-oxygen system at low oxygen pressures and high temperatures are badly needed for predicting the behavior of metallurgical systems.

MATERIAL, APPARATUS AND PROCEDURE

The experiments were carried out in graphite

crucibles using previously prepared metal and slags. The carbon-saturated iron, made by melting ingot iron in graphite crucibles, contained only incidental traces of chromium and silicon. Desired chromium contents in the initial metal for a run were obtained by charging chromium metal of 99+% purity in weighed proportion with the iron. The lime-silica-alumina slags all contained 16% Al_2O_3 and covered a range of CaO/SiO_2 ratios. They were prefused from C. P. oxides in graphite crucibles. About half of the homogenized melt was poured off for use as chromium-free slag, and chromium was added to the remaining portion as Cr_2O_3. The analyses of the slags are given in Table 1.

Table 1. Analyses of Initial Slags

Designation	CaO %	SiO₂ %	Al₂O₃ %	ΣCr %	Cr²⁺ %	$\frac{Cr^{2+}}{\Sigma Cr}$
1530	30.7	53.6	16.3	0.00	—	—
1535	35.5	48.6	16.5	0.00	—	—
1540	40.8	42.8	16.3	0.00	—	—
1545	45.3	38.8	16.1	0.00	—	—
1530C	29.6*	51.6*	15.7*	3.74	3.59	0.96
1535C	34.3*	47.0*	16.0*	3.28	2.92	0.89
1540C	39.6*	41.5*	15.8*	2.94	2.73	0.93
1545C	44.2*	37.9*	15.7*	2.33	2.30	0.99

* Calculated from analysis of corresponding primary slags to adjust for dilution by the chromium addition.

Some observations made during slag preparation are significant. When the chromic oxide was added, there was an immediate and vigorous reaction evolving carbon monoxide. This was accompanied over a period of several minutes by a rapid color change from bright green to deep blue. Following this behavior, there persisted a much slower bubbling reaction. The chromium-bearing slag had to be poured under a protective nitrogen atmosphere to prevent reoxidation at the surface to the green chromic oxide. This behavior suggested the reduction of chromic to chromous ion, and the analyses recorded in Table 1 showed that 89 to 99% of the total chromium content was Cr^{2+}. The divalent chromium was determined by dissolving the slag in a sulfuric-hydrofluoric acid mixture containing a known excess of sodium vanadate, back-titrating, and calculating the chromium equivalent of the oxidant consumed. No minute investigation of the chromous/chromic ratio was made because it appeared from our results together with those of Körber and Oelsen[5] that the chromium could be considered as virtually all divalent for the purposes of this study.

The final design of the apparatus and the experimental procedure were developed to solve problems that arose during initial runs in stationary graphite crucibles. A carbide formed on the crucible wall from reaction between the slag and the graphite.

This carbide was identified by X-ray diffraction as Cr_7C_3 with faint traces of Cr_3C_2. It was found that the carbide was not a true equilibrium phase for the system under study, as its solubility in the metal had not been exceeded. Once formed at the slag-graphite interface, however, it persisted because its solubility in the slag was virtually nil and transport to the metal was exceedingly slow. The carbide reaction could not be followed independently because of sampling difficulties, and it would have interfered with both kinetic and distribution studies. A solution was found in the use of a rotating crucible,[12] by which means the slag was confined in a "crucible" created of the liquid metal by centrifugal action while the metal itself remained in contact with graphite to assure carbon saturation. No free carbide was found in any runs with the rotating crucible.

The design of the furnace is shown schematically in Fig. 1. The crucible and its thermal insulation

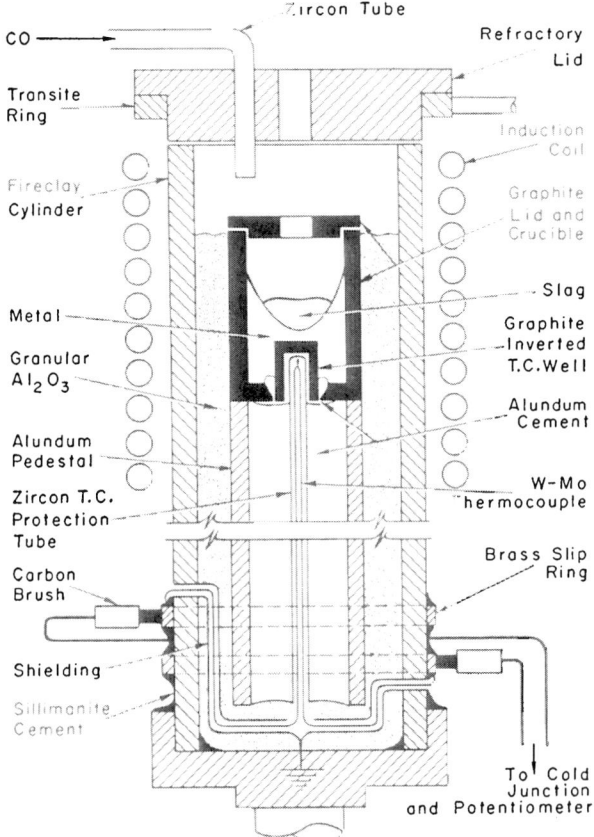

Fig. 1. Schematic cross section of rotating crucible furnace.

were supported so that they could be rotated at a controlled rate (275–345 rpm) about a vertical axis within a stationary induction coil. The most notable feature was the inverted thermocouple that was necessary to obtain reliable temperature measurement without disturbing the slag-metal interface during the run. (Variable emissivities of the slags made optical pyrometer readings unreliable.) The shielded leads from the *W/Mo* thermocouple were connected to the cold junction by means of brass slip rings and carbon brushes. Careful checks showed that this arrangement introduced no errors. The inverted thermocouple wells were machined separately from the crucibles so that they could be positioned very accurately to reproduce the "geometry" or thermal environment of the hot junction from run to run. Thermocouple calibrations and checks on reproducibility were made by thermocouples immersed in the metal from above during dummy runs. Temperature was controlled within ±5° C during experimental runs. The furnace was flushed continuously by CO that was vented and burned at the sampling hole at a fast enough rate to prevent effectively inward diffusion of air, so the oxygen pressure above the slag was determined by reaction between graphite and CO at crucible temperature and the prevailing atmospheric pressure.

The rotating crucible not only eliminated the problem of carbide formation but offered another advantage, as well as several disadvantages. A large mass of metal was needed to form the paraboloid of revolution, but only a small volume of slag could be contained in this cup out of contact with graphite at reasonable rates of rotation. As a result of the large difference in masses of slag and metal, the metal concentrations remained practically constant during substantial changes in slag composition, which greatly simplified the mathematical treatment of results. Opposed to this were the disadvantages that the small amount of slag seriously limited the number of samples that could be taken without a major decrease in slag weight and that the area of the slag-metal interface could not be rigidly controlled.

The experimental procedure was to melt 1000–1200 grams of metal under CO in the rotating crucible. The desired temperature was approached slowly to avoid overshooting, as kish would have been rejected if subsequent cooling were necessary. The metal was held at temperature for 10 min, and then 12 grams of prefused slag (−20 mesh) was added into the center of the metal cup by means of a silica funnel. The slag melted in less than 30 sec, and zero time was taken as the moment that the slag was added. Slag samples were taken at predetermined time intervals by an aspirator with a graphite tip designed to limit the sample to about 0.5 gram. The samples were cooled in the CO atmosphere above the crucible before being exposed to air. The metal samples were taken immediately after the slag samples if needed. In the kinetic runs, the apparent diameter of the top of the slag button was measured by means of a pair of calipers maintained at a fixed sighting distance; these measurements were made immediately after the release of a CO bubble and just before taking a slag sample. The distribution studies were carried out in a similar fashion except that they required a differ-

ent sampling schedule and no measurement of slag diameter.

KINETICS OF CHROMIUM REDUCTION

The kinetic study consisted of 15 runs of 20 to 22 min duration with slag samples taken at approximately 4 min intervals. The metal contained virtually no chromium, and the initial slag compositions were those listed with the "C" designations in Table 1. The temperature range from 1500 to 1650° C was covered in 50–deg increments. The results have been presented in detail elsewhere[4] and will be only summarized here.

If all of the chromium in the slag is assumed to be divalent, as discussed earlier, the reaction under study may be written as follows:

$$Cr^{2+}_{(slag)} + O^{2-}_{(slag)} + [C] \rightleftharpoons [Cr] + CO \ (g) \quad (1)$$

Tests of the experimental data for order of reaction showed that they best fit first order with respect to chromium, as opposed to 1/2, 3/2, and second order. This is consistent with equation 1 though not sufficient evidence to prove that the process is controlled by chemical reaction rate. If the process is treated by chemical kinetics, it is permissible to write $a_C = a_{CO} = 1$ and to take a_{Cr} as very low and nearly constant because of the experimental conditions selected. The rate of the back-reaction from right to left can therefore be neglected until the reaction nears completion. Furthermore, the activity of oxide ions available for reaction is assumed to depend upon the total composition of the slag rather than the "CrO" content alone, so it presumably does not change much during a given experiment. Variation of oxide-ion activity with slag basicity can be incorporated into composition dependence of the rate constant. Hence, the rate equation for equation 1 may be simplified to:

$$\frac{-d(Cr)}{dt} = Ak'(Cr) \quad (2)$$

where A is the interface area and k' is an experimental coefficient. The integrated form of this first-order expression is

$$\ln \frac{(Cr)_0}{(Cr)_t} = Ak't \quad (3)$$

where $(Cr)_0$ is the concentration of chromium in the slag at zero time.

Values of the product Ak' were determined from the slopes of the lines fitted by least squares to plots of $\ln (Cr)_0/(Cr)_t$ versus time for each run. The values ranged from 1.4 to 11.6 × 10⁻² min⁻¹ with a mean of 5.5 × 10⁻². There was no consistent trend with temperature or with slag composition.

There is a great deal of uncertainty about the area of the slag-metal interface, A. Only a diameter of the projected area in a plane normal to the axis of rotation could be measured. The surface of revolution up to any given diameter can be calculated for slag-free metal from the equation of the paraboloid. The shape of the metal cup is altered by the presence of slag, however, depending on the amount and density of the slag and surface-tension effects. An exact calculation is impossible at present. Furthermore, the projected area varied erratically during a run, and from run to run, because of the entrapment, growth, and escape of CO bubbles as well as from apparent changes in surface energies. For this reason, it seemed wisest to use only diameter measurements that were made immediately after the escape of a bubble and near the end of the run when the rate of gas evolution had slowed down, in order to minimize the effect of enclosed bubbles. Efforts to determine the area by freezing a heat under rotation were unsuccessful because of distortion of the interface during solidification. The theoretical areas for slag-free metal paraboloids corresponding to the measured diameters were calculated; these values are admittedly somewhat low and uncertain but are the best approximation available.

The Ak' products may be divided by the estimated areas and multiplied by the slag weight/100 to obtain a coefficient in terms of mass of chromium transferred per unit area, rather than change of concentration. The coefficients calculated in this way gave an average value of

$$k \cong 1 \times 10^{-3} \text{ gram Cr min}^{-1} \text{ cm}^{-2} (\%\text{Cr})^{-1}$$

for transfer of chromium from slag to carbon-saturated iron. This value may be considered as a mass-transfer coefficient or a chemical rate constant depending on the rate-controlling mechanism of the process.

The values of k for individual runs scattered over about an order of magnitude. Much of this can probably be attributed to inability to control or allow for variations in effective interface area, as already noted. The fact that no systematic trend for k to vary with either slag composition or temperature emerged above the scatter is interpreted to mean that the process is not very sensitive to slag composition or temperature. For the first-order process of sulfur transfer in an almost identical system, Chang and Goldman[13] showed that the rate of sulfur transfer from slag to metal was also essentially independent of slag composition and only slightly temperature dependent, with an average value of 0.8×10^{-3} gram S min^{-1} cm^{-2} (% S)$^{-1}$. The rate of FeO reduction from the slag, while of different order, is also nearly independent of temperature.[14] The results on the rate of chromium transfer from slag to metal are therefore consistent with related experiments even though their precision is lower than was desired.

SLAG-METAL DISTRIBUTION OF CHROMIUM

Preliminary work showed that as equilibrium was approached, practically all of the chromium was reduced from the slag. The initial chromium content of the metal was increased to about 14% in an effort to obtain chromium concentrations in the slag that were significant compared with the analytical accuracy of ±0.05% chromium.

It proved to be impossible to reach a constant slag-metal distribution ratio for chromium for reasons that will become clear. Figure 2 shows the

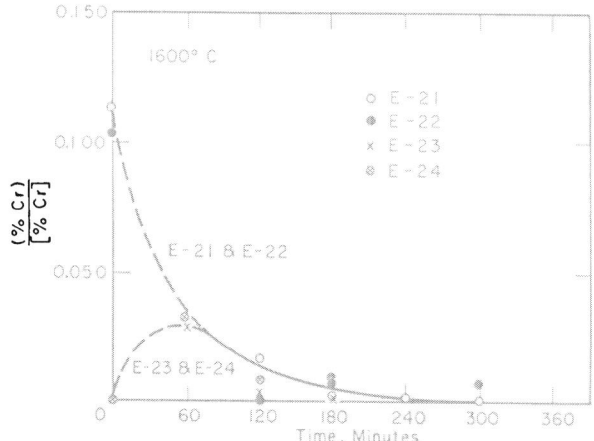

Fig. 2. Chromium distribution as a function of time, 1600° C.

typical behavior when an attempt was made to approach equilibrium from both sides, i.e., starting with high and low chromium concentrations in a slag for a given metal composition. When the chromium started high, it failed to level out at a finite concentration in the slag, and when it started at zero, it rose to a pronounced maximum and then fell in the same way as the high initial chromium. Essentially zero chromium was reached for all slags in about 3 hr.

The explanation of the failure to reach a constant chromium distribution ratio in these experiments was to be found in the concurrent reduction of silica from the slags. This reaction had been negligible during the brief kinetic runs with relatively high initial CrO in the slag, but it became appreciable during the 5 to 6 hr distribution runs, amounting to as much as 35% of the original slag weight. The reaction was accentuated under the experimental conditions used in this work because the large mass of metal acted as a sink that remained at nearly zero silicon content. Under these conditions, equilibrium with respect to silicon could not have been reached until the silica activity in the slag had been reduced to a very low value. The experiments were started with high-silica slags in order to permit a large drift of silica content during the run.

The results could be correlated by plotting the chromium distribution ratio as a function of the simultaneous silica content of the slag in the continuously changing system, using data taken from the late stages of the runs where the curves for initially high and low chromium concentrations coincided. Figure 3 is a combined plot of a number of points from each of the seven runs. The lime-silica ratios corresponding to the silica contents (at constant Al_2O_3/CaO ratio) are shown on the upper abscissa scale. It is evident that the chromium distribution ratio is essentially zero for lime-silica ratios above 1.2, and it rises to a value of only about 0.04 for basicity down to 0.6. This relation is not very temperature-sensitive, as points for both 1500 and 1600° C fall in the same band. Much of the scatter can be attributed to the fact that the chromium contents of the slag were not high compared with the limit of accuracy of chemical analysis.

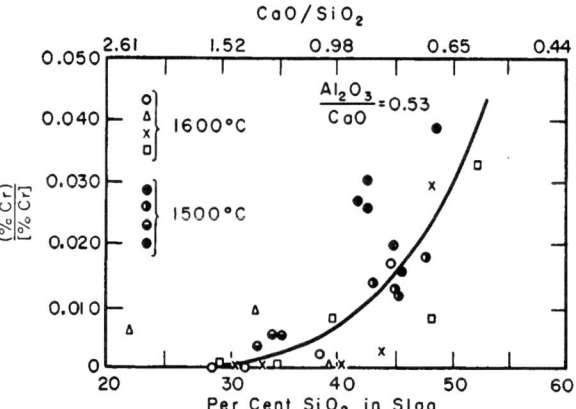

Fig. 3. Relation of chromium distribution to simultaneous silica content of slag.

This general behavior is best explained in terms of relative departure from equilibrium or what may be called *effective oxygen pressures* for the various reactions. Molecular rather than ionic representation will be used to avoid later complication by silicate structure. The process of chromium distribution may be written as:

$$2CrO_{(slag)} \rightleftharpoons 2[Cr] + O_2 \qquad (4)$$

$$K_{Cr} = \frac{a_{Cr}^2}{a_{CrO}^2} \cdot p_{O_2} \qquad (5)$$

$$p_{O_2} = K_{Cr} \cdot \frac{a_{CrO}^2}{a_{Cr}^2} \qquad (6)$$

The oxygen pressure defined by equation 6 is that which would have to exist for the system as a whole to maintain a given chromium distribution ratio as the equilibrium value. If the system is initially put together with a relatively high CrO content in the slag and low [Cr], this constitutes a comparatively high state of oxidation. This situation can not continue to exist in a vented system with excess carbon, as oxygen will be removed according to equation 1, with corresponding decrease in the ratio a_{CrO}^2/a_{Cr}^2 until the oxygen pressure approaches that in equilibrium with graphite and CO at 1 atm.

An identical situation exists with respect to the silica, for which the following reactions apply:

$$SiO_{2(slag)} \rightleftharpoons [Si] + O_2 \qquad (7)$$

$$p_{O_2} = K_{Si} \frac{a_{SiO_2}}{a_{Si}} \qquad (8)$$

If the prevailing activity ratio a_{SiO_2}/a_{Si} is higher than would be in equilibrium with the oxygen pressure established by other reactions, silica will tend to be reduced. In those runs of Fig. 2 where the initial CrO in the slag was low, SiO_2 was at first reduced by Cr in the metal with consequent increase in the (CrO)/[Cr] ratio. This continued until the oxygen pressures defined by equations 6 and 8 were equal. Thereafter, both CrO and SiO_2 were reduced simultaneously by carbon, so that a definite relation between the chromium distribution ratio and the silica content of the slags was established as shown in Fig. 3.

An essential part of the picture is the slowness of reduction of silica from slags by carbon.[15] A rough evaluation of this rate was made for runs which had no chromium in the original metal (to avoid reduction of silica by chromium instead of carbon). The data were not sufficient to establish the order of the reaction, so first order was assumed arbitrarily to calculate a rate constant comparable with that for chromium. The rate of reduction of silica was found to be about 20 to 60 times slower than that for chromous oxide, depending on temperature.

The upper curves of Fig. 2, with high initial CrO, represent conditions where the original oxygen pressure set by chromium was higher than that corresponding to the silica-silicon ratio. The CrO was reduced at its comparatively rapid rate by carbon until the oxygen pressure reached that corresponding to the silica activity. From this time on, CrO could not be reduced any more rapidly than silica, and the reduction of CrO could only keep pace with the slower rate of silica reduction as already noted.

The condition where the reduction of CrO and SiO_2 by carbon is proceeding simultaneously can be represented by setting equations 6 and 8 equal:

$$p_{O_2} = K_{Cr} \frac{a_{CrO}^2}{a_{Cr}^2} = K_{Si} \frac{a_{SiO_2}}{a_{Si}}$$

This may be rearranged to give

$$\frac{a_{CrO}^2}{a_{Cr}^2} = K_{SiCr} \frac{a_{SiO_2}}{a_{Si}} \qquad (9)$$

If it is assumed that activity coefficients for CrO, [Cr] and [Si] do not change greatly over limited composition ranges, then a simplification can be made:

$$\frac{(\% CrO)^2}{[\% Cr]^2} = K'_{SiCr} \frac{a_{SiO_2}}{[\% Si]} \qquad (10)$$

Equation 10 predicts a variation of the chromium distribution ratio with silica in the slag, at nearly constant [Si], such as was found in this study. It is interesting to note that essentially the same relation was observed by Körber and Oelsen,[5] who worked with unit silica activity in the slag and varying silicon in the metal. A similar line of reasoning can be used [4] to derive the index numbers found empirically by Faust[6] for titanium, $(SiO_2)/[Si] = K_{SiTi} (TiO_2)/[Ti]$; by Wentrup, Maetz, and Heller[7] for manganese, $(SiO_2)/[Si] = K_{SiMn} (MnO)^2/[Mn]^2$; and by Oelsen and Maetz[8] for sulfur distribution between iron and acid slags of essentially constant silica activity, $K_{SiS} = (S)/[S][Si]^{\frac{1}{2}}$. The same principles will reconcile the kinetic effects in the desulfurization of iron by slags of early additions of manganese or silicon[9] as contrasted with late additions of MnO or SiO_2.[10]

Thermodynamics dictates that a system cannot be at equilibrium for one component alone but must come to equilibrium for all components simultaneously. When reactions proceed at characteristically different rates, the slowest one becomes controlling for the entire system. The slowness of silica reduction in carbon-saturated systems gives this reaction an unique role that has long been recognized empirically without rational explanation.

This study of chromium distribution could have been made one of equilibrium by working to the equilibrium silicon content in the iron for desired slag basicities. The determination of the equilibrium silicon would, however, have been very tedious, and the chromium distribution ratios would have been even lower than those observed. The results might not have been as interesting as the ones obtained.

From the practical viewpoint, the results of this study suggest that substantially complete recovery of chromium in the metal is to be expected in the iron blast furnace or other processes making high-carbon iron as long as there is a significant reduction of silicon, at least for metal contents up to 14% Cr and slags having lime-silica ratios above 0.6.

SUMMARY

The reduction of chromium from lime-silica-alumina slags by carbon-saturated iron is first order with respect to chromium. The reaction rate is about 1×10^{-3} gram min^{-1} cm^{-2} (% Cr)$^{-1}$ and is substantially independent of temperature in the range from 1500 to 1650° C and of slag basicity in the range from 0.6 to 2.5 lime-silica ratio. The chromium in the slag is divalent under the conditions studied.

The carbide Cr_7C_3 is formed by slag-graphite reaction by a different mechanism at a slower rate. This carbide is nearly insoluble in the slag but is soluble in carbon-saturated iron up to at least a 14% chromium content.

A constant slag-metal distribution ratio for chromium was not obtained because of simultaneous reduction of silica from the slag. In the changing system at times when silica reduction was controlling, the ratio (% Cr)/[% Cr] ranged from zero at lime-silica ratios above 1.2 to only 0.04 at a basicity ratio of 0.6. This and related observations have been explained in terms of equivalent oxygen pres-

sures and the slowness of reduction of silica by carbon, which was 20 to 60 times slower than chromium reduction under the conditions studied.

For practical purposes, the recovery of chromium in the metal should be nearly complete under strongly reducing conditions similar to those in the hearth of a blast furnace.

References

1. H. M. Chen and J. Chipman, *Trans. Am. Soc. Metals,* 38, 20 (1947).
2. N. J. Grant, E. C. Roberts, and J. Chipman, *Trans. A.I.M.E.,* 200, 145 (1954).
3. D. C. Hilty, W. D. Forgeng, and R. L. Folkman, *Trans. A.I.M.E.,* 203, 253 (1955).
4. C. W. McCoy, *The Behavior of Chromium in Slag-Metal, Systems under Reducing Conditions,* Unpublished doctoral thesis, Carnegie Institute of Technology, Pittsburgh, Pa., 1956.
5. F. Körber and W. Oelsen, *Mitt. Kaiser-Wilhelm-Inst. Eisenforsch. Düsseldorf,* 17, 231 (1935).
6. E. Faust, *Arch. Eisenhüttenw.,* 12, 361 (1939).
7. H. Wentrup, H. Maetz, and P. Heller, *Arch. Eisenhüttenw.,* 20, 139 (1949). (Brutcher Translation 2543.)
8. W. Oelsen and H. Maetz, *Arch. Eisenhüttenw.,* 16, 283 (1942/43).
9. K. M. Goldman, G. Derge, and W. O. Philbrook, *Trans. A.I.M.E.,* 200, 534 (1954).
10. N. J. Grant, O. Troili, and J. Chipman, *Trans. A.I.M.E.,* 191, 672 (1951). Cf. also, p. 666.
11. C. G. Maier, "Sponge Chromium," *U. S. Bur. Mines Bull.,* No. 436 (1942).
12. E. P. Barrett, W. F. Holbrook, and C. E. Wood, *Trans. A.I.M.E.,* 135, 73 (1939).
13. L. C. Chang and K. M. Goldman, *Trans. A.I.M.E.,* 176, 309 (1948).
14. W. O. Philbrook and L. D. Kirkbride, *J. Metals,* 8, 351 (1956).
15. J. C. Fulton and J. Chipman, *Trans. A.I.M.E.,* 200, 1136 (1954).

Discussion

CHIPMAN pointed out the similarity between the reaction of reduction of chromous oxide to that of desulfurization in a carbon-saturated iron system. He said that this study will also be helpful in understanding the reactions for sulfur removal. RICHARDSON proposed that chromium might be present in the slag in several valence forms, as a neutral atom, in the chromous form, and perhaps in the chromic form, and inquired if this was studied. PHILBROOK said that analytical results indicated that 89 to 99% of the chromium was present as the chromous ion. Some samples oxidized somewhat on cooling, so that it was reasonable to assume that all the chromium was present in the chromous state. However, the analytical method would not have shown the possibility that some neutral chromium would have reacted with chromic ion to form the chromous ion.

LANGENBERG pointed out that the removal of slag samples for analysis might seriously affect the kinetic system being studied, so that conditions would change appreciably from the beginning of the run to the end. PHILBROOK said that the total amount of slag taken in sampling was approximately 25% of that originally charged. As a consequence, greatest reliance was placed on results from the early part of the run.

C. W. Sherman

P. Vallet

Presiding

Section 5

Kinetics and Slag-Metal Reactions

by L. S. Darken

Kinetics of Metallurgical Reactions with Particular Reference to the Open Hearth

In talking about the physical chemistry of metals to chemists with a minimum knowledge of metals, I find it rather difficult to bring out the fact that although the chemistry of metals is certainly genuine chemistry, it involves several points of marked difference from the better-known branches of chemistry. Perhaps foremost among these is the absence of the molecule in the metallic state. After hearing that in solid and liquid metals we do not have entities corresponding to the molecules by which the freshman was introduced to chemistry and with which the organic chemist is perennially concerned, many people are disposed to regard the whole subject rather skeptically and are not at all inclined to listen sympathetically to any further departures from what might be considered orthodox chemistry.

However, the point on which I should like to elaborate here is a second way in which the chemistry of metals differs considerably from the traditional chemistry which most of us learned as undergraduates and even as graduates. This pertains to the subject of reaction rates. The difference in kinetic behavior is very closely related to the absence of the molecule.

To illustrate the classical type of reaction-rate experiment, let us recall one from a laboratory course. In this experiment, acetic anhydride is mixed with water; after a few seconds of shaking, mixing is fairly complete, as judged by the disappearance of the striation lines which were initially quite visible. With a little more shaking, the solution is perfectly clear and homogeneous. After this, we transfer the solution to a Dewar flask, insert a Beckman thermometer, and sit down to follow the rate of the reaction $Ac_2O + H_2O \rightarrow 2AcOH$ by observing the temperature rise as a function of time. These measurements are customarily taken throughout the usual laboratory period of two hours, with perhaps a few others later in the day.

This leisurely sort of reaction rate never occurs in homogeneous metals. For example, if liquid iron and liquid silicon are mixed, a substantial amount of heat is evolved in a great hurry, as fast as mixing occurs, and the thought of leisurely following this homogeneous reaction rate as one does for acetic acid is quite grotesque. The theoretical counterpart or interpretation of this difference lies in the nature of the metallic state. The great mobility of electrons in metals or, alternatively, the rapidly resonating two-electron covalent bond (of Pauling) gives a model in which every atom is bonded to each of its nearest neighbors; if nearest neighbors are changed, the electrons or bonds change just as rapidly. This statement is just a rephrasing of the general thesis that the metallic state is nonmolecular in nature. The mobility of the bonds in the metallic state is to be contrasted with their relative durability in such well-known molecules as the common gases or acetic anhydride. Reactions in the latter examples are usually pictured in terms of molecular collisions, only a very small fraction of which are successful in changing bonds to produce reaction. Clearly, the picture is very different for the metallic state, as we have no molecules, and the mobility of bonds is very high. This, of course,

Dr. Darken is Associate Director, E. C. Bain Laboratory, United States Steel Corporation, Monroeville, Pennsylvania.

means that the whole subject of homogeneous reaction rates in the usual sense, and the usual treatment in terms of monomolecular, bimolecular, and so on, may be dismissed entirely in the consideration of a homogeneous metal.

A consequence of this, or really another way of saying this same thing, is that local equilibrium prevails in the metallic state. By this is meant that any reaction occurring in measurable time in a small (infinitesimal) volume unit is accompanied by small (infinitesimal) change in the molal free energy. In other words, a large over-all free-energy change takes place in a large number of small steps. It is readily seen that such is not the case for the example of the formation of acetic acid which may be taken as typical of a large group of nonmetallic reactions; the formation of H_2O or HI from the gaseous elements might also be mentioned as typical. Here, in contrast to a metal, it is evident that within a small volume element a molal free-energy change of many kilocalories takes place as reaction proceeds. Or in still other words, the unit elementary step in proceeding from reactant through activated complex to product involves much energy.

The obvious question at this point is what mechanism of reaction do we have in the metallic state. The equally obvious answer is that we have (1) diffusive processes in the single-phase regions, and (2) heterogeneous reactions. Obviously, the unit jump which is the elementary step of the diffusive process involves a rather small over-all free-energy change; even for a rather high concentration gradient this change would be unlikely to be over one calorie per gram atom. Schematically, the typical energetic difference between the elementary step in a molecular and in a diffusive process may be represented as shown in Fig. 1.

Having concluded, or perhaps more properly postulated, that the diffusive process is the dominant mechanism for the dissipation of free energy in a single-phase region of solid or liquid metal, it is tempting and, I believe, profitable to extend this concept of local equilibrium to a large number of heterogeneous reactions of metallurgical interest. Herein is the clue to one rather fruitful meaning of the expression "high-temperature chemistry"—chemistry in the temperature region where local

equilibrium prevails. From this viewpoint, the earlier statement concerning local equilibrium in metallic systems implies that the usual temperatures of metallurgical interest may be considered "high" —steel, for example, might be considered at "high temperature" down through the usual heat-treating range. This concept, of course, carries certain implications; for example, that aside from surface-energy considerations during pearlite growth in an Fe—C alloy, austenite, ferrite and cementite cannot co-exist in any given volume element. This "doctrine" of local equilibrium may apparently be applied fruitfully in a wide variety of reactions of metallurgical interest, and in such cases it will be observed that reaction-rate problems are reduced to problems in (1) nucleation, and (2) diffusion. It is not in any sense intended to imply in connection with local equilibrium that a volume element is necessarily in the lowest possible free energy state with respect to all conceivable phases, but only with respect to those that are actually present. Thus, all volume elements of a molten metal might be supersaturated with respect to a particular gas at 1 atm pressure, providing no gas phase is actually present in contact with the metal; for example, if the metal is surrounded by slag. Many other metastability phenomena, as in the Fe—C and Fe—N systems, readily come to mind; these are in no way incompatible with the prevalence of local equilibrium.

It must be admitted that direct experimental evidence of the prevalence of diffusion-controlled reactions is not as extensive and conclusive as might be wished in the temperature ranges of metallurgical interest. In view of the extensive treatment Carl Wagner included in this series, I shall not attempt to explain a long list of phenomena in these terms but shall merely mention a remark that he made to me to the effect that after he has finished a diffusion problem and thinks, "Ah, now I shall get on to something else," he usually finds that the next problem involves diffusion, too. Instead, I should like to mention one case which does not appear to be controlled by boundary-layer diffusion—this to illustrate the limits of applicability. The example I have in mind involves the nitrogenization of gamma iron by an atmosphere of gaseous N_2 (plus 1% of H_2 to protect against oxide). It is well estab-

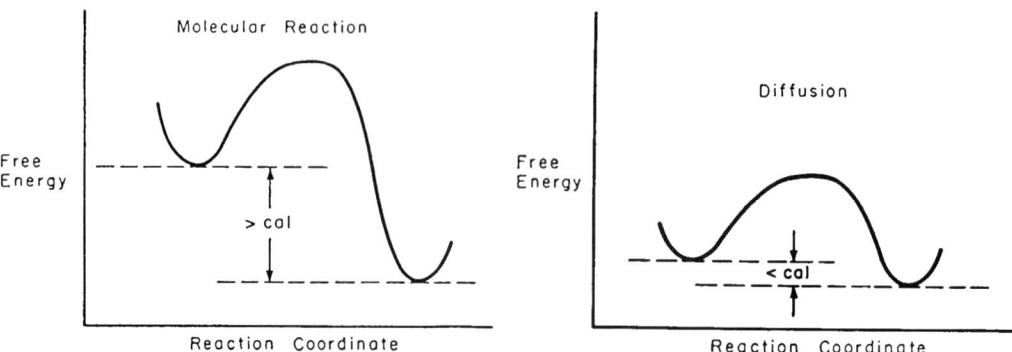

Fig. 1. Energy differences for elementary steps in a molecular reaction and in diffusion.

lished that the solubility of N_2 gas is 0.024%, yet, as shown in Fig. 2, the surface of the metal does not rise to this value even after considerable lapse of time. I think it is quite evident in this case that the chemical potential (or activity) of nitrogen suffers a rather abrupt change within the few atom layers in the immediate vicinity of the surface. That is, local equilibrium can hardly be said to prevail here; a small-volume element including a portion of the interface is not in equilibrium. The reason here seems rather clear: the process of the

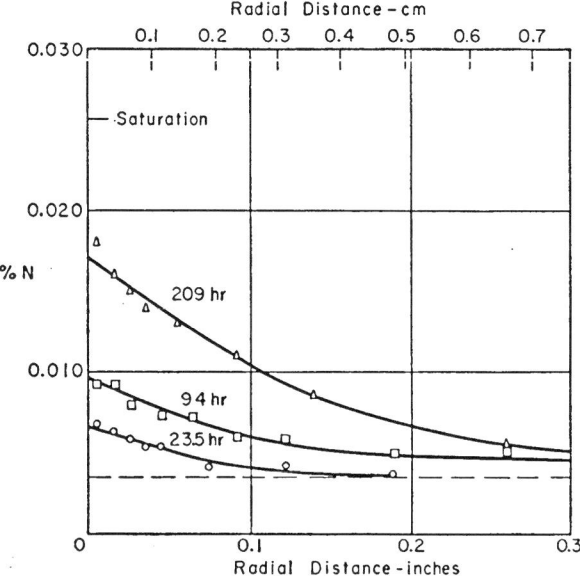

Fig. 2. Penetration curve of nitrogen in iron rod at 955° C. [From Darken and others, Trans. A.I.M.E., *191*, 1174 (1951).]

transition of nitrogen from N_2 gas to gamma iron is a complex one involving at some stage the breaking of the covalent N—N bond, possibly in the steps (1) absorption of N_2 molecules on the surface, (2) dissociation to atoms on the surface, and (3) inward diffusion. This type of difficulty is not particularly surprising in this and similar cases; "high temperature" for this reaction might reasonably be expected to lie above the temperature region where the dissociation of gaseous N_2 is appreciable. Certainly at metal-metal and metal-slag boundaries where no covalent bonds are involved, we would not anticipate bond-breaking difficulties. (A possible exception in the transfer of silicon from, or to, the partially covalent silicate ion or network should be mentioned, though it seems more likely that in slags near 1600° C dissociation would be extensive.)

KINETICS OF THE OPEN HEARTH

The fundamental idea of the control of the primary open-hearth reaction, oxidation of carbon, in terms of (1) the heat-requirement problem and (2) boundary-layer diffusion, has been discussed by the author in BOHS.[1] The first topic will be summarized only very briefly here, but the second will be expanded somewhat.

Metallurgical Heat Requirement

From the practical viewpoint, a major, if not the major factor in determining the production rate of the open hearth is the amount of the metallurgical heat requirement and the rate at which this can be supplied by the fuel within the limitations of the refractory materials.

In a typical heat, the fuel furnishes about four million Btu per ton of steel produced. The heat required for the metallurgical processes is not easily defined, but it ranges downward from about one million Btu, depending not only on the charge but on the manner of calculation. In any event, this requirement appears small in comparison with the heat furnished by the fuel and invites the erroneous conclusion that the metallurgical heat requirement is of little consequence compared to the heat losses. In fact, the truth lies near the opposite viewpoint, in that the difficulty of transferring heat to the bath is so great that, except in duplex practice, this becomes the limiting factor, and rather small variations in its net heat requirement are of considerable significance.

The two major factors that must be understood to realize the practical significance of the metallurgical heat requirements are (1) the availability to the bath of only high-temperature heat, and (2) the slowness of transfer of heat to the bath, occasioned mainly by the insulating effect of the slag.

As reported in BOHS, a statistical survey of several hundred heats was conducted. In this were considered separately such factors as nature of the charge, relative effect of cold pig versus hot metal, effect of air versus ore oxidation, and variations of amount of charged limestone. The evidence seems fairly convincing that there is a statistical relation between the heat requirement of the metallurgical reactions and production rate. It appears that a 1% increase in the metallurgical heat requirement is accompanied in general by a drop in the production rate of about 0.5 to 0.8%. The conclusion that this is more than a casual relationship can hardly be avoided. It is further estimated that out of the 3- to 5-hr variation in charge-to-tap time (usually found in any large group of heats of similar type in one shop) about 1.5 to 2.5 hr may be attributed to those variables that affect the heat requirement. Roughly one-half of the variability in production rate appears to be attributable to variability in the metallurgical heat requirement.

Viewed from the other side, the practical problem in conventional practice is predominantly one of maintaining a high heat flow from the flame. This flow is closely related to the maintenance of a deep-seated boil rate which in turn depends not merely on an intense flame but especially on an adequate supply of air for both fuel and bath requirements —the latter being an important factor. We shall return to some aspects of this later.

Kinetics of the Carbon Boil

The principal chemical reaction of the open hearth is, of course, that of carbon (initially in the metal) with oxygen (initially in gas and ore) to form gaseous oxides of carbon. As oxygen, from air or ore, must first enter the slag, the first problem in understanding the mechanism is a determination of the steps by which oxygen from the slag combines with carbon to form carbon monoxide.

Let us first consider conditions at the slag-metal interface. In accord with the previous discussion, it is assumed that local equilibrium prevails at this interface and, hence, that the oxygen content of the metal here is that corresponding to equilibrium with the slag (deferring until later the details of the argument that we need not be troubled in this connection by gross compositional change of the slag in this vicinity). It is further assumed, as verified by spoon samples taken at different depths, that the main mass of metal is nearly uniform in oxygen content and, hence, that the rate-controlling step in the vicinity of this interface is diffusion of oxygen through a relatively thin boundary layer in the metal. The flux of oxygen is then (by Fick's first law), $\frac{D([O]_{SE} - [O])}{100 \, \Delta l} \rho$ where [O] is the oxygen content in weight percentage of the main bath, $[O]_{SE}$ is the corresponding percentage of oxygen at the interface, Δl is the film thickness, and ρ is the density of the metal. Since, at steady-state boil, each atom of oxygen transferred reacts with one atom of carbon, the rate of carbon consumption per unit area is 12/16 of this. Each unit mass of carbon removed per unit cross section decreases the percentage of carbon by $100/l\rho$, where l is the bath depth. Hence, the rate of carbon drop is

$$-\frac{dc}{dt} = \frac{12D([O]_{SE} - [O])}{16l\Delta l} \qquad (1)$$

Thus, we may compute the rate of carbon drop for the open hearth provided we know or estimate the quantities on the right side of this equation. The diffusivity, D, of oxygen in liquid steel has not been measured directly, and there is considerable range in the reported value of the diffusivity of various elements in liquid iron. The apparently careful work of Morgan and Kitchener[2] gives $D = 5 \times 10^{-5}$ cm²/sec for cobalt and 7×10^{-5} for carbon. The earlier work considered in BOHS, as well as some later, led, for several elements, to values of 10^{-4} or higher. For the present purpose a rounded value for the diffusivity of oxygen in liquid steel of 10^{-4} cm²/sec is taken. The normal depth, l, of a quiet bath is taken as 13.5 in. or 34 cm (an erroneous value of 76 cm is given on page 614 of BOHS—this corresponds to the actual height during boil). An average value of $[O]_{SE} - [O]$ during normal boil conditions is taken from the extensive work of Brower and Larsen[3] as 0.04. The thickness, Δl, of the diffusion layer remains a major uncertainty, but

we may tentatively try the value 0.003 cm taken in BOHS as typical of that of many boiling liquids. These figures when inserted in equation 1 give for the rate of carbon drop under "normal" boil conditions

$$-\frac{dc}{dt} = \frac{12}{16} \times 10^{-4} \times \frac{0.04}{34 \times 0.003} = 3 \times 10^{-5} \, \%/sec$$

$$\text{or } 0.11 \, \%/hr \qquad (2)$$

which, considering the many uncertainties involved (including the fact that this figure should be multiplied by the ratio of actual slag-metal-interface area to the nominal area) is in astonishingly good agreement with observed rates of 0.12 to 0.18% per hour. A similar calculation for conditions following heavy oreing gives a rate about five times as great—again in agreement with practice.

The rate of approach to manganese (or sulfur) equilibrium may be computed from the same postulates, thus giving as the rate of manganese transfer

$$\frac{d\Delta[Mn]}{dt} = \frac{D}{l\Delta l} \Delta[Mn] \qquad (3)$$

where $\Delta[Mn]$ is the difference in the percentage of manganese between the two sides of the diffusion boundary layer and the other symbols have the same meaning as previously except that D represents now the diffusivity of manganese (assumed very close to that of oxygen). Integration between limits from $t = 0$ gives

$$\Delta \log \Delta[Mn] = \frac{Dt}{2.3l\Delta l} \qquad (4)$$

For 90% approach to equilibrium the left side is equal to one and the time t_e to achieve this is found, using the same numerical values, to be

$$t_e = \frac{2.3l\Delta l}{D} = \frac{2.3 \times 34 \times 0.003}{10^{-4}}$$

$$= 2300 \text{ sec} \approx 40 \text{ min} \qquad (5)$$

This is in very good agreement with the observed time of about one-half hour to achieve or reachieve substantial equilibrium after an addition to the furnace. It will be noticed that, if this observed time is used to evaluate $\Delta l/D$ empirically, a slightly smaller value is obtained than that previously estimated; this value could have been used to compute the rate of carbon drop and gives even better agreement.

Bubble Formation and Growth

Although the picture or model of the open-hearth kinetics as developed so far seems very encouraging and satisfactory in at least a semiquantitative sense, it is by no means a complete picture. We must next consider how the oxygen, once it has entered the metal from the slag, leaves the metal as oxides of carbon; the relation between the two processes must also be considered. It is frequently said, as an introduction to chemical kinetics, that the reactants may proceed to products by a great many reaction paths; of these (in favorable cases) one is predominant and is, hence, the one upon which to focus attention; further, that, of the many steps in this path, one is usually significantly slower than the rest and, thus, is the rate-controlling step. Unfortunately, the fundamental open-hearth reaction cannot be treated by this simplified procedure. The main reason is that only a very small portion of the total oxygen entering the metal can be stored there at any given time and, hence, that it must leave at substantially the same rate at which it enters.

In accord with the earlier general discussion, it is apparent that the mechanism by which oxygen leaves the metal is bound up with the nucleation of CO bubbles and the subsequent growth of these by diffusion of oxygen and carbon through a boundary layer in the metal adjacent to the bubble. During the major portion of the boil, the carbon content is much higher than that of oxygen, and, therefore, the diffusion of oxygen rather than that of carbon limits the rate. This is readily seen by noting that the two fluxes are substantially equal (to form CO), and hence, by Fick's law (and assuming the diffusivities approximately equal) the change in % C across the diffusion layer is approximately equal to the change in % O. As the latter is small, a few hundredths of 1%, it follows that the former is also small and, hence, that the % C at the interface is substantially that of the main mass of metal—in contrast to the situation for oxygen.

A great mass of evidence for liquids, including the experiments of Körber and Oelsen for steel, indicates that spontaneous nucleation of gas bubbles does not occur in homogeneous liquids for degrees of supersaturation comparable to that of CO in the open hearth (except for the condition known as frothing).

The normal deep-seated carbon boil undoubtedly originates on the bottom. The most reasonable and only obvious place of origin of the bubbles is at the crevices, cracks, and pores in the bottom of the furnace. Here entrapped gas serves as the starting point or nucleus. The condition for initial growth is that the virtual CO pressure (proportional to the product of [% C] by [% O]) exceed the sum of: atmospheric pressure; the ferrostatic pressure; and the capillary overpressure, $2\gamma/R$, where γ is the surface tension and R is the bubble radius determined by the nature of the crevice. This aspect has been mentioned in the literature several times, perhaps most recently by Vallet,[4] who shows a chart for the radius of the equilibrium bubble as a function of the product of % C and the "excess" % O; this ranges from 2 to 10 microns for degrees of supersaturation usually observed in the open hearth. Bubbles exceeding this minimum size grow by

transport of carbon and oxygen from the metal to the gas phase, breaking loose and then rising. As each tears away, it leaves a small tail of gas behind in the crevice, somewhat analogous to the case of water dripping from a faucet. This tail then serves as the start of the next bubble.

During the growth and rise of the bubble through the metal, we again assume that local equilibrium prevails and now focus attention on the diffusion boundary layer in the metal adjacent to the bubble. (The possibility that the principal diffusion layer might be in the gas may be dismissed immediately because of the fact that the gas is nearly pure CO + CO_2 and no diffusion is required for the reaction to proceed.) The flux of oxygen here, again by Fick's law, is the product of its diffusivity and concentration gradient; the total rate of transfer is equal to this product times the bubble area. At steady-state boil this rate of oxygen transfer must equal that across the slag-metal boundary; thus,

$$\frac{DS([O]_{SE} - [O])}{\Delta l} = \frac{DS'([O] - [O]_{CE})}{\Delta l'} \quad (6)$$

where S signifies surface or interface area, the prime indicates a quantity referring to the gas-metal interface and $[O]_{CE}$ indicates the oxygen content of the metal at the gas-metal interface (carbon equilibrium). As D is identical on the two sides, and Δl is probably very close to $\Delta l'$, and as S' is probably only slightly greater than S, it follows from this (as well as being verified by experience) that $[O]$, the observed oxygen content of the bath, must be nearly midway between $[O]_{SE}$ and $[O]_{CE}$ and cannot closely approximate either. In other words, we cannot say that the rate of boil is controlled predominantly by either interface, but rather jointly by both.*

The over-all picture thus developed for the oxidation of carbon in the open hearth is that of two nearly equal and closely coupled diffusion boundary layers. They provide a resistance to the passage of oxygen, such that the over-all rate is changed by altering conditions at either layer.

Foaming or Frothing

Occasionally, there develops, in the open hearth, a very undesirable condition known as foaming or frothing. Under this condition, the slag is puffed up, apparently by a large number of very small bub-

* An estimate may be made from the ratio $S/S' \cong 3$, which is obtained from the fact that a boiling bath is roughly twice as deep as a quiet one. This assumption, in conjunction with those of the equality of Δl and $\Delta l'$ and an average bubble size of 5 cm, gives

$$[O]_{SE} - [O] = 3([O] - [O]_{CE}) \quad \text{or} \quad [O] = 1/4[O]_{SE} - 3/4[O]_{CB}$$

The fact that this gives [O] lower (more near carbon equilibrium) than observed by Larsen and Brower suggests that it would be of interest to determine $[O]_{SE}$ experimentally by equilibrating, in the laboratory, several open-hearth metals and slags. Possibly the estimate made of the gas content of the boiling bath is high.

bles. Usually there is no indication of a deep-seated boil; the heat flow to the bath and the rate of carbon drop are very low; the furnace is not functioning. Although there is no pretense of adequately explaining the nature of this difficulty here, two possibly associated phenomena may be mentioned: (1) in the absence of boil, sufficient supersaturation may occasionally build up on the upper surface of the metal to lead to spontaneous bubble formation in the top boundary layer of the metal; as these bubbles escape the metal almost immediately, they do not have time to grow to sufficient size to pass rapidly through the slag; (2) if the metal is quiet, bubbles rising from the bottom will not recirculate and will reach the top rapidly as small bubbles without time to grow. In fact, a calculation from Stokes' law indicates that a bubble rising through quiet metal would achieve a size of only 1 mm by the time it rose to the slag.

The Boundary Layer in the Slag at the Slag-Metal Interface

In the previous discussion no mention was made of the barrier to diffusion presented by the boundary layer in the slag adjacent to the metal. In view of the fact, discussed at length by Wagner, that, in general, we should take account of the boundary layer in each phase at a liquid-liquid interface, let us next consider this layer. The only type of information we have obviously related to this is that pertaining to heat flow.[5] Data on the thermal conductivity of slag are not available; however, it may very roughly be taken as about the same as for glasses, namely, about 12 Btu per hr per sq ft per ° F for 1 in. The average rate of heat input to the bath is about 24,000 Btu per sq ft per hr. The sharp temperature rise in the slag immediately above the metal is about 100° F.[5] These figures give a boundary-layer thickness $\Delta l = 12 \times 100/24000 = 0.05$ in. or 0.12 cm. This is considerably greater than the diffusion boundary layer of the metal ($\Delta l = 0.003$ cm) and would give us serious cause for concern that the prior treatment for oxygen transport is deficient in neglecting the resistance of the boundary layer in the slag, were it not for the point stressed by Wagner that the boundary-layer concept is complex, and that the thickness for heat flow is quite different from that for diffusion. From these concepts, we may estimate the ratio of the two thicknesses as proportional to the cube root of the ratio of the diffusivity to the thermal diffusivity; thus, as the thermal diffusivity is about 10^{-3} cm²/sec and the diffusivity is about 10^{-6} cm²/sec, the cube root of the ratio is about 1/10 or the diffusion boundary layer in the slag has a thickness of approximately 1/10 of 0.12 or 0.012 cm. This is only slightly greater than that of the metal (0.003 cm). Equating the flux through the two boundary layers (using the superscript M for metal and S for slag), we now find

$$D^M \frac{\Delta[O]^M}{\Delta l^M} = \frac{D^S}{4} \frac{\Delta[Fe]^S}{\Delta l^S} \qquad (7)$$

the factor of 4 arising from the ratio of equivalent weights. As D^S is about 1/10 of D^M, it is seen that $\Delta[Fe]^S$ is expected to be about 160 times $\Delta[O]^M$ or about 4%. This is not a vanishingly small quantity in comparison to the iron content of the slag, and it is reasonable to suppose that it could be detected by carefully designed experiments.

If this very rough reasoning is correct, there is some reason to believe that the slag immediately in contact with, and in equilibrium with, the metal is slightly less oxidizing than the main portion of slag. Though this may be of little consequence so far as carbon oxidation is concerned, it may be of considerable importance in conjunction with equilibria involving oxygen, such as manganese, sulfur, and phosphorus distributions. It has generally been considered that the distribution with respect to these, attained in the open hearth, corresponds to the equilibrium at complete equilibrium of the main mass of slag and metal. Possibly this idea needs revision. Also, looking a little farther, it may be possible to utilize or alter the slag gradient to achieve desired effects.

SUMMARY AND CONCLUSIONS

In this general review of chemical kinetics at high temperatures, the idea of local equilibrium is developed and exploited. The exceedingly short life of atomic configurations in this region leads to the general abandonment of the classical viewpoint. Instead, most reaction-rate problems at high temperature may be resolved to nucleation problems and diffusion problems. The diffusion problems frequently involve diffusion through thin boundary layers. The reactions of the open hearth are treated in these terms.

We may, at least tentatively, accept the following as the result of direct observation of the open hearth:

(a) During boil, the metal rises so that its depth in extreme cases approaches twice that of a quiet bath. This leads to the immediate conclusion that the volume of gas bubbles in the bath at any given instant may equal the volume of the metal itself. Recalling that a sphere inscribed in a cube occupies about one-half the volume of the cube, we arrive at a picture of bubbles so numerous that in extreme cases they are almost tangent to each other.

(b) The size of the observed bubbles at the slag surface commonly ranges up to the size of one's fist. With some interpretation, it is not unreasonable to conclude that the main gas evolution is in bubbles about a centimeter or, perhaps, a few centimeters in diameter.

(c) The rate of carbon drop during steady-state boil is commonly about 0.15% C per hr. From this, it is readily computed that each cc of metal liberates about 0.04 cc of CO per sec.

(d) A steady-state boil exists. By this is meant that once boil starts it tends to persist. It does not tend to stop itself nor does its rate tend to increase very rapidly.

By combining the first and third observations that the metal contains up to its own volume of gas but is producing this at the rate of only 1/25 its own volume per sec, we arrive at the conclusion that on the average a given portion of gas may be in the metal for as long as 25 sec. The mean bubble life is thus many seconds. In contrast, the computed time of rise of a bubble through otherwise static liquid steel is a small fraction of a second. To reconcile this gross discrepancy without violating the second observation, it seems necessary to postulate a much more violent turbulent motion of the bath than is usually assumed. That is, the bubbles must be swirled from top to bottom perhaps hundreds of times before escaping. To sweep a large bubble down, the metal in its vicinity must be moving at an absolute velocity greater than the relative velocity of the bubble. The local velocity of the metal necessary to carry a bubble of centimeter diameter down is difficult to estimate but is indeed quite high.

As auxiliary evidence, the relation between size and time for a bubble may be computed from the rate of diffusion through the boundary layer. This gives the bubble radius in centimeters to be roughly one-quarter the time in seconds. Thus, the observed bubble size of a few centimeters is consistent with the duration of stay (many seconds) and the concept of boundary-layer diffusion.

As a further check, we may estimate the bubble size from the total amount of bubble surface (several times the cross section of the bath) and total bubble volume. This also gives a mean bubble diameter of several centimeters.

Thus, the various listed observations are consistently correlated—the only surprising implication being the high velocity of the turbulent agitation.

The general picture of the boil at steady state may be formulated by starting with this turbulence. The turbulence, in conjunction with the properties of the metal as viscosity and diffusivity, then may be regarded as primarily responsible for the thickness of the boundary films. These, in turn, control the rate of passage of heat and of oxygen and, hence, the rate of evolution of CO. Also, intimately connected with this is the rate of generation of CO bubbles, which is connected with cracks, crevices, and surface tension. The rate of evolution of CO determines the amount of turbulence. This brings us back to the arbitrary starting point. It is quite clear that a balance is struck—i.e., a definite rate of boil is established—such as to satisfy all these requirements simultaneously. Thus, no one factor can be said to determine the rate of boil—all must be considered jointly.

References

1. *Basic Open Hearth Steelmaking*, Second Edition, Physical Chemistry of Steelmaking Committee, A.I.M.E., New York, 1951.
2. D. W. Morgan and J. A. Kitchener, *Trans. Faraday Soc.*, 50, 51 (1954).
3. T. E. Brower and B. M. Larsen, *Trans. A.I.M.E.*, 172, 164 (1947).
4. P. Vallet, *Iron and Steel*, 28, 463 (1955).
5. B. M. Larsen, "A New Look at the Nature of the Open-Hearth Process," NOHC Special Report Series No. 1, A.I.M.E. (1956).

Discussion

TAYLOR reviewed briefly the question of whether the oxygen content in an active open-hearth bath could be obtained better from a poured test or a bomb test. He felt that experience and experimental work favored the latter. As a result he was concerned about the oxygen values used by Darken in the calculations.

DARKEN noted that this is a perennial question concerning the bomb test, and his feeling was that the character of the bomb sampling shifts the composition of the steel away from what it was when the sample was taken. The bomb tends to create a local boil as the metal goes over the edge of the cold mold where the gas bubbles are nucleated. The steel tends to come to equilibrium as a result of these gas bubbles forming. TAYLOR replied that a good bomb sample should be slagged, and as a result the metal would flow over only about one-half inch of cold steel. Under these conditions he doubted if there was more boiling than obtained when pouring through air and into an open mold.

TENENBAUM said that there were three major periods in the carbon boil that Darken should consider. The first period comes early in the heat and is the ore boil. In this condition the gas volume probably approaches the metal volume, and Darken's assumed centimeter size of a bubble seems a reasonable one. The 0.15% per hr carbon drop appears reasonable. The second stage must be passed through if one is to make steel with a carbon content under 0.10%. It is generally reached around 0.20% C. At that point the slag, which has previously been mushy as a result of solid particles and which contains many gas bubbles, suddenly becomes quite hot. The bubbles and the mushiness disappear. That is to say, the bubbles continue to rise through but they do not remain in the slag. At this point the metal volume visibly drops, indicating a decrease in the amount of gas entrapped in the metal phase. During this condition and without oreing or use of oxygen, one gets a fairly rapid rate of carbon drop close to 0.60% per hr, and the gas volume is roughly 25% of the metal volume. The third stage arises between 0.05 and 0.06% C. At this point the gas volume becomes very small because the rate of carbon drop is low. Here the metal volume is probably not very much different from what one would get with a stagnant bath. One might estimate 10% of the metal volume as gas. The second stage, being a steady condition, might lead to more realistic figures for the calculations made by Darken.

DERGE suggested that Darken's calculations might be made more realistic if one considered the problem of hindered settling as a result of the metal phase being at least 50% gas. This large gas volume would tend to hinder the upward movement of individual bubbles. DARKEN said that he had not included this, although he felt that it would be reasonable to do so. PHILBROOK inquired about the assumption of several different film thicknesses for the chemical processes going on simultaneously. He visualized the film as a thin layer through which matter and heat are transferred by diffusive processes. He felt that the juggling of film thickness for different reactions in different materials being transferred was not compatible with the considerations for heat and fluid-momentum transfer. WAGNER stated that for different processes such as mass transfer and heat exchange there would be a different film thickness to be used in each case because of the different dimensionless numbers which are incorporated in the characteristic equations.

by P. Vallet

Theory of Degassing of the Open-Hearth Bath during Carbon Boil

During the carbon reaction in an open-hearth furnace, the metal is no longer in direct contact with the furnace gases but is protected by a layer of molten slag.

The residual nitrogen and hydrogen contents of the open-hearth steel originate in the furnace charge or in the various additions (moisture, for instance, so far as hydrogen is concerned). The following paragraphs will show that these elements are partially eliminated by the evolution of CO, which is generated in large quantities during the carbon reaction.

Even if some nitrogen and hydrogen reach the metal through the slag layer by diffusion, it is admissible to accept that such a process is more than compensated by the elimination just mentioned.

GENERAL PRINCIPLE

Large quantities of a gas, g, dissolved in a liquid metal can be removed by bubbling through the metal a stream of another gas, g', less soluble than g in the metal, or less objectionable for the use of the final product. A certain amount of g diffuses into the bubbles of g' to bring up its partial pressure equal to, at most, that corresponding to an equilibrium between the metal and g at the same temperature, as defined in the preceding paragraph.

Each bubble of g' removes only a small amount of g, but it is clear that a very great number of g' bubbles can eventually carry away a sizable volume of g.

Dr. Vallet is affiliated with the Institut de Recherches de la Sidérurgie, Saint-Germain-en-Laye, France.

W. GELLER'S THEORY

Only the conclusions of the theory will be given.[1] Suppose that p_{N_2} is the pressure of the nitrogen atmosphere which is in equilibrium with the nitrogen-saturated metal. When a carbon monoxide bubble, initially free from nitrogen, rises from the hearth through the metal, it has a tendency to take up nitrogen. If the metal and the bubble had time to reach equilibrium, the partial pressure of nitrogen in the bubble would be equal to p_{N_2}. In the reverse hypothesis, where equilibrium is not attained, the partial pressure of nitrogen would be less than p_{N_2} and could be represented by the product $f p_{N_2}$, f being less than unity. It will be seen that the first hypothesis is a limiting case of the second where $f = 1$.

Geller's argument leads to the following equation:

$$\frac{K_N p}{f}\left(\frac{1}{\%\,[N]_0} - \frac{1}{\%\,N}\right) + \%\,[N]_0 - \%\,[N]$$
$$+ 2.33\,(\%\,[C]_0 - \%\,[C]) = 0 \quad (1)$$

In this equation, $\%\,[N]_0$ and $\%\,[C]_0$ represent the initial weight percentages of nitrogen and carbon and $\%\,[N]$ and $\%\,[C]$ represent the weight percentage of nitrogen and carbon at any other stage of the process. Here, p is the total pressure in the bubble; K_N is the equilibrium constant for the dissolution of nitrogen in iron as expressed by the equation:

$$N_2\,(g) = 2[N] \quad (2)$$

This constant is defined by the following equation:

$$K_N = \frac{[\% \, N]^2}{p_{N_2}} \qquad (3)$$

Were the nitrogen content of the metal assumed to be low, it will be seen that K_N is numerically equal to the square of the weight percentage of the nitrogen dissolved in iron when the nitrogen pressure in equilibrium with the metal is 1 atm. At 1600° C, it is found that experimentally $K_N = 0.0016$.

Equation 1 has been simplified by Geller:[1]

$$\% \, [N] - \% \, [N]_0 = \Delta N$$

$$\% \, [C] - \% \, [C]_0 = \Delta C$$

This leads to the following equation:

$$\frac{K_N p}{f} \frac{\Delta N}{\% \, [N]_0 \, \% \, [N]} - \Delta N = 2.33 \Delta C \qquad (4)$$

When $f = 1$, the equation is the same as Geller's except for a few details. The coefficient f is, in any case, identical with what Geller has called the "yield or efficiency of nitrogen removal" (represented by the Greek letter η (in the original memorandum). In his general equation, he has not introduced η as has been done here with f but has considered it only afterward.

Equation 1 shows that $\% \, [C]$ is related to $\% \, [N]$ by a conventional algebraic formula of the type $y = ax + b + c/x$. The curve representing the variations of $\% \, [N]$ as a function of $\% \, [C]$ is therefore a hyperbola. Only one arc of this curve is of interest, i.e., the arc corresponding to values of $\% \, [C]$ and $\% \, [N]$, which are lower than $\% \, [C]_0$ and $\% \, [N]_0$, respectively.

The variations ΔN and ΔC are negative since $\% \, [N]$ and $\% \, [C]$ decrease constantly. Geller[1] has represented the variations of $\% \, [N]$ not as a function of $\% \, [C]$ but as a function of $\% \, [C]_0 - \% \, [C] = -\Delta C$, i.e., the proportion of carbon removed during the operation. The theoretical curve corresponding to equation 4 would also be an arc of a hyperbola having the horizontal axis as an asymptote and starting from the point having as coordinates $\Delta C = O$ and $\% \, [N]_0$.

This curve is plotted for $f = 1$ by Geller using the data of Debuch,[2] and Bardenheuer and Thanheiser.[3] All the experimental curves lie above the "theoretical curve" as defined by Geller. This result effectively confirms that f cannot be equal to unity.

As Geller has pointed out, experimental curves are hyperbolic in form but drop less abruptly than the theoretical curve (namely, $f = 1$). An empirical value can be found for f corresponding to determined experimental values for $\% \, [N]$, $\% \, [N]_0$, $\% \, [C]$, $\% \, [C]_0$, as equation 4 is of the first degree in f. Moreover, considering the accuracy of the experiments, there is every reason to believe that the new theoretical curve using the value of f deduced from the extreme values of $\% \, [N]$ and $\% \, [C]$ during a given operation must be very close to the

experimental curve. In this respect, equation 4, in which f has an adequate value, can be accepted as satisfactory.

In the experiments of Debuch cited by Geller, the nitrogen content decreased by about 0.002 to 0.004% and the carbon content by about 0.2 to 0.3%. The parameter f is found to have a mean value of 0.18.

Discussion of Geller's Theory

This theory has the disadvantage of not explaining why the value of f is, in fact, so low as 0.18. Moreover, this theory effectively assumes that a stream of CO is blown through the metal in the same way as an inert gas such as argon not originating in the metal.

The CO bubbles, which seem to form mainly on the hearth, are initially very small and expand as they rise, due to the continuous accumulation of carbon monoxide produced by the carbon-removal reaction. The fact that the partial pressure of nitrogen in the bubble is less than that corresponding to equilibrium could be adequately explained by the fact that the rate of transport of CO across the gas-liquid interface is larger than the rate of transport of nitrogen.

SUGGESTED THEORY

A given single gas bubble j located near the slag-metal interface in an open-hearth furnace, at the moment t, will be found to contain CO and some nitrogen. During an infinitesimal moment of time dt, it will receive infinitely small amounts of CO and nitrogen.

It has been shown[4] that the speed of carbon removal due to bubble j is given by the following equation:

$$\frac{7}{3} \frac{dm_{Cj}}{dt} = k_1 s_j \, \% \, [C] \, \Delta[O] \qquad (5)$$

in which m_{Cj} represents the weight of carbon contained in the bubble j as CO, and s_j is the surface area of the bubble; $\Delta[O]$ is the excess oxygen defined by Larsen,[5] i.e., the difference between the actual oxygen content of the bath and that which would be found if the carbon and oxygen dissolved in the iron were in equilibrium with CO at 1 atm; k_1 is the coefficient of the reaction rate of carbon removal.

The diffusion rate of nitrogen into the bubble j during the same time dt can be determined in the same manner. By applying the conventional kinetic theory to the reaction represented by equation 2 read from right to left—as has been done in the similar case of the carbon reaction—it can be assumed that nitrogen diffuses into the bubble at a rate proportional to its surface area and to the square of the concentration of nitrogen contained in the metal, which has a tendency to dissolve into the metal; it can be assumed also that the dissolution rate is proportional to the surface area of the bubble and to the concentration of nitrogen in the bubble, measured by the partial pressure of p_{Nj}. The over-all rate of increase of the mass m_{Nj} of nitrogen contained in the bubble is therefore the difference between the two preceding rates. This can be expressed by:

$$\frac{dm_{Nj}}{dt} = (k_N\,[\%\,N]^2 - k_N'p_{Nj})s_j \qquad (6)$$

where k_N is the kinetic coefficient of the nitrogen-removal reaction and k_N' is that of the reverse reaction.

If the bubble were in equilibrium with surrounding metal with respect to nitrogen, it would not receive any nitrogen, and the first member of equation 6 would be zero. This would give

$$k_N\,[\%\,N]^2 = k_N'p_{Nj} \qquad (7)$$

which is another way of writing the law of mass action already expressed by equation 3. A comparison of equations 7 and 3 shows that

$$\frac{k_N'}{k_N} = K_N \qquad (8)$$

By this relationship, equation 6 may be expressed as

$$\frac{dm_{Nj}}{dt} = k_N s_j\,([\%\,N]^2 - K_N p_{Nj}) \qquad (9)$$

It has been seen that, at 1600° C, $K_N = 0.0016m$ and the nitrogen pressure in the bubble is certainly very low. If the metal had a nitrogen content of 0.010%, the corresponding equilibrium pressure would be about 0.06 atm, but the partial pressure of nitrogen in the bubble is certainly lower. As a rough estimate, the second term of the parentheses in equation 9 can be neglected, and the equation then becomes

$$\frac{dm_{Nj}}{dt} = k_N s_j\,[\%\,N]^2 \qquad (10)$$

If each side of equation 5 is divided by its counterpart in equation 10, there results

$$\frac{7}{3}\frac{dm_{Cj}}{dm_{Nj}} = \frac{k_1\,\%\,[C]\,\Delta O}{k_N\,[\%\,N]^2} \qquad (11)$$

The infinitesimal masses of carbon and nitrogen dm_{Cj} and dm_{Nj} which enter bubble j originate in the metal. They are therefore equal and opposite in sign to the infinitely small variations dm_C and dm_N in the masses of carbon and nitrogen m_C and m_N dissolved at the moment t in the bath. Moreover, the ratio of these variations is equal to the ratio of the infinitely small variations of the relative weighted percentages of these elements dissolved in the metal. Equation 11 can therefore be written as follows:

$$\frac{7}{3}\frac{d\,\%\,C}{d\,\%\,[N]} = \frac{k_1\,\%\,[C]\,\Delta[O]}{k_N\,[\%\,N]^2} \qquad (12)$$

It can be seen that the surface of the bubble is no longer important, since s_j is eliminated from the equation. Moreover, this equation is quite different from Geller's differential equation. To integrate this equation, new hypotheses have to be made.

It can be assumed that the probability of the encounter of two atoms dissolved in iron is independent of the nature of these atoms. If so, the ratio of the kinetic coefficients k_1/k_N will be equal to the ratio of the exponential function of Arrhenius' law, corresponding to the carbon-removal reaction and to the nitrogen-removal reaction, respectively:

$$\frac{k_1}{k_N} = \exp\left(\frac{E' - E}{RT}\right) \qquad (13)$$

E' being the apparent activation energy of the nitrogen-removal reaction and E that of the carbon reaction. It can be seen that if E' and E were equal the temperature would have no effect on the process. Then the exponential is equal to unity. The situation would be approximately the same if E' were of the same order of magnitude as E, considering the slight temperature increase of the bath (100° C, according to Geller), whereas T is about 1800° K.

If it be assumed that the excess of oxygen remains approximately constant, integration will give, as a first approximation:

$$\frac{3}{7}\cdot\ln\left(\frac{\%\,[C]}{\%\,[C]_0}\right) = \exp\left(\frac{E' - E}{RT}\right)\frac{\%\,[N] - \%\,[N]_0}{\%\,[N]_0\,\%\,[N]} \qquad (14)$$

When $\%\,[C]$ does not differ much from $\%\,[C]_0$, this equation can be somewhat simplified by using Geller's notation as above and by substituting for $\ln(1 + \Delta C/\%\,[C]_0)$ the first term of its Taylor's development in series. This gives

$$\frac{7}{3}\frac{\Delta C}{\%\,[C]_0} = \exp\left(\frac{E' - E}{RT}\right)\Delta[O]\frac{\Delta N}{\%\,[N]_0\,\%\,[N]} \qquad (15)$$

The results of the experiments thus far are, on the whole, too incomplete to enable a quantitative checking of equations 14 or 15. Usually the final nitrogen and carbon contents of the metal alone have been determined; sometimes the final temperatures have been measured as well, but never (so it seems) the values of $\Delta[O]$. Although these values appear in the equations, the integration effected has no meaning unless $\Delta[O]$ has remained constant, and it is necessary to make sure of this.

However, in Debuch's experiments[2] cited by Geller,[1] the final carbon content of the metal would seem to be about 0.10%. If so, it may be expected that $\Delta[O]$ is about equal to the mean value indicated by Larsen[3] excluding the periods in which additions are made, that is to say, 0.020%.

The use of equation 14 for the heats 2, 3, 4, 7 to 10, 12, 13, and 15 of Debuch, gives the quotient

$$\frac{k_N}{k_1 \Delta[O]}$$

with an average value of 21.9 for the 10 heats. From equation 13, this gives

$$21.9 = \frac{1}{\Delta[O]} \exp\left(\frac{E - E'}{RT}\right)$$

If [O] is taken as $= 0.020$, and $T = 1900°$ K (or $1627°$ C), then

$$\log 0.438 = \frac{E - E'}{1900R} \log e$$

It has been estimated that E is approximately, 29,000 cal.[4] In addition, $R/\log e = 4.575$. Thus,

$$1.64 \times 1900 \times 4.575 = 29{,}000 - E'$$

that is,

$$-3100 = 29{,}000 - E'$$

and E' has a value of about 32,000 cal, which is very close to the apparent activation energy of the carbon-removal reaction.

The same quotient has also been calculated for the heats 2 to 15 of Debuch for the fusion period (between initial charging and complete fusion) during which the temperature remained about constant. A value of 48.9 was found to be an average for the 14 determinations. It was found, with [O] $= 0.02\%$, that the exponential appearing in equation 13 would be very close to unity. This would mean that $E' = E$ and E' would be approximately equal to 29,000 cal, a value very close to the preceding, in view of the approximations involved.

The low value obtained for this apparent activation energy indicates that chemical kinetics is not rate controlling, for as Goodeve[6] has indicated, a reaction with so low an activation energy should proceed immeasurably fast at 1500 to 1600° C. The process is clearly controlled by diffusion rates as in the case of carbon elimination.

References

1. W. Geller, *Stahl u. Eisen*, **64**, No. 1, 10 (1944).
2. K. Debuch, Unpublished doctoral thesis, Techn. Hochschule, Aachen, 1941.
3. P. Bardenheuer and G. Thanheiser, *Mitt. Kaiser-Wilhelm-Inst. Eisenforsch. Düsseldorf*, **17**, 133 (1935).
4. P. Vallet, *Rev. mét.*, **51**, 709 (1954); *Iron and Steel*, **28**, 463 (1955).
5. B. M. Larsen, *Trans. A.I.M.E.*, **145**, 67 (1941).
6. C. F. Goodeve, *Discussions Faraday Soc.*, No. 4, 9 (1948).

Discussion

DARKEN discussed at some length the problem of rationalizing the calculated rate at which nitrogen is removed from a metal bath by the carbon boil to the observed rate. He pointed out that the carbon-oxygen reaction could be explained in terms of the diffusion process through the boundary layer. Because conditions for the diffusion of carbon, oxygen, and nitrogen must be similar at the bubble interface, the nitrogen transfer rate can be estimated from the observed results of the carbon-oxygen reaction. Assuming the limiting condition to be that the inside surface of the gas bubble is in equilibrium with nitrogen in the gas, then one could estimate how far from equilibrium the nitrogen content of the bubble would be as a result of the diffusion layer. The difference turns out to be very, very small. With a bath with 0.004% [N], the nitrogen content of the bubble at equilibrium should be 1% N_2. In general, the observed nitrogen content of the bubble is only approximately one-third of the amount computed from the foregoing argument. He surmised that the reaction was slow because of covalent bonding. CHIPMAN also suggested that the lack of nitrogen removal by the carbon boil may be due to the replenishment of nitrogen from either slag or atmosphere at a regular rate, as has been observed for hydrogen. DARKEN added that this argument assumed a nitrogen solubility in the slag, a point on which there is no good information.

SHANAHAN asked (in Darken's Fig. 2) if the saturation value of nitrogen at the surface was not observed because of the necessity of taking a finite sample size. If an exceedingly thin sample from the surface could have been taken, it might have been found that the surface layer was in equilibrium with the nitrogen atmosphere, as far as nitrogen content was concerned. DARKEN replied that the shape of the curves was such as to indicate no irregularity in the diffusion coefficient right up to the absorption layer which would be 10 angstroms in from the surface. It would be only there that a sharp upturn in the nitrogen content would be possible, and he pointed out that this absorption layer was actually part of the reaction system.

RICHARDSON pointed out that a very high activation energy of 60 to 80 kcal was inherent in Darken's assumption that chemical-reaction kinetics were limiting in the removal of nitrogen. The value is somewhat higher than is normally observed at 1600° C for reactions involving nitrogen.

by J. C. Fulton
and J. Chipman

Kinetic Factors in the Desulfurization
of Pig Iron by Blast-Furnace Type Slags

INTRODUCTION

The desulfurization of molten pig iron by blast-furnace type slags has been the subject of a number of investigations. It has been shown that desulfurization can be accelerated by the addition of deoxidizing agents such as Si,[1] Mn, or Al.[2] Conversely, deceleration of sulfur transfer occurs when oxide reduction[3, 4] is in progress. The very slow approach to equilibrium in the reduction of SiO_2 from the slag maintains a prolonged hindrance to sulfur transfer. In order to be able to study desulfurization in the absence of the competing silica reduction, an investigation of silicon equilibrium between metal and slag was made.[5] With the aid of the silicon equilibrium data, Ramachandran, King, and Grant[6] have measured the rate of sulfur transfer and the quantity of carbon monoxide evolved under conditions of minimum interference from side reactions. These authors concluded that the reaction is electrochemical in nature.

The transfer of sulfur from metal to slag involves the reduction process:

$$[S] + 2e^- = S^{2-} \qquad (1)$$

This reaction can proceed only when electrons are supplied by a concurrent oxidation such as:

Dr. Fulton is Associate Director of Research, Allegheny Ludlum Steel Corporation, Brackenridge, Pennsylvania. Dr. Chipman is Professor of Metallurgy, Massachusetts Institute of Technology, Cambridge, Massachusetts.

The authors wish to thank the American Iron and Steel Institute for support of this work. They are also indebted to Mr. Donald L. Guernsey and his associates for the careful analytical work reported.

$$[C] + O^{2-} = CO + 2e^- \qquad (2)$$

$$[Si] + 2O^{2-} = SiO_4^{4-} + 4e^- \qquad (3)$$

$$[Fe] = Fe^{2+} + 2e^- \qquad (4)$$

At low concentrations of carbon or silicon or at moderate concentrations of FeO, the electrode potential due to reactions 2, 3, or 4 may become so small that reaction 1 is stopped or severely slowed. In blast-furnace practice, reaction 2 provides the driving force for desulfurization. The silicon content of the metal is normally so low that the reverse of reaction 3 occurs, and this process, competing with reaction 1, results in a rather slow transfer of sulfur. Similarly, the presence of Fe^{2+} (as FeO) in the slag may, through reversal of reaction 4, place a low limit on the desulfurizing power of the slag.

It was the purpose of this study to simplify the problem by using a slag-metal system which was already essentially at equilibrium with respect to reactions 3 and 4 so that the over-all process consisted mainly of the combined reactions 1 and 2; that is:

$$[S] + [C] + O^{2-} = CO \text{ (g)} + S^{2-} \qquad (5)$$

EXPERIMENTAL PROCEDURE

As in previous studies,[7, 8] the reaction was carried out in a graphite crucible fitted with a graphite stirrer and operating in an atmosphere of carbon monoxide. The design was modified as shown in Fig. 1. The new stirrer minimized disturbance of the slag-metal interface where there was merely a revolving cylinder of 13/16 in. O.D. inside of a crucible 2 1/4 in. I.D. Stirring speeds from 12 to

500 rpm were used. Temperature control and sampling procedure were the same as previously described.[7]

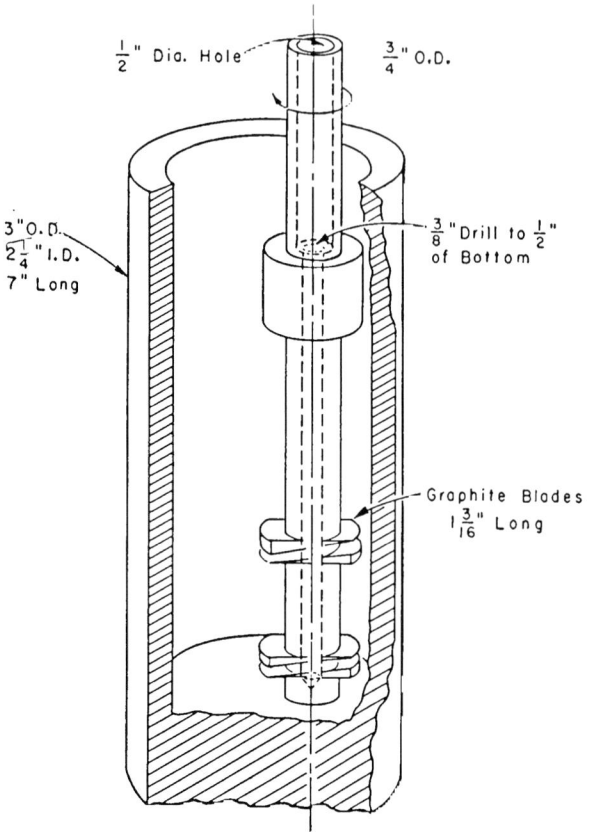

Fig. 1. Schematic drawing of graphite crucible and stirrer used in kinetic experiments.

In making a run, 400 grams of metal were charged, and about one hour was permitted for graphite saturation and thermal equilibration of the metal bath. Enough sulfur to give an initial concentration of 0.80% was added as iron sulfide about 10 min before the slag. An old graphite stirrer was used to homogenize and saturate the metal bath prior to the actual experimental period. At the start of the run, a freshly machined graphite stirrer was put in so as to define more accurately the geometric conditions in the system. A sample of the metal was taken 2 to 5 min before adding the slag.

A premelted master slag sized to −6/+20 mesh was used in order to expedite fusion. The initial composition of the slag in weight percentage was CaO 39.0; SiO₂ 40.0; Al₂O₃ 12.6; MgO 8.4. Two hundred grams were used in each experiment except 223, in which the amount was 400 grams. Zero time was taken when the slag was observed to be fused. The lag between the time of addition and "zero" time was narrowed by slowly mixing the slag with the stirrer. Normally 5 min were required to add and heat the slag to the experimental temperature. The temperature was held at 1500 ± 10°

throughout the run. Samples of the metal were obtained at 0, 10, 20, 40, 60, 120, and 180 min.

RESULTS AND DISCUSSION

The results of nine experiments in this investigation are summarized in Table 1. The metal in the charge contained 6.0 ± 0.1% Si for all runs except 229 where no silicon was added to the ingot iron. The silicon remained substantially constant throughout all experiments except 229, where it increased from 0.06 to 0.21%.

Table 1. Summary of Experimental Results

Experiment No.	Stir Rate (rpm)	Conditions	Slope	δ_S' (cm)
224	100	Normal	−0.034	0.033
227	12	Normal	−0.027	0.042
228	35	Normal	−0.023	0.049
230	100	Normal	−0.024	0.047
231	500	Normal	−0.066	0.017
234	0	Normal	−0.011	0.102
223	100	Double slag wt	−0.037	0.030
233	100	CO through bath	−0.026	—
229	100	No Si charged	−0.004	—

The experimental data have been analyzed according to the two-film hypothesis recently presented by Gordon and Sherwood [9] and as adapted to steelmaking reactions by Wagner. The symbols necessary are defined as follows:

C_S' = concentration of sulfur in the bulk metal at any time, moles/cm³

$C_S'^*$ = equilibrium concentration of sulfur in the metal

D_S' = diffusion coefficient of sulfur in the metal, cm²/min

δ_S' = effective boundary-layer thickness for sulfur in the metal, cm

A = area of slag-metal interface, cm²

V = volume of metal, cm³

t = time, min

If we assume that transfer of sulfur from the bulk metal to the phase boundary controls the rate of the over-all reaction, then the expression for the transfer of sulfur from the metal is:

$$V \frac{dC_S'}{dt} = -\frac{D_S'}{\delta_S'} A (C_S' - C_S'^*) \qquad (6)$$

or

$$\frac{2.3 d \log (C_S' - C_S'^*)}{dt} = -\frac{D_S' A}{\delta_S' V} \qquad (7)$$

Noting that the concentration of sulfur may be expressed in any units, we may substitute for C_S', yielding:

$$\frac{d \log ([\% S] - [\% S^*])}{dt} = -\frac{D_S' A}{2.3 \delta_S' V} \qquad (8)$$

Thus, in a plot of log ([% S] − [% S*]) versus time, the slope is equal to $-D_s'A/2.3\delta_s'V$. In a system of simple geometry such as used in these experiments, we can use the fact that $V/A = L$ where L is the bath depth in cm. Thus,

$$\delta_s' = -\frac{D_s'}{2.3 \times L \times \text{slope}} \qquad (9)$$

The equilibrium sulfur content has been calculated from the work of Hatch and Chipman.[8] Their curves indicates 0.013% S at equilibrium in all experiments except 223 where 0.008% S is found.

Effect of Stirring Rate

In Fig. 2, log {[% S] − [% S*]} has been plotted versus the time in minutes for the first seven ex-

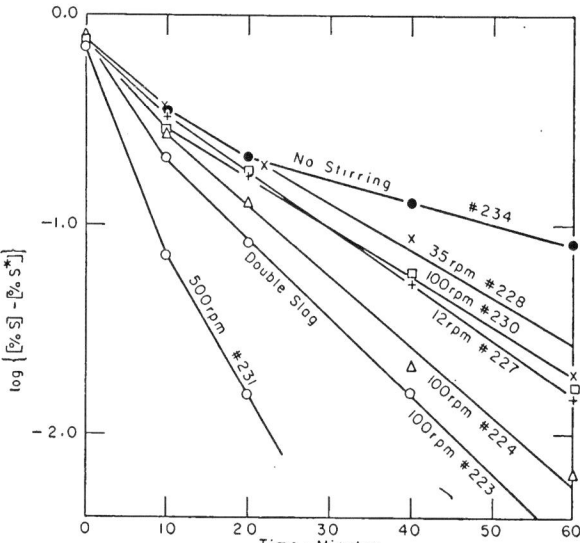

Fig. 2. Influence of stirring rate on the rate of desulfurization at 1500° C.

periments. Runs 224 and 230 indicate that the reproducibility of the experimental method is not very good; these two experiments were identical in every respect. In view of the lack of a definite trend between the runs at 12, 35, and 100 rpm, it seems evident that only drastic changes in stirring rate are important. That is, run 234 with no stirring seems to be somewhat slower than those at 12, 35, and 100 rpm. Likewise, the run at 500 rpm definitely shows a faster reaction rate. Experiment 223 with double the normal slag volume was only slightly faster than normal runs at the same stirring rate. The early portion of each run is nonlinear, due in all probability to increased slag-metal interfacial area and added stirring in the rapid stage of gas evolution.

The slope of the linear part of each rate curve is shown in Table 1. The effective film thickness is also tabulated, based on a value of $D_s' = 6.0 \times 10^{-3}$ cm²/min, taken from the work of Holbrook, Furnas,

and Joseph.[10] While the boundary-layer thicknesses appear to be of about the normal order of magnitude, they cannot be considered accurate because of the unknown influence of gas evolution.

The effect of the evolution of carbon monoxide is not well understood. Bubbling probably contributes to stirring rate and may affect the interfacial area. Moreover, the difficulty of carbon-monoxide bubble nucleation has been suggested many times as being a possible reaction barrier. Experiment 233 was designed to shed some light on these questions. This experiment was conducted in exactly the same fashion as 224 and 230 except for the method of flushing carbon monoxide through the system. In the normal runs, carbon monoxide entered the system through a 1/4-in. I.D. graphite tube which directed gas down on the slag surface. In experiment 233, a similar graphite tube of 1/4-in. I.D. by 1/2-in. O.D. with a closed end was used to bubble carbon monoxide through the metal and slag. The gas flow rate was 100 cm³/min through 4 1/16-in. diam holes which were drilled through the side of the graphite tube at about 1/2 in. from the bottom. This position also corresponded to 1/2 in. up from the bottom of the liquid metal which was 7/8 in. deep in a normal run without the graphite tube immersed. In Fig. 3 the results are com-

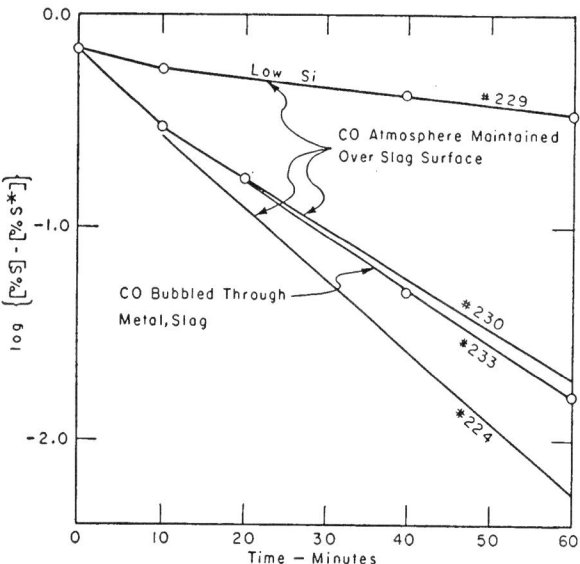

Fig. 3. Influence of CO bubbling and of metal composition on desulfurization at 1500° C.

pared with two normal runs. It is evident that bubbling had no effect on the rate.

Gas composition might also be considered in carbon-monoxide evolution mechanics. It is worth noting that very fast rates of desulfurization have been observed by Trentini, Wahl, and Allard,[11] who used nitrogen to blow powdered lime through tuyères into a converter-type vessel. The success of the process suggests that the carbon-monoxide sink provided by the nitrogen bubbles might be a factor.

Effect of Silicon Content of Metal

The work of Grant, Troili, and Chipman[1] has clearly demonstrated that desulfurization is hindered by simultaneous reduction of silica from the slag. A repetition of this type of experiment was considered desirable in this investigation in order to provide a direct comparison of the chemical and physical factors associated with sulfur transfer under identical experimental conditions. In experiment 229, no silicon was added to the charge. Comparison with normal runs is shown in Fig. 3. It is evident that the absence of silicon from the metal and the consequent silica reduction from the slag are much more effective in decreasing the rate of sulfur transfer than the complete omission of stirring. Chemical factors thus outweigh in importance the diffusion factor, and it is only when the former are closely controlled that the latter can be evaluated.

SUMMARY

Experiments have been made to help clarify the mechanism of sulfur transfer between liquid pig iron and blast-furnace type slags. Except for one run with no stirring, all experiments were made under conditions of forced convection by means of a graphite stirrer. With 6% Si in the metal to maintain equilibrium with SiO_2 in the slag, the transfer rate is slightly affected by stirring rate. To this extent, the process, under constant chemical conditions, is diffusion controlled.

Bubbling CO through the bath was without effect.

Omission of Si from the charge causes a drastic slowing of the reaction, indicating the overriding importance of chemical factors in determining the rate of transfer.

References

1. N. J. Grant, O. Troili, and J. Chipman *Trans. A.I.M.E.,* 191, 672 (1951).
2. K. M. Goldman, G. Derge, and W. O. Philbrook, *Trans. A.I.M.E.,* 200, 534 (1954).
3. N. J. Grant, U. Kalling, and J. Chipman, *Trans. A.I.M.E.,* 191, 666 (1951).
4. N. J. Grant, J. W. Dowding, and R. J. Murphy, *Trans. A.I.M.E.,* 197, 1451 (1953).
5. J. C. Fulton and J. Chipman, *Trans. A.I.M.E.,* 200, 1136 (1954).
6. S. Ramachandran, T. B. King, and N. J. Grant, *Trans. A.I.M.E.,* 206, 1549 (1956).
7. J. C. Fulton, N. J. Grant, and J. Chipman, *Trans. A.I.M.E.,* 197, 185 (1953).
8. G. G. Hatch and J. Chipman, *Trans. A.I.M.E.,* 185, 249 (1949).
9. K. G. Gordon and T. R. Sherwood, *Chem. Eng. Prog. Symposium Ser.,* No. 10, 15 (1954).
10. W. R. Holbrook, C. C. Furnas, and T. L. Joseph, *Ind. Eng. Chem.,* 24, 993 (1932).
11. B. Trentini, L. Wahl, and M. Allard, *Trans. A.I.M.E., J. Metals,* 9, 1133 (1957).

by W. A. Fischer

The Deoxidation of Iron Melts

with Silicon

Whereas until now the equilibrium conditions of the deoxidation by silicon have been treated very profoundly and frequently, the literature contains only isolated references to the chronological progress of this reaction. In the following discussion a first attempt will be made to widen our knowledge of this problem. The questions which we have posed ourselves and have endeavored to answer are as follows:

In what manner is the deoxidation by silicon of an iron melt, being in equilibrium with the atmosphere (air) and the crucible, dependent on:

(1) The amount of silicon added.
(2) The temperature of the melt.
(3) The size of the furnace.
(4) The agitation of the melt.
(5) The kind of the crucible lining.

In addition, it was intended to examine what kind of inclusions are formed during the different periods of time and in the different crucibles during the course of the reaction.

EXPERIMENTAL METHOD

For carrying through the experiments, samples were sucked from the iron melts after the addition of the silicon at intervals of 1/2 min each, and the changes of the oxygen and silicon contents were followed until the initial contents present before the silicon addition had been reached again. The

Dr. Fischer is affiliated with the Max-Planck-Institut für Eisenforschung.

This paper is an extract of the thesis of M. Wahlstev, Technische Hochschule, Aachen, Germany, 1956 (prepublication).

experiments were carried through in a high-frequency furnace of a capacity of 4 kg, using crucibles made of silica, aluminum oxide, magnesium oxide, lime, and fluorspar at temperatures of between 1530 and 1700° C. The influence of the furnace size was examined in high-frequency furnaces lined with silica and having capacities of 4, 50, and 300 kg.

The silicon additions amounted to 0.05 to 1.5%. In order to determine the separation time of the primary products segregating from the melt immediately after the silicon addition, we proceeded in the following way: for each sample the product $K'_{Si} = [\% \ Si][\% \ O]^2$ was obtained from the previously determined silicon and oxygen contents, and this product was compared with the equilibrium value K_{Si} according to Gokcen and Chipman[1] for the temperature involved. Plotting these two values against the time results in two curves which at first differ considerably from each other but meet each other after a definite period of time has elapsed. This period will be considered as the separation time of the primary products of reaction.

In order to study the inclusions present in the samples, metallographic examinations were carried out. The amount and composition of the inclusions were determined by means of electrochemical isolation and by a microanalysis of the isolated products.

PROGRESS OF DEOXIDATION PROCESS

In Fig. 1 are plotted the initial contents of oxygen and silicon as well as the contents remaining after separation of the primary products of melts that have been carried through in silica crucibles. The scattering points result from the inaccuracy in the deter-

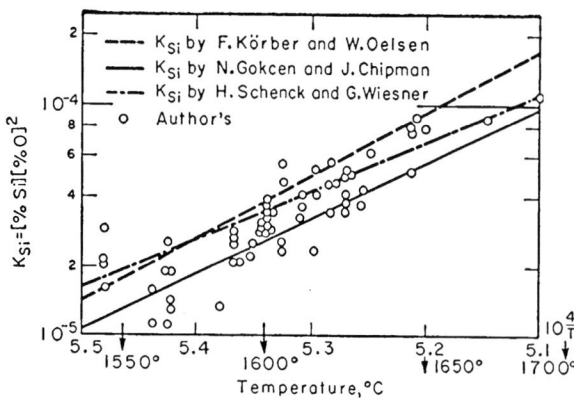

Fig. 1. Silicon-oxygen equilibrium in liquid iron at temperatures between 1530 and 1700° C (silica crucibles).

mination of the rather small silicon contents, i.e., between 0.002 and 0.006%. Under these circumstances, the agreement with the values obtained by other research workers may be considered as being quite satisfactory.

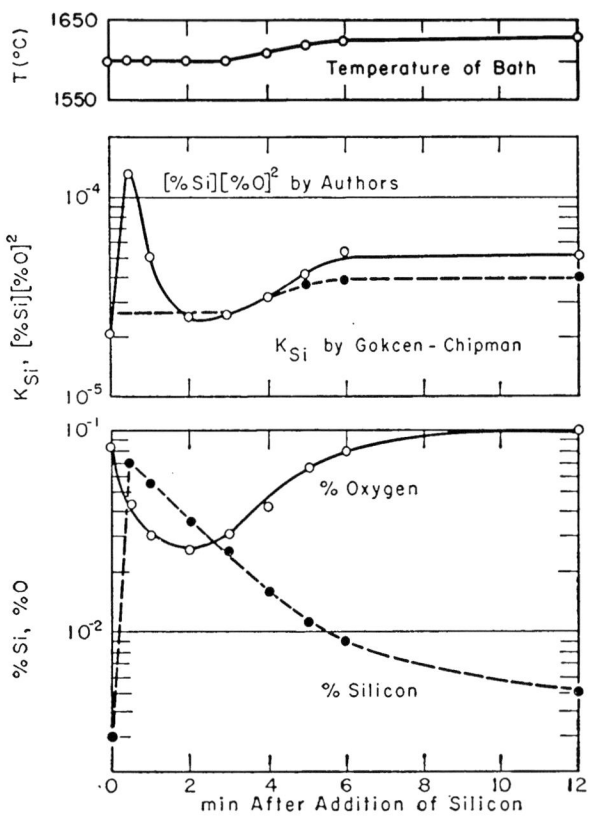

Fig. 2. Run of heat after addition of 0.75% Si in a 4-kg silica crucible at 1600° C.

Figure 2 shows the progress of the deoxidation at 1600° C in acid 4-kg crucibles after an addition of 0.15% Si to the melt. Immediately after the addition, the silicon contents attain a maximum value corresponding to about half of the amount of silicon added. They fall off then, as the time of the experiment increases, and after about 10 min they again reach their initial content of 0.003 to 0.004%

before the addition of silicon. The oxygen contents decrease within 2 min after the silicon addition from 0.08% to 0.026% and then rise again to their initial contents. The K'_{Si} values, having been calculated from the silicon and oxygen contents as determined, attain the K_{Si} values of Gokcen and Chipman[1] after a period of 2 min, which corresponds to the defined separation time of the primary products in this crucible. During the further progress of the melt, both values remain equal although the oxygen and silicon contents are undergoing big changes. The reoxidation, therefore, takes place only on the surface of the melt and proceeds by diffusion of the silicon in the melt toward the surface. There is practically no incorporation in the melt of the deoxidation products fromed in this process.

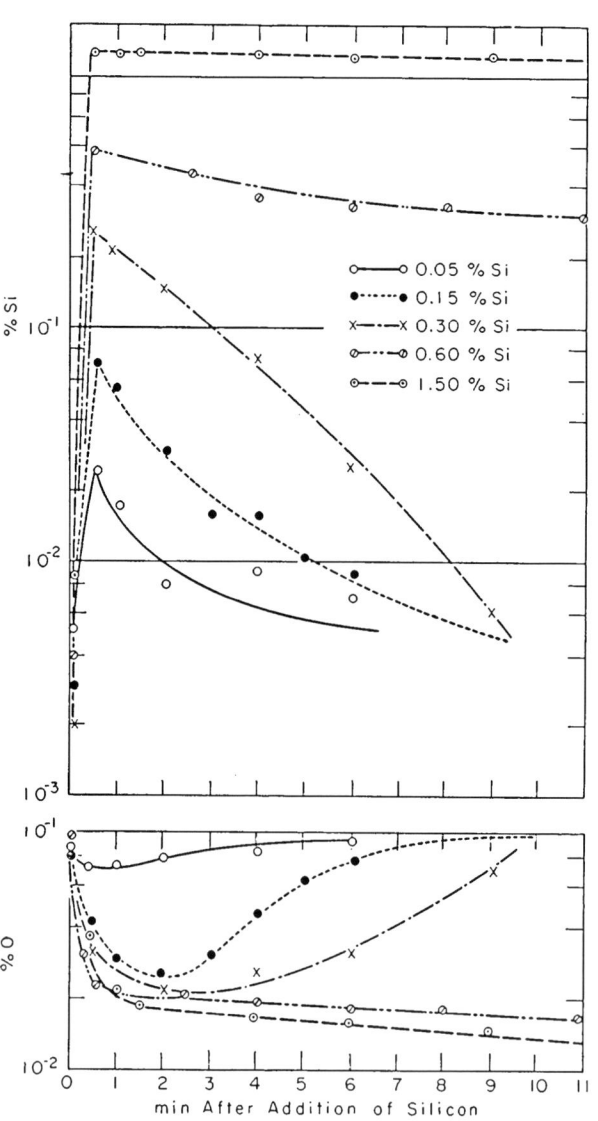

Fig. 3. Run of deoxidation after addition of various quantities of silicon at 1600–1630° C.

In Fig. 3 the chronological progress of the silicon and oxygen contents is represented for the same test conditions after an addition of silicon in amounts varying from 0.05 to 1.5%. The uniform distribution of the silicon is accomplished in all melts within a period of 1/2 min. The lowest value of the oxygen contents of the various melts decreases with increasing silicon additions. The time

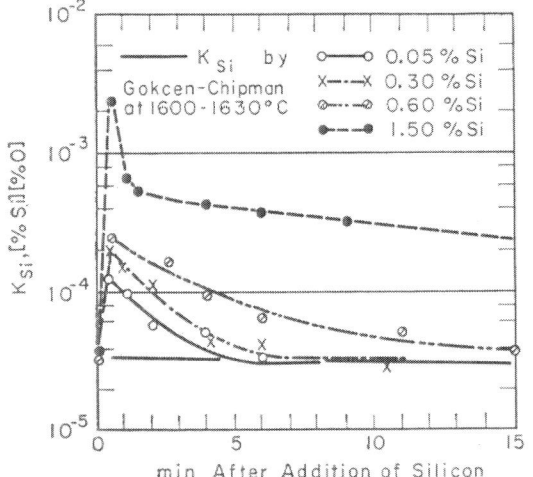

Fig. 4. Variations of values of [% Si] [% O]² after addition of 0.05–1.5% Si at 1600–1630° C (4-kg silica crucible).

required for the segregation of the primary products increases considerably with increasing silicon additions, as is shown in Fig. 4. At 1.5% Si, it can no longer be measured, since the curves of the K'_{Si} and

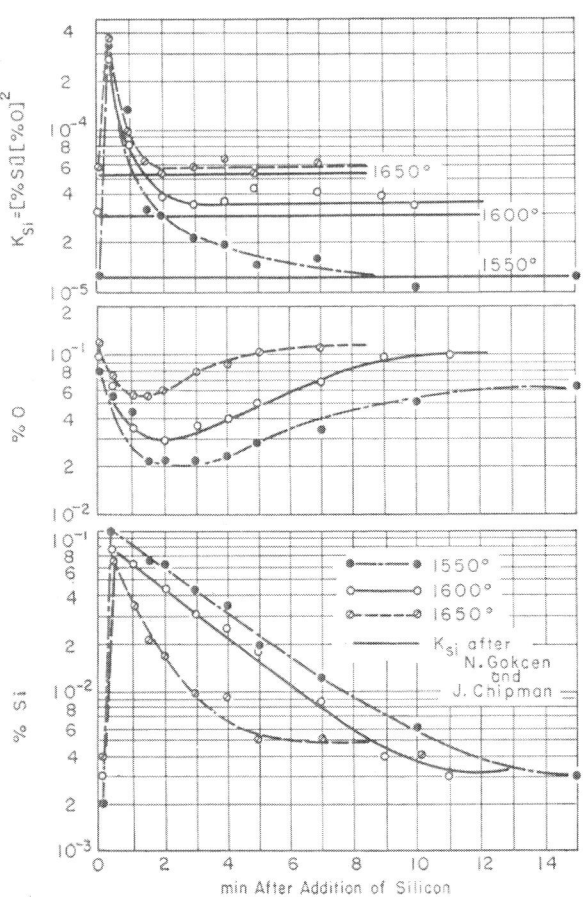

Fig. 5. Separation of silicate deoxidation product as influenced by temperature after 0.15% Si addition.

K_{Si} values no longer touch each other within the experimental time. The investigations carried through by Gokcen and Chipman[1] show that up to 15% Si in the iron melt no effect of the amount of the silicon on the K_{Si} values can be observed. The obstructed segregation of the primary products, as observed, cannot be attributed, therefore, to an activity of the dissolved silicon changing with the concentration.

The segregation rate is dependent on the temperature in the sense that it increases as the temperature increases. Figure 5 shows this fact in the example of the melt which has been deoxidized with 0.15% Si.

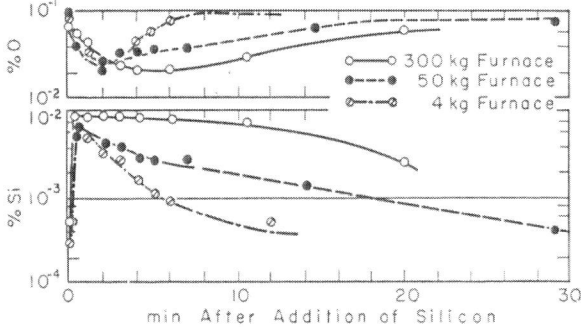

Fig. 6. Influence of the volume of furnace on the deoxidation run after addition of 0.15% Si at 1600° C.

In Fig. 6 the progress of the silicon and oxygen contents is represented after an addition of 0.15% Si at about 1600° C in high-frequency furnaces of different sizes of 4, 50, and 300 kg of bath weights. The time required for a uniform distribution of the silicon is equal to less than 1/2 min in all of the furnaces. As shown in Fig. 7, there is a linear relationship between the burning loss rate of the

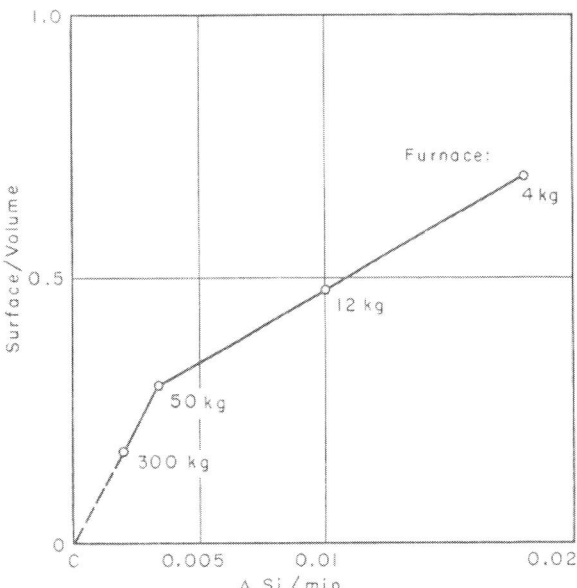

Fig. 7. Influence of the quotient surface/volume of the heat on the velocity of silicon oxidation at 1600° C.

silicon and the ratio between the bath surface and the bath volume, with the exception, however, of the 4-kg melt. Regarding the duration of the segregation of the primary products, however, this relationship no longer holds true. With 2 min, it is of equal magnitude in the 4-kg furnace as well as in the 50-kg furnace and increases to 5 min only in the 300-kg furnace. The main factors of influence here are probably the diameter of the crucible and the agitation rate of the melt.

The extraordinarily strong effect which the agitation of the melt exerts on the segregation of the deoxidation products may be seen in Fig. 8, in

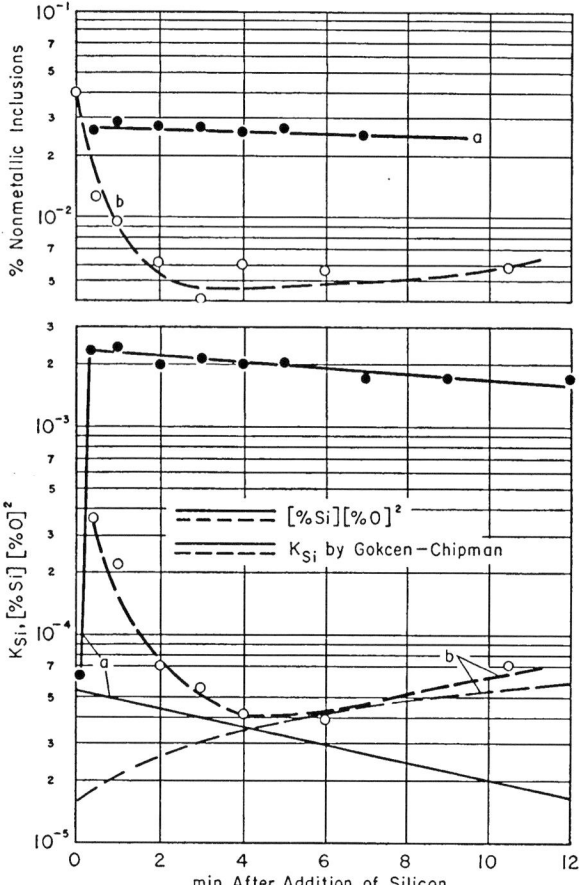

Fig. 8. Run of deoxidation after addition of 0.75% Si. (a) furnace current switched off; (b) furnace current switched on.

which the deoxidation process in the 300-kg high-frequency furnace with an agitated melt (furnace switched on) is compared with a quiet melt (furnace switched off). Whereas in the agitated melt the segregation of the primary products requires only about 5 min, in the quiet melt practically no segregation whatever occurs within the experimental time.

The effect of agitation on the segregation of oxides has been studied by electrochemical isolation of the non-metallic inclusions from samples. With the furnace switched on, the amount of inclusions goes through a minimum value which coincides in time with the segregation of the primary

products; with the furnace switched off, only a very small decrease of the amounts of inclusions takes place as the time of experiment increases. The effect of the crucible lining on the segregation time of the primary products is shown in Fig. 9 in the

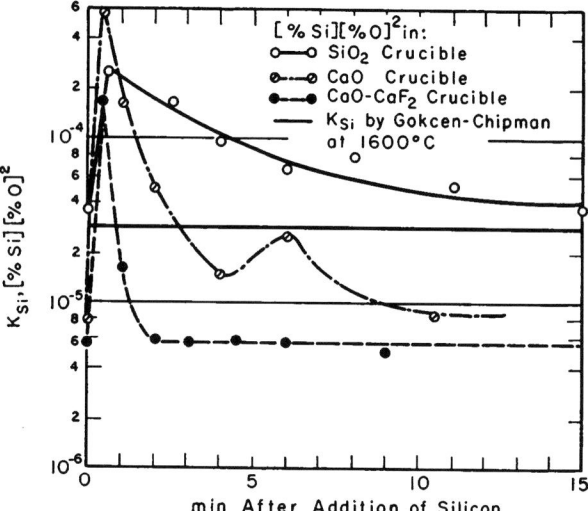

Fig. 9. Influence of the lining on the silicon-oxygen equilibrium after addition of 0.6% Si at 1600° C.

example of 4-kg melts after deoxidation with 0.6% Si at about 1600° C. Whereas in the silica crucible the segregation of the primary products requires about 16 min, it already takes place in the lime crucible within 3 to 4 min, and in the fluorspar crucible, within 1 to 2 min. As is shown in the figure, the K'_{Si} values are, furthermore, considerably smaller than the values given by Gokcen and Chipman.[1] Because of the smaller activity of the silicic acid in the lime silicates formed as slags, the constant of the silicon deoxidation in these crucibles is smaller by approximately one decimal power.

DEOXIDATION PRODUCTS

In conclusion, I should like to refer to some other observations made on the deoxidation products obtained during the reaction. Here the following kinds should be distinguished:

(1) Products resulting from melts before the silicon addition.

(2) Products resulting from melts with primary segregations.

(3) Products from melts without primary segregations.

Figure 10 shows inclusions of a melt made in a silica crucible before the silicon addition. These are siliceous wüstite inclusions whose content of silicic acid decreases with decreasing size of the inclusions, as may be recognized distinctly from the ratio of the bright wüstite to the darker fayalite.

In the other crucibles, the iron melts before the silicon addition have less than 0.001% silicon by

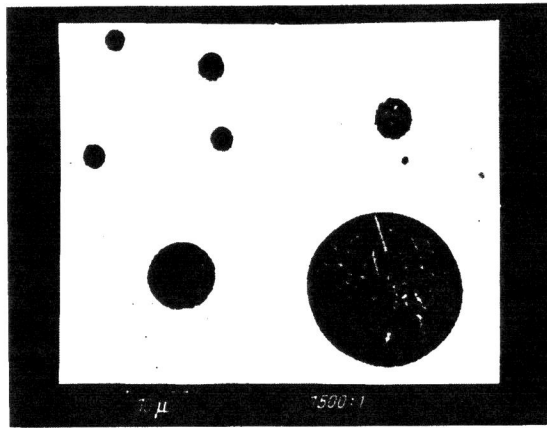

Fig. 10. Duplex g'obule of wüstite and wüstite-fayalite eutectic.

chemical analysis. According to these low contents, there is practically no more silicic acid to be found in the inclusions of these melts. As may be seen in Fig. 11, the inclusions consist of wüstite inclusions

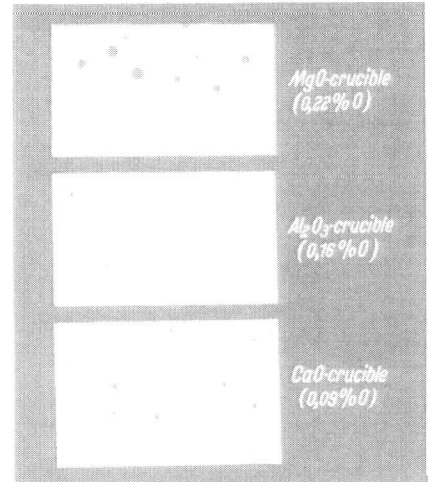

Fig. 11. Wüstite inclusions from various crucibles.

which are homogeneous or, rather, only very slightly cored. The MgO, CaO and Al₂O₃ contents of these inclusions are also below the limit of the determining power of the analytical methods used. (After

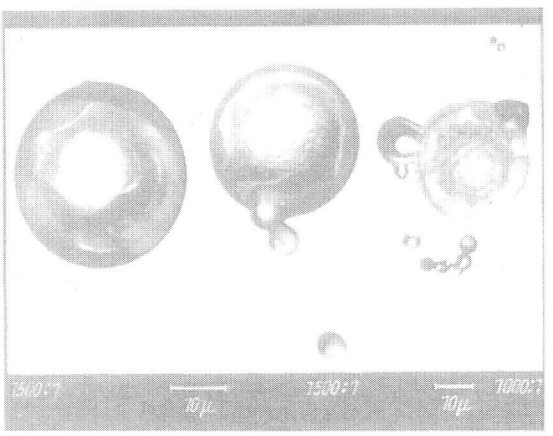

Fig. 12. Immiscible glassy silicate.

the addition of silicon only vitreous inclusions with different FeO content are found in all melts, the color of which is accordingly lighter or darker.)

Particularly numerous are the inclusions which, according to the miscibilty gap, in the system FeO—SiO₂, are cored. Figures 12 and 13 represent dif-

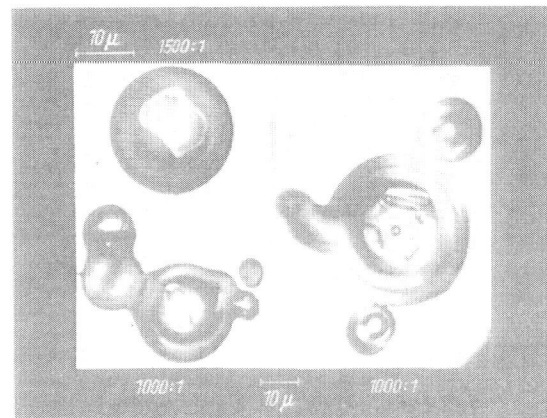

Fig. 13. Immiscible two-phase glassy silicate.

ferent stages of these cored glasses. Those cored glasses will be present in the melt immediately after the silicon addition. The inclusions of the melts originating from the switched-off furnace are on the average bigger than those from the agitated melts. Here, furthermore, often rather strange intermediate shapes occur if several inclusions are fusing together. Figure 14 shows the process of

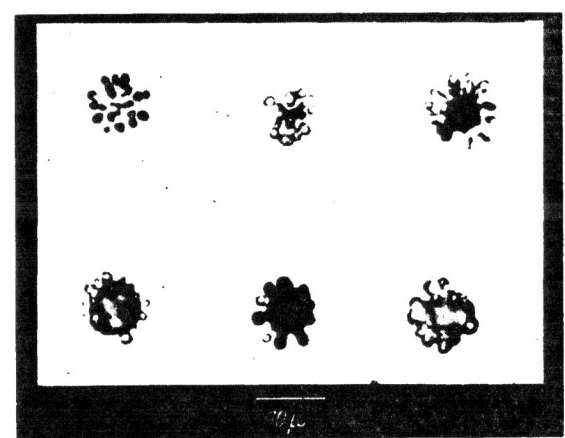

Fig. 14. Formation of glassy inclusions.

the formation of inclusions during the cooling of the melts.[2] A great number of small inclusions grow together into bigger inclusions. A clear distinction between the primary inclusions segregating out immediately after the silicon addition and the secondary inclusions formed during cooling and solidification, however, could not be made.

References

1. N. A. Gokcen and J. Chipman, *J. Metals*, **4**, 171 (1952).
2. C. A. Zapffe and C. E. Sims, *Trans. A.I.M.E.*, **154**, 192 (1943).

Discussion

FETTERS expressed a strong interest in this type of study which he referred to as involving "transients." That is, because of chemical reactions and changing conditions during deoxidation of steel, the actual final state of the steel is somewhere in between two equilibrium conditions, and we have what one might call a transient deoxidation product.

ELLIOTT observed that the experimental results indicate that there may be two separate problems: that of forming silica from silicon and oxygen, and that of floating the nonmetallics out of the system. He inquired whether the experimental procedures used were able to show the extent to which either of these processes controlled the observed change in oxygen and silicon content. FISCHER replied that this situation is under study. But they had observed that even in an unstirred melt the segregation rate was very, very slow.

HILTY said that an experiment of his in the past had indicated that there was a finite time required for getting silicon and oxygen atoms together to form SiO_2. He also stated that his work in this area had indicated that the size of inclusions found in steel was strongly influenced by the temperature of the bath from which the sample was taken and the rate at which the sample was cooled. This observation was verified by Elliott and Larson. He also reported that the steel sample which had appreciable silicon and oxygen contents before sampling oftentimes would not show any inclusions under very high magnifications with an optical microscope. If this sample was heated to just below the melting point, inclusions would appear in the microstructure.

In answer to a question by Olette, FISCHER said that they had successfully melted pure iron in a pure lime crucible and had obtained very low oxygen contents of around 0.001%. It was possible to hold this melt in contact with the lime crucible for over 2 hr at approximately 10^{-7} atm pressure with no pickup of oxygen by the melt from the crucible being noted.

P. Vallet

G. Derge

Presiding

Section 6

Reaction Rates in Iron and Steelmaking Processes

by T. B. King
and S. Ramachandran

Electrochemical Nature of Sulfur Transfer in the System Carbon-Saturated-Iron—Slag

Reactions between two liquids such as a slag and an alloy often prove relatively easy to study if knowledge of equilibrium conditions is sought. The dimensions of the system are not important except for experimental convenience. Reaction kinetics in such systems are much more difficult to study. The rate of the over-all reaction may be readily measured, but the results are likely to be peculiar to the experimental system used.

In the most general case the over-all reaction will involve a sequence of steps including both mass transfer and chemical reaction. There may well be no single step sufficiently slow that it essentially controls the over-all reaction rate.

Mass transfer in slag and alloy is effected by diffusion and convection. In most experimental systems convection cannot be eliminated, and it is better merely to make the conditions of convection reproducible from one experiment to another. This is most readily accomplished by mechanical stirring or induction heating. However, in such cases the conditions of convection defy specification by dimensionless parameters such as have proved useful in less complex flow studies. Thus the "scal-

ing-up" of kinetic studies on slag-metal systems would seem to be impossible.

This does not mean that such studies are unfruitful. Rather, the information to be sought is not quantitative information on the rate of slag-metal reactions but a more complete knowledge of reaction mechanisms. This is not provided by equilibrium studies but may be important practical knowledge.

Such an approach may be illustrated by an investigation of the reaction by which sulfur is transferred from carbon-saturated iron to oxide slags. The complexity of this reaction may be inferred from the work of Grant[1] and Derge[2] and their co-workers who have shown that the rate of the reaction is considerably influenced by "deoxidizers" such as silicon and aluminum. Nevertheless it is possible to learn something of the reaction mechanism from suitable experiments.

The apparatus used here has been described previously.[3, 4] Essentially it comprises an induction-heated carbon crucible containing slag and metal which have been brought together in the liquid state and are stirred by rotation of the crucible against a stationary paddle. The speed of rotation is such that natural convection is unimportant; therefore, the temperature distribution is not critical; the carbon crucible prevents induction stirring of the metal. Slag and metal volumes are maintained constant for each experiment. Thus, convection conditions are reproducible. Provision is

Dr. King is Professor of Metallurgy, Massachusetts Institute of Technology; Dr. Ramachandran is associated with the Allegheny Ludlum Steel Corporation.

The authors wish to acknowledge the assistance of Dr. C. Wagner with the theoretical treatment. The financial assistance of the American Iron and Steel Institute is also gratefully acknowledged.

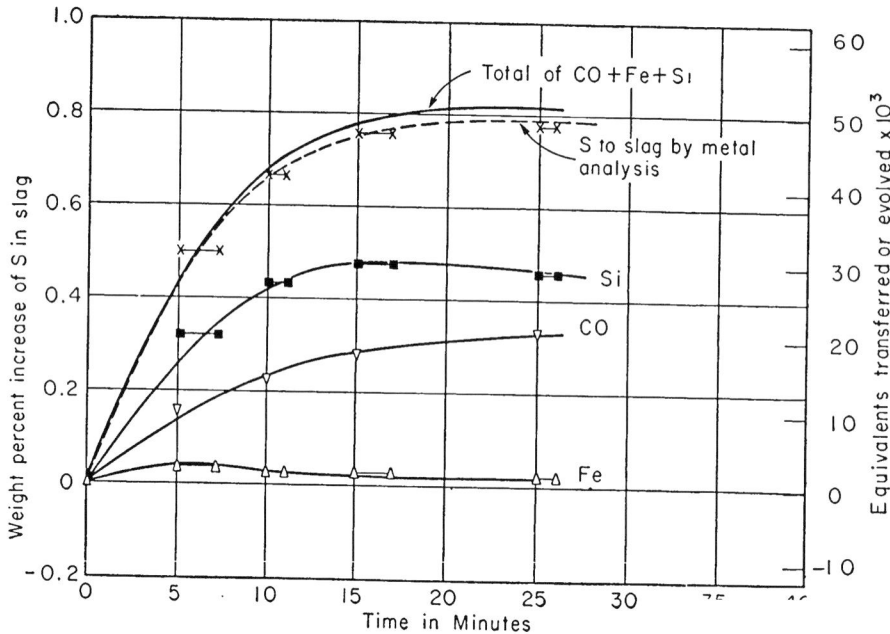

Fig. 1. Equivalents of S, Fe, and Si transferred from metal to slag and equivalents of CO evolved. Run S12: temperature 1502° C; Slag 48% CaO, 21% Al₂O₃, 31% SiO₂; initial [Si] 0.38%; equilibrium [Si] 0.5%.

made for sampling slag and metal and for measurement of the rate of CO evolution, which accompanies sulfur transfer, by continuous pressure recordings. Chemical analyses of samples are made to complete the information obtainable. The analyses are complete in the sense that a mass balance for all reacting elements except carbon and oxygen can be made for each sampling time.

The results of eight runs with this apparatus are given in Figs. 1 to 8. Either lime-silica-alumina or lime-alumina slags were used, and the carbon-saturated iron contained about 1.0% sulfur and silicon ranged from a trace to 0.38%. In one run the iron contained 0.53% aluminum. The stoichiometric relations are best shown by plotting as functions of time the numbers of chemical equivalents of elements transferred from iron to slag or, in the case of carbon monoxide, the number of equivalents evolved. There were some difficulties in analysis of slags for sulfur, and for consistency, the amount of sulfur transferred to the slag has been calculated from the more accurate metal analysis.

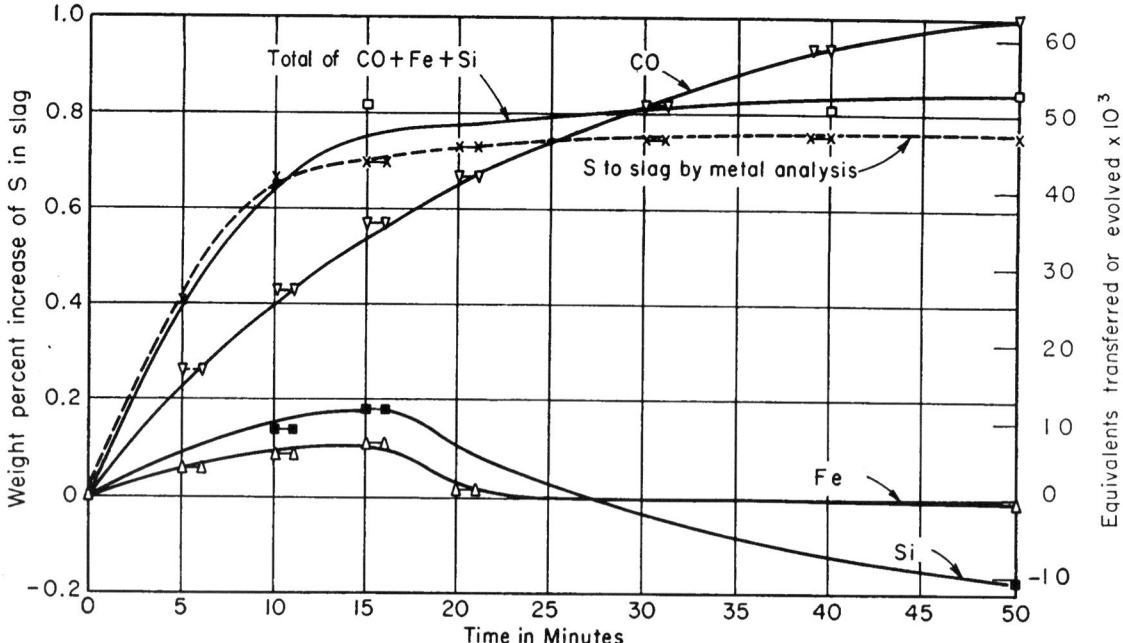

Fig. 2. Equivalents of S, Fe, and Si transferred from metal to slag and equivalents of CO evolved. Run K10: temperature 1505° C; slag 48% CaO, 21% Al₂O₃, 31% SiO₂; initial [Si] 0.1%; equilibrium [Si] 0.5%.

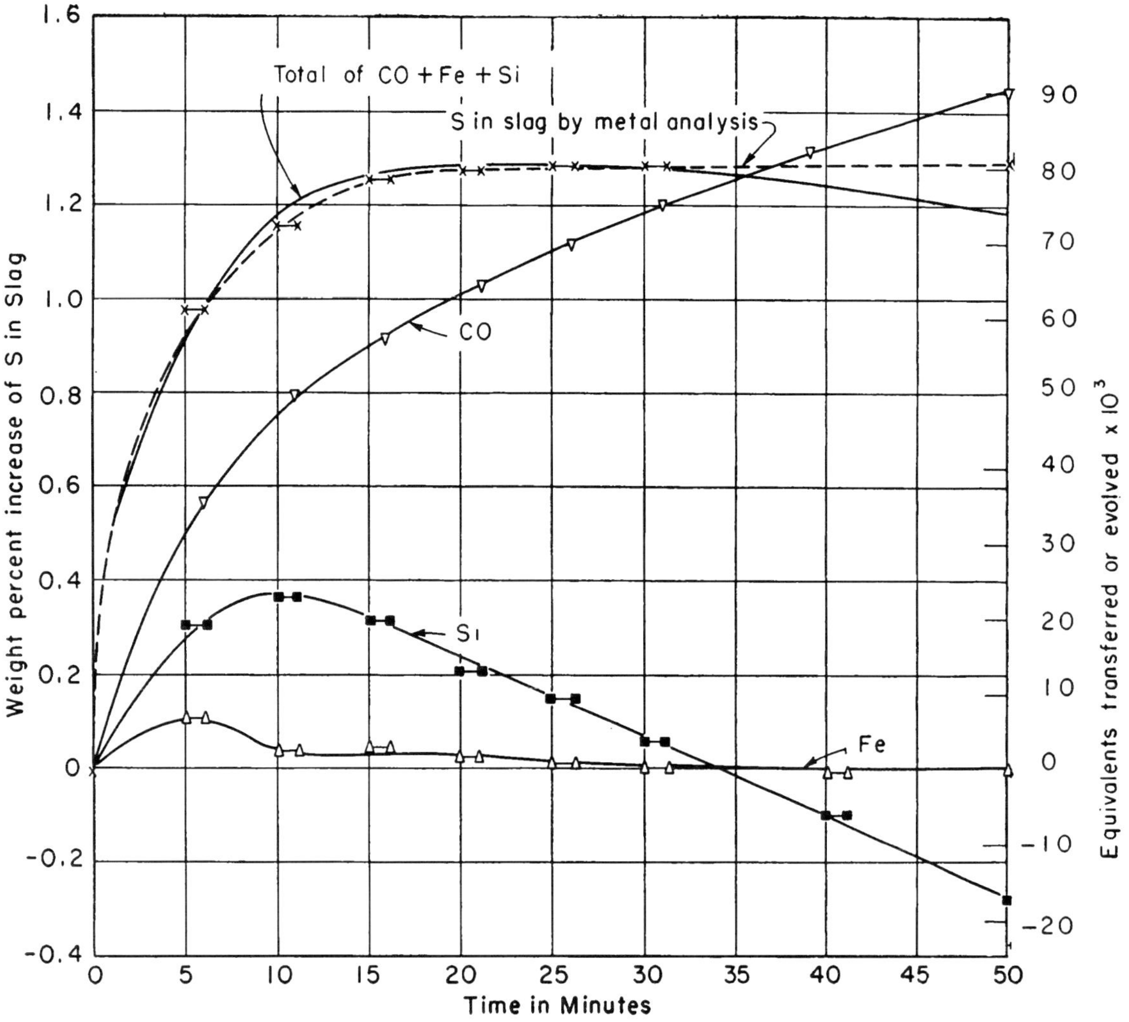

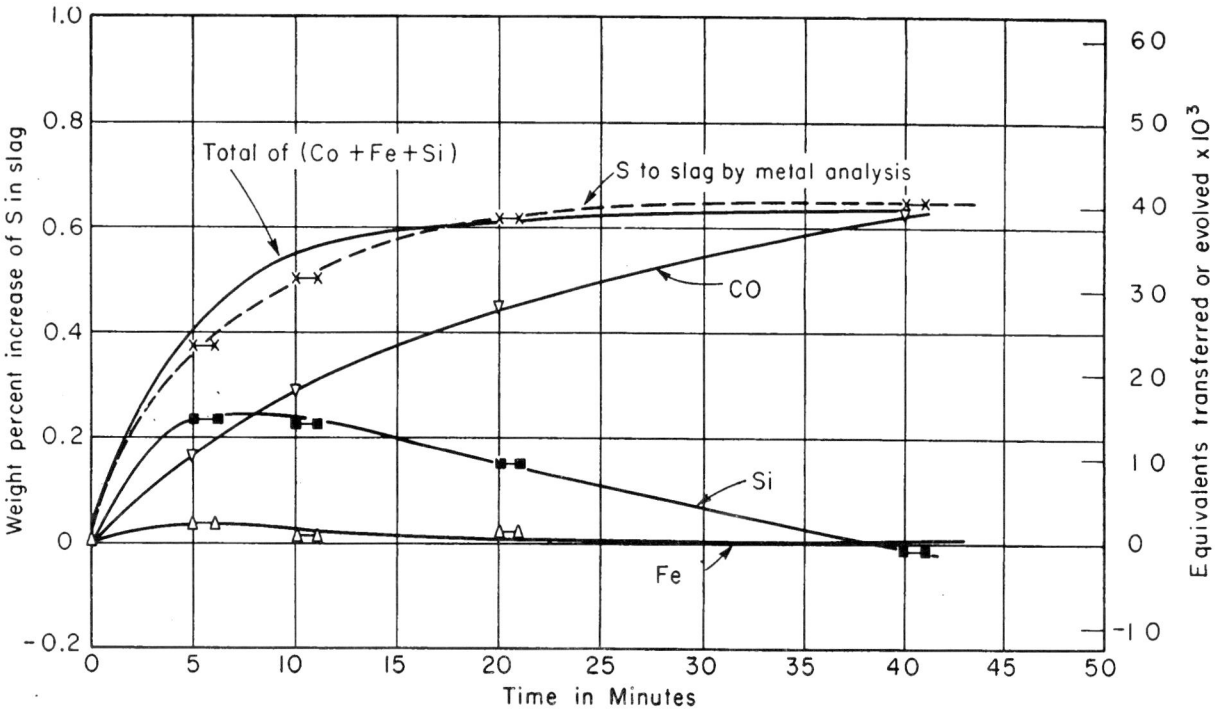

Fig. 3. Equivalents of S, Fe, and Si transferred from metal to slag and equivalents of CO evolved. Run K8: temperature 1505° C; slag 48% CaO, 21% Al₂O₃, 31% SiO₂; initial [Si] 0.16%; equilibrium [Si] 0.5%.

Fig. 4. Equivalents of S, Fe, and Si transferred from metal to slag and equivalents of CO evolved. Run S29: temperature 1486° C; slag 40% CaO, 16% Al₂O₃, 44% SiO₂; initial [Si] 0.65%; equilibrium Si 12%.

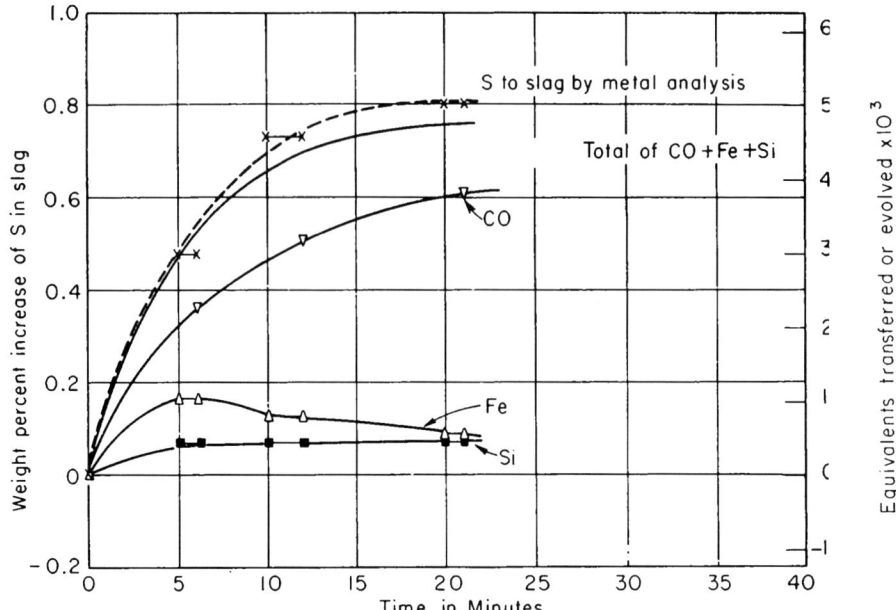

Fig. 5. Equivalents of S, Fe, and Si transferred from metal to slag and equivalents of CO evolved. Run S14: temperature 1546° C; slag 50% CaO, 50% Al₂O₃; initial [Si] 0.03%.

The essential point of interest is that the amount of CO evolved is not equivalent to the amount of sulfur transferred to the slag. In the simplest case, for example, in run S26 where a lime-alumina slag and an essentially pure iron-sulfur alloy were used, oxidation of iron also takes place in the initial stages, and finally iron is reduced from the slag when most of the sulfur has been transferred. When the metal contains silicon or aluminum, these are also oxidized in the initial stages, and the amount of CO evolved is correspondingly smaller.

Even though the initial silicon content of the iron is well below that corresponding to equilibrium with silica in the slag, graphite, and CO at 1 atm pressure, for example in run S29, oxidation of silicon takes place in the early stages. Within the limits of experimental error the total number of equivalents of sulfur transferred to the slag at any time is equal to the sum of the numbers of equivalents of carbon monoxide, iron, and silicon, or aluminum.

The stoichiometric relations between the various

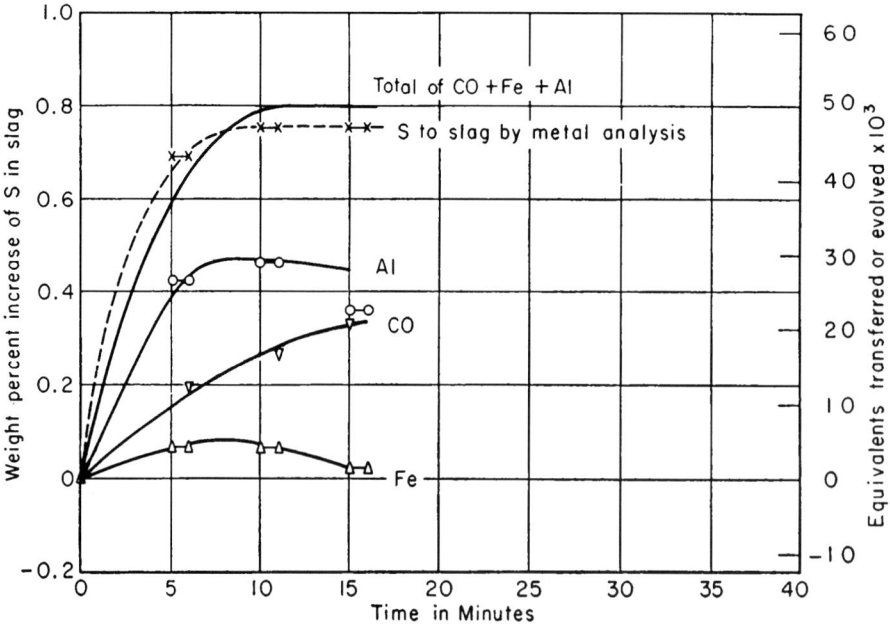

Fig. 6. Equivalents of S, Fe, and Al transferred from metal to slag and equivalents of CO evolved. Run S15: temperature 1550° C; slag 50% CaO, 50% Al₂O₃; initial [Al] 0.53%.

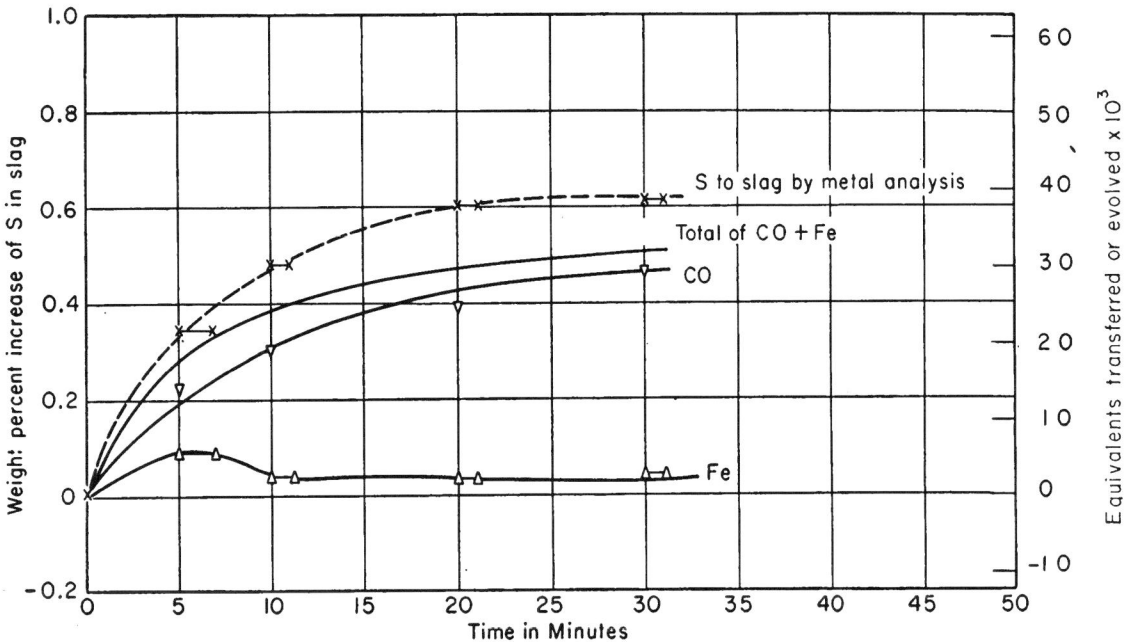

Fig. 7. Equivalents of S and Fe transferred from metal to slag and equivalents of CO evolved. Run S26: temperature 1590° C; slag 45% CaO, 55% Al₂O₃.

simultaneous reactions may be represented as follows:

$$2\dot{n}_S = 2\dot{n}_{CO} + 2\dot{n}_{Fe} + 4\dot{n}_{Si} + 3\dot{n}_{Al} \qquad (1)$$

where $\dot{n}_i$ is the rate of transfer of i from metal to slag or the rate of CO evolution in moles per second. The sign is reversed for slag-to-metal transfer. Relation 1 is to be expected from the requirement that electroneutrality between the ionic and metallic solutions be maintained.

Without regard to the detailed mechanism, we may represent the simultaneous reactions taking place in the early stages as follows,

$$[S] + [C] + O^{2-} = S^{2-} + CO \qquad (2)$$

$$[S] + Fe = S^{2-} + Fe^{2+} \qquad (3)$$

$$2[S] + [Si] + 2O^{2-} = 2S^{2-} + SiO_2 \ (slag) \qquad (4)$$

The oxidation of silicon to SiO_2 is indicated be-

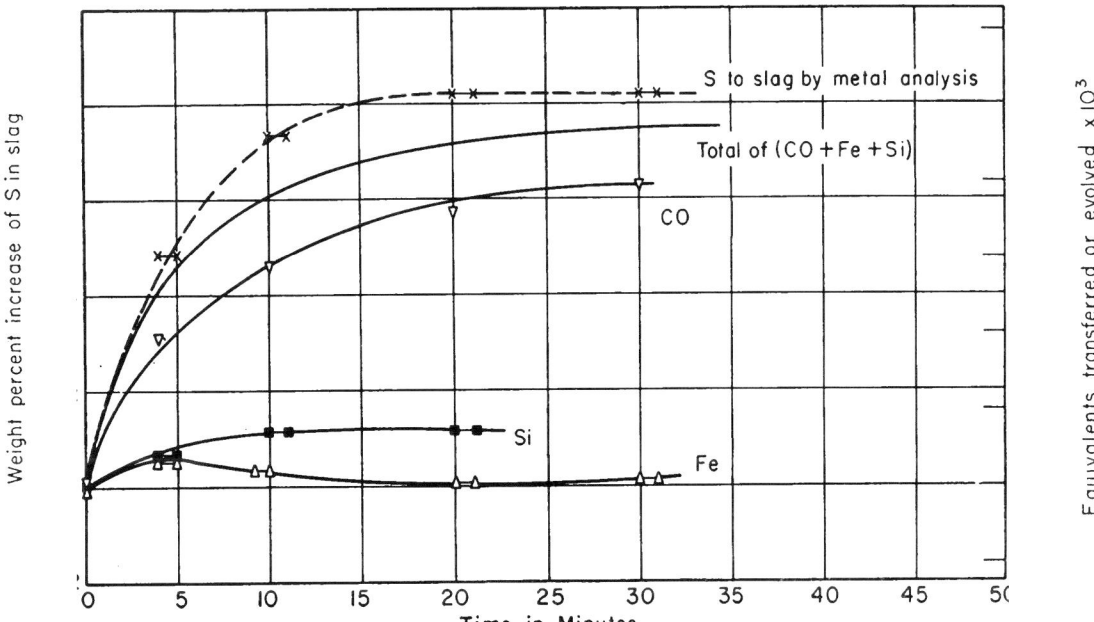

Fig. 8. Equivalents of S, Fe, and Si transferred from metal to slag and equivalents of CO evolved. Run S27: temperature 1604° C; slag 40% CaO, 60% Al₂O₃; initial [Si] 0.05%.

cause the ion involved in these slags is not known with certainty.

In the later stages, reactions 3 and 4 may be replaced by the following:

$$[C] + Fe^{2+} + O^{2-} = CO + Fe \qquad (5)$$

$$2[C] + SiO_2 \text{ (slag)} = 2CO + [Si] \qquad (6)$$

which can be obtained by combination of reaction 2 with 3 and 2 with 4, respectively.

Reactions between slag and metal necessarily involve (a) transport of reactants to the slag-metal interface, (b) reaction at the interface, and (c) transport of products away from the interface. Transport occurs by a combination of diffusion and convection processes, but in a well-stirred system there are appreciable concentration gradients only in regions close to the interface.

In the present case it is not possible to indicate whether any one of these steps is sufficiently slow so that it essentially controls the over-all reaction rate. For example, the concentration of sulfur in the slag is, in the early stages, small compared with its concentration in the metal. We may therefore be tempted to assume that transport of sulfur from the slag-metal interface to the bulk slag controls the reaction rate. The rate of transport is expressed as follows:

$$\frac{\dot{n}_S}{A} = \frac{D_S''}{\delta_S''} (C_S''^* - C_S'') \qquad (7)$$

where D_S'' is the diffusion coefficient of sulfur in the slag, δ_S'' is the effective thickness of the diffusion boundary layer for sulfur, $C_S''^*$ and C_S'' are, respectively, the interface and bulk concentrations of sulfur in the slag, and A is the nominal phase-boundary area.

The assumption of transport control implies that phase-boundary reactions are at equilibrium and that therefore the interface concentrations are equilibrium concentrations. The further assumption that transport of sulfur in the slag is rate controlling implies that the concentrations of other reactants at the phase boundary do not differ appreciably from their bulk concentrations. Therefore, $C_S''^*$ can be expressed in terms of bulk concentrations. For the equilibrium between iron and sulfur according to reaction 3,

$$K_3 = \frac{C_S''^* \cdot C_{Fe}''^*}{C_S'^* \cdot C_{Fe}'^*} \qquad (8)$$

From which, bearing in mind the above assumptions, we have:

$$C_S''^* = K_3 \frac{C_{Fe}' \cdot C_S'}{C_{Fe}''} \qquad (9)$$

Comparison of the early stages of runs S14 and S26, in which the slags have similar viscosities, shows that the rate of sulfur transfer is much higher in run S14 in spite of the fact that $C_S''^*$ is, from equation 9, much lower. Obviously transport of sulfur

in the slag is not the exclusive rate-controlling process.

Similar considerations applied to other possible rate-controlling processes show that the experimental results cannot be interpreted satisfactorily in terms of a single rate-controlling transport process. We may therefore recognize three possibilities:

(a) Mixed transport control.
(b) Control by the rate of a phase-boundary reaction.
(c) Mixed control by phase-boundary reactions and transport processes.

According to Darken's views,[5] the first should be the most likely possibility because absolute-reaction-rate theory indicates that for homogeneous and simple heterogeneous reactions at high temperatures the approach to equilibrium at the reaction site is extremely rapid compared with the rates of transport to and from this site. Darken has called this "local equilibrium." We may now examine the present results according to this viewpoint. This will require some knowledge of the equilibrium constants for the relevant reactions. Data are available on reactions involving CO evolution; since a third phase is thereby introduced, such reactions should first be considered more closely with regard to the site of nucleation of CO.

The observations of Brower and Larsen[6] on CO evolution during the boil in open-hearth furnaces coupled with experiments of Körber and Oelsen[7] suggest that nucleation of CO takes place mainly on the furnace bottom. Darken[5] has therefore suggested that the rate-controlling process in evolution of CO during the boil is oxygen transport from the slag-metal interface to the bulk metal and from the bulk metal to CO bubbles. Under these conditions the rate of CO evolution cannot be higher than the maximum possible rate of transport of oxygen to the metal. The rate of oxygen transport to the bulk metal is given by

$$\frac{\dot{n}_O}{A} = \frac{D_O'}{\delta_O'} (C_O'^* - C_O') \qquad (10)$$

the symbols having the same significance as previously. Since D_O'/δ_O' is not known, there is no possibility of making an absolute estimate of $\dot{n}_O$. However, Ramachandran, King, and Grant[4] have shown that in this present series of experiments the ratio $\dot{n}_O/\dot{n}_S$ may be estimated on the reasonable assumption that D_O'/δ_O' and D_S'/δ_S' are not very different.

The calculation was made for desulfurization by lime-alumina slags so that the stoichiometric relation 1 simplified to

$$\dot{n}_S = \dot{n}_{CO} + \dot{n}_{Fe} = \dot{n}_{Fe} - \dot{n}_O \qquad (11)$$

The maximum value of the ratio $\dot{n}_O/\dot{n}_S$ was shown to be 0.061. Hence, according to relation 11, $\dot{n}_S \cong \dot{n}_{Fe}$. The experimental results, as may be verified from Fig. 1 to 8, show, however, that $\dot{n}_S$ is certainly not equal to $\dot{n}_{Fe}$ but rather that $\dot{n}_S$ and $\dot{n}_{CO}$ are of the same order.

Most of the CO evolution which is observed must therefore take place by a mechanism which is not limited by the rate at which oxygen can be transferred to the bulk metal from the slag-metal interface.

It should be noted that the case Darken has treated is different in one important respect. According to Ramachandran, King, and Grant,[4] the maximum value of $(C_O'^* - C_O')$ is, converting to weight percentage, 0.014%, whereas Darken estimates that, for the much more oxidizing conditions in the open hearth, the concentration difference should normally be about 0.04%; this is not the maximum value, which could be as high as 0.20%.

Alternative mechanisms for CO evolution which do not involve transport of oxygen to the bulk metal require reaction between oxygen in the slag and carbon at the slag-metal phase boundary (or the slag-crucible boundary). If CO is also nucleated, there the reaction occurs at a line contact among the three phases and might be expected to be slow. However, the reaction is similar to hydrogen evolution at a metallic electrode in an aqueous solution. We may therefore consider that the reaction involves discharge of oxygen ions at the slag-metal phase boundary with the formation of CO dissolved in the slag, which then diffuses to bubbles nucleated at the phase boundary. The solubility of CO molecules in the slag is not known, although it may be expected to be low.

There remains the possibility that the reaction at the slag-metal interface involves solution of carbon in the slag as carbonate ions. Slags which have been contained in graphite crucibles do contain about 1% carbon, but it is difficult to conceive of all this carbon as being in solution.

Based on the assumption that CO evolution is indeed similar to cathodic evolution of hydrogen, the rate should be proportional to the area of the slag-metal interface. Oxygen ions are discharged at this interface so that transport of carbon to the interface is necessary. Hence the rate of CO evolution must be equal to the rate at which carbon arrives here (regardless of what the actual rate-controlling step may be), or

$$\frac{\dot{n}_{CO}}{A} = \frac{D_C'}{\delta_C'}(C_C' - C_C'^*) \qquad (12)$$

If transport of carbon in the metal were the rate-controlling step (which is highly unlikely), then $C_C'^*$ would become an equilibrium carbon concentration. Similarly, over-all transport control of the complete sulfur-transfer reaction would mean also that $C_C'^*$ is an equilibrium concentration. It is therefore of interest to examine whether equilibrium at the interface is actually reached. The over-all CO evolution reaction may, in view of the above discussion, be written as in reaction 5, for which the equilibrium constant is

$$K_5 = \frac{p_{CO} \cdot C_{Fe}'^*}{C_C'^* \cdot C_{Fe}''^* \cdot C_O''^*} \qquad (13)$$

According to the results of Hatch and Chipman[8] the equilibrium iron oxide content of a slag in equilibrium with graphite, iron, and CO at 1 atm pressure is about 0.03%. This is the equilibrium

$$C_{graphite} + O^{2-} + Fe^{2+} = CO + Fe \qquad (14)$$

Assuming that, in the present case, the concentration of carbon at the interface, $C_C'^*$, is not very different from that of the bulk metal, C_C', which is the saturation value, we may therefore expect that the iron oxide content of the slag at the interface should not be higher than 0.03%. In fact, the bulk iron oxide content reaches 0.2 to 0.4%, and the interface concentration must be still higher. Hence, unless we make the assumption that $C_C'^*$ is an order of magnitude lower than C_C' or that the supersaturation of CO in the slag is greater than 10 atm, both of which are unlikely assumptions, we must conclude that equilibrium is not reached and that therefore the reaction is not exclusively transport controlled. It would seem that the slow step in the reaction may be discharge of oxygen ions at the slag-metal interface, implying a rather large activation energy for this process.

Reactions between ionic and metallic phases such as are being considered here may be considered to take place by direct collisions of the reactants at the same site or, alternatively, by consecutive electrochemical partial reactions, the sites of which need not coincide, since electrons may flow through the metal from the anodic to cathodic sites.

On this basis the simultaneous reactions in sulfur transfer may be written:

	Initial Stages	Later Stages
Anodic Reactions	$[C] + O^{2-} = CO + 2e^-$	$[C] + O^{2-} = CO + 2e^-$
	$Fe = Fe^{2+} + 2e^-$	
	$\frac{1}{2}[Si] = \frac{1}{2}Si^{4+} + 2e^-$	
	$\frac{2}{3}[Al] = \frac{2}{3}Al^{3+} + 2e^-$	
Cathodic Reactions	$[S] + 2e^- = S^{2-}$	$[S] + 2e^- = S^{2-}$
		$Fe^{2+} + 2e^- = Fe$
		$\frac{1}{2}Si^{4+} + 2e^- = \frac{1}{2}[Si]$

As has already been noted, cathodic and anodic processes are equivalent, and there is no net flow of current. It may also be noted that the discharge of oxygen ions can take place at the slag-crucible interface in addition to the slag-metal interface. The former process is a true "local-cell" action such as occurs in metallic corrosion. Electrons furnished by the anodic process may flow to the metal phase through the graphite crucible to be available for cathodic reactions.

The present experiments may now be examined to see if a decision can be made regarding the relative contributions of the electrochemical mechanism and the direct reaction. The simplest experimental conditions are those in which lime-alumina slags and iron free from silicon and aluminum are used. Only the initial stages will be considered so that the back reaction may be neglected. It has already been shown that at least partial control of the rate by a phase-boundary reaction exists, and we will further assume that the rate is considerably less than that for exclusive transport control. Under these conditions the concentrations of reactants at the interface are not very different from their bulk concentrations, and therefore, concentration polarization is neglected in what follows.

In the electrochemical mechanism, it is reasonable to presume that the anodic evolution of CO is the slow step and that the other electrode reactions, anodic oxidation of iron and cathodic oxidation of sulfur, are in equilibrium. Essentially, we have an iron electrode in contact with the slag, and therefore, the electrode potential, E, can be expressed in terms of the concentration of iron ions in the slag, C_{Fe}''. According to Nernst's formula,

$$E = E_0 + \frac{RT}{2F} \ln C_{Fe}'' \tag{15}$$

where E_0 is the standard potential for the hypothetical state of unit concentration of iron ions in the slag.

The rate of CO evolution is related to the current density, J_{CO}, for this anodic process as follows

$$\frac{\dot{n}_{CO}}{A} = \frac{J_{CO}}{2F} \tag{16}$$

J_{CO} is related to the electrode potential by an equation of the following type,[9]

$$J_{CO} = k_{CO} \cdot C_C' \cdot C_O'' \cdot \exp\left[2(1-\alpha)\frac{EF}{RT}\right] \tag{17}$$

where k_{CO} is the rate constant for CO evolution as an anodic process, C_C' and C_O'' are the concentra-tions of carbon and oxygen ions in metal and slag, respectively. In equation 17 it is assumed that the reaction is between carbon in the metal and oxygen ions in the slag; α is related to the activation process at the anode and has a value of about 0.5. Thus

$$\frac{\dot{n}_{CO}}{A} = \frac{k_{CO}C_C' \cdot C_O''}{2F} \cdot \exp\left[2(1-\alpha)\frac{EF}{RT}\right] \tag{18}$$

Substituting equation 15 into 18 we have

$$\frac{\dot{n}_{CO}}{A} = \frac{k_{CO} \cdot C_C' \cdot C_O''}{2F}$$
$$\times \exp\left[\frac{2F(1-\alpha)}{RT}\left(E_0 + \frac{RT}{2F}\ln C_{Fe}''\right)\right] \tag{19}$$

which may be put into the form

$$\frac{\dot{n}_{CO}}{A} = \text{const } C_{Fe}''^{(1-\alpha)} \tag{20}$$

The rate of CO evolution should then be proportional to a fractional power of the concentration of iron ions in the slag. However, the analyses for iron in the slag are quite difficult, and the results are not considered accurate enough for an evaluation of equation 20. Under the present assumptions concerning which electrode reaction is the slow step, we would expect equilibrium between iron and sulfur at the metal-slag interface, and hence, from equation 8, bearing in mind that interface concentrations are approximately equal to bulk concentrations,

$$C_{Fe}'' = K_8 \cdot C_{Fe}' \cdot \frac{C_S'}{C_S''} \tag{21}$$

With the substitution of equation 21 in equation 20, with C_{Fe}' constant and with the metal-slag-interface area assumed constant,

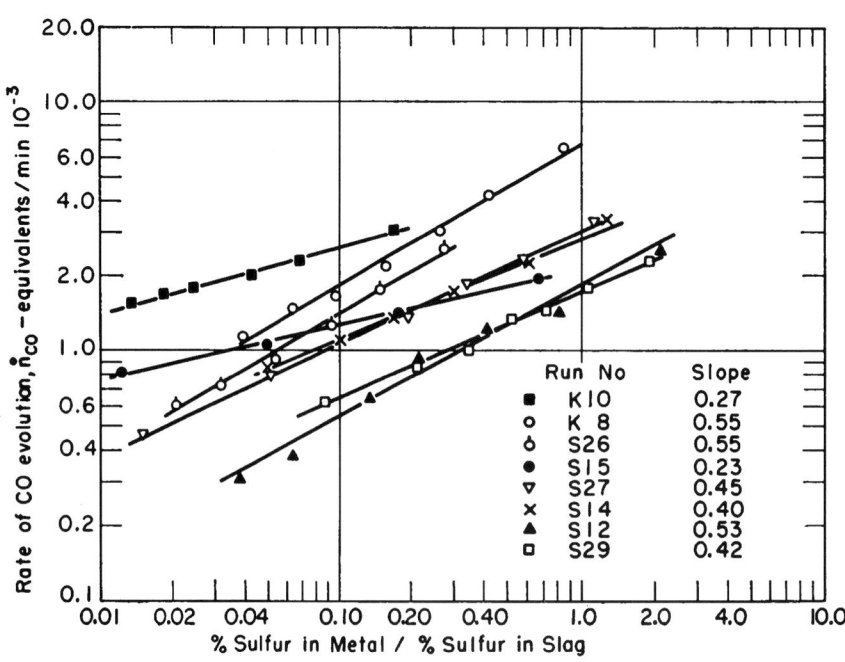

Fig. 9. Plot of rate of CO evolution, $\dot{n}_{CO}$, versus (% sulfur in metal/% sulfur in slag) in logarithmic coordinates. Slopes are indicated.

$$\dot{n}_{CO} = \text{const} \left(\frac{C_S'}{C_S''} \right)^{(1-\alpha)} \qquad (22)$$

This relationship has been tested by obtaining the slopes of the CO evolution curves from Figs. 1 to 8 and plotting these as functions of the ratio $\%[S]/\%S_{slag}$ on logarithmic coordinates. As Fig. 9 shows, the points fall on straight lines within the accuracy of such data. The slopes of these lines, with two exceptions, vary from 0.4 to 0.55. It is not clear why the slopes for runs K10 and S15 should be about half this value. Run S15 involved aluminum oxidation, but run K10 was not significantly different in initial conditions from run K8. The expected slope of such plots is approximately 0.5, since this is a likely value for $(1 - \alpha)$. These results are therefore not inconsistent with the electrochemical mechanism suggested. However, in view of the approximations involved in the theoretical treatment and the experimental inaccuracies, which are magnified by the necessity for taking slopes of CO evolution curves, it is not claimed that this constitutes a proof that the electrochemical mechanism predominates. Further experiments are clearly necessary.

References

1. N. J. Grant, O. Trioli, and J. Chipman, *Trans. A.I.M.E.*, 191, 672 (1951); N. J. Grant, U. Kalling, and J. Chipman, *Trans. A.I.M.E.*, 191, 666 (1951).
2. G. Derge, W. O. Philbrook, and K. M. Goldman, *Trans. A.I.M.E.*, 188, 1111 (1950); K. M. Goldman, G. Derge, and W. O. Philbrook, *Trans A.I.M.E.*, 200, 534 (1954).
3. S. Ramachandran, T. B. King, and N. J. Grant, "Proc. Blast. Furn. Conf.," *A.I.M.E.*, 14, 338 (1955).
4. S. Ramachandran, T. B. King, and N. J. Grant, *Trans. A.I.M.E.*, 206, 1549 (1956).
5. L. S. Darken and R. W. Gurry, *Physical Chemistry of Metals*, McGraw-Hill Book Co., New York, 1953.
6. T. E. Brower and B. M. Larsen, *Trans. A.I.M.E.*, 172, 137 (1947).
7. F. W. Körber and W. Oelsen, *Naturwissenschaften*, 23, 462 (1935).
8. G. G. Hatch and J. Chipman, *Trans. A.I.M.E.*, 185, 274 (1949).
9. H. Eyring, S. Glasstone, and K. Laidler, *J. Chem. Phys.*, 7, 1053 (1939).

Discussion

DARKEN expressed a skepticism that King first rule out the diffusion mechanism by assuming that the CO evolution occurred by bubble formation in the metal. He felt this to be a straw man because most people would find it difficult to visualize how such a bubble would nucleate. This is especially true in view of the low oxygen level in the system being treated. He also pointed out that one should look to the diffusion layer in the slag at the slag-metal interface as a limiting condition because of the very low iron content of these slags. He expressed skepticism concerning the reaction involving carbon in the metal, oxygen in the slag, and CO in the gas phase. This type of reaction was particularly difficult for him to visualize because of the very small interfaces involving three phases which could exist. He felt that a two-phase reaction would be more reasonable and plausible. KING replied that he did not propose that the reaction was exclusively phase-boundary controlled. Instead, it was partially phase-boundary controlled and partially diffusion controlled. In spite of Darken's arguments, KING also felt that it should not be too great an improbability that a bubble could nucleate at the slag-metal interface. The activation energy need not be too high.

WAGNER interposed that he felt that there was an appreciable solubility of CO in the slag. To support this viewpoint, he cited the data presented by Richardson to the effect that neutral metal atoms are also soluble in slags. He reviewed the proposition that local cell action might be controlling. There is a possibility at the slag-graphite interface that oxygen ions from the slag combine with carbon from the crucible to form CO, and the electrons from this reaction are delivered by the graphite crucible to the site of the reaction with sulfur. He also said that he felt that the nucleation problem was not a serious one in the system and cited the low overvoltage necessary for hydrogen evolution on a platinum electrode. KING pointed out that in the treatment of the problem the possibility of local cell action has been considered, but the experimental work which was reported did not illuminate this matter. However, currently, Dr. Baak is studying this problem. He has been conducting an experiment in which sulfur transfer can be accomplished only by local cell action. A sodium borate slag and a metal phase of silver containing sulfur are contained in an alumina crucible. Contact with the slag and metal phases is made by means of graphite and nickel rods, respectively. Current flows through the rods only when they are short-circuited, and the amount of current is closely related to the number of equivalents of sulfur transferred.

VALLET said that he felt that both the diffusion and the nucleation problems had to be accounted for in treating this mechanism. PHILBROOK pointed out that the general results obtained in the experiments reported were similar to those that had been reported several years ago.[*] It seems quite certain that nucleation of the bubbles does not occur in the metal or at the metal-crucible interface. He proposed that it was fundamentally simpler to express the reactions as follows:

$$[Fe] + [S] \rightleftharpoons Fe^{2+} + S^{2-}$$

This was suggested by Darken in the discussion of the earlier paper to which Philbrook had reference. The equation says that an electron exchange occurs when iron and sulfur leave the melt. Being reversible, it is an oxidation-reduction process. If the ferrous-iron concentration is high, equilibrium for the reaction is reached. Other elements like silicon and manganese can also participate. Consequently one can write the same type of equation for each of them. Alternatively, one could say that silicon has an effect by reducing the ferrous iron in the slag. He also proposed the reaction:

$$Fe^{2+} + [C] \rightleftharpoons (C^{2+}) + [Fe]$$

If there is any O^{2-} in the slag nearby, it could react with C^{2+} to form a CO bubble. This might be equivalent to what Dr. Wagner calls the solubility of CO in the slag. To him, this was a more plausible explanation

* G. Derge, W. O. Philbrook, and K. M. Goldman, *Trans. A.I.M.E.*, 188, 1111 (1950).

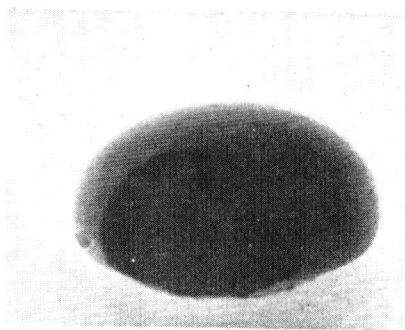

Fig. 10. A drop of liquid iron saturated with carbon and no sulfur under a blast-furnace slag with 1% S and CaO/SiO₂ = 1.35. The interface is convex. Interfacial tension about 790 dyne cm⁻¹. Graphite crucible.

than that which requires the discharge of an oxide ion which is very stable as compared to some of these other ions that are participating.

PHILBROOK went on to say that the discussion had not touched on other aspects of desulfurization that he con-

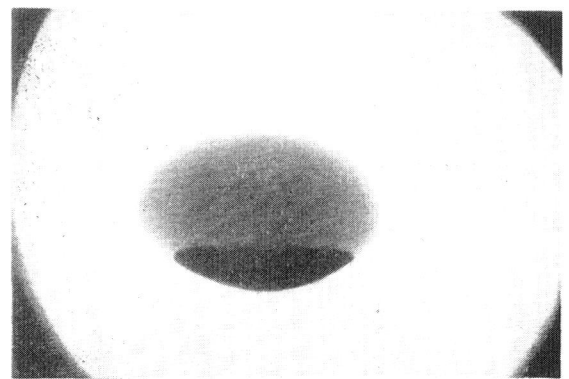

Fig. 11. A drop of liquid iron saturated with carbon and 0.7% sulfur under a blast-furnace slag with 1% S and CaO/SiO₂ = 1.35. The interface is plane. No CO bubbles. Interfacial tension less than 5 dyne cm⁻¹. The quenched slag is brownish near the interface. Graphite crucible.

sidered to be of importance: (a) the effect of slag composition, and (b) the role of interfacial behavior. It is well known that desulfurization is more rapid in acid slags than it is in basic slags. One might explain this by the argument that acid slags have a higher concen-

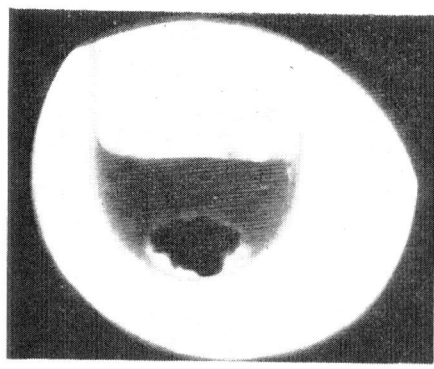

Fig. 13. A drop of liquid iron with 3% carbon and 0.7% sulfur under a blast-furnace slag presaturated with sulfur (CaS), 20 min after pouring. The drop is of an irregular shape and is almost surrounded with a large CO bubble. Alumina crucible.

tration of Fe²⁺ ions which would tend to retard desulfurization. Also, the strong coordination between silicon and oxygen in the basic slags would tend to displace the reaction

$$[Si] + 2[S] \rightleftharpoons Si^{4+} + 2S^{2-}$$

to the right. The surface behavior of sulfur has been studied by Kozakevitch and at the Carnegie Institute of Technology. Both studies show that sulfur is strongly adsorbed on the surface of liquid iron. The slag measurements were made in an argon atmosphere and showed that those slags having a high desulfurizing power had the lowest concentrations of sulfur adsorbed on the surface. These slags also give the most rapid desulfurization. Thus, it may be that conditions governed by the slag and metal surfaces may be of great importance.

PHILBROOK continued his comments by noting that one of the unique phenomena in desulfurization is that gas bubbles rising from within the melt (not at a crucible wall) carry droplets of iron up into the slag and leave these droplets on the surface of the slag. This suggests that a gas bubble also carries a surrounding film of iron, and indicates the possibility of a reaction involving the slag-gas bubble-iron film.

KOZAKEVITCH presented five X-ray pictures (Figs. 10, 11, 12, 13, and 14) of the contact between liquid slag and liquid metal as influenced by the sulfur content of the slag and metal and carbon content of the metal. The persistent and violent reaction which is accom-

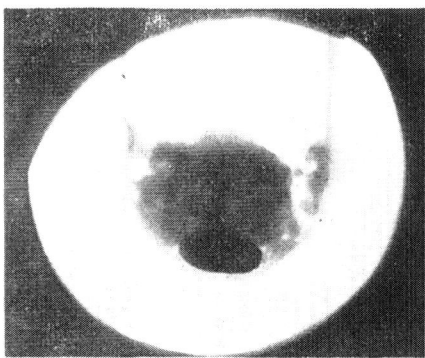

Fig. 12. A drop of liquid iron with 3% carbon and 0.7% sulfur under a blast-furnace slag presaturated with sulfur (CaS), 10 min after pouring. Alumina crucible.

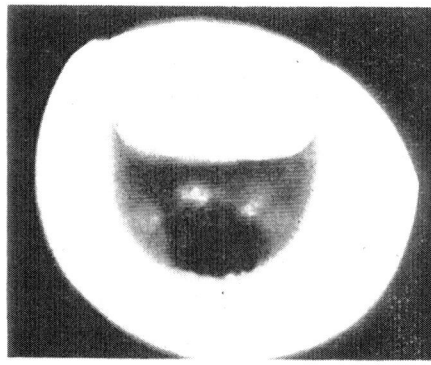

Fig. 14. A drop of liquid iron with 3% carbon and 0.7% sulfur under a blast-furnace slag presaturated with sulfur (CaS), 30 min after pouring. Very violent reaction at the interface, the metal begins to be emulsified. Alumina crucible.

panied by gas evolution is evident with the high-carbon iron and CaS-saturated slag. KING commented that the markedly different physical conditions between Kozakevitch's experiment and those reported in the paper made it difficult to compare factors that might be rate controlling.

In concluding the discussion, KING commented that they had proposed a mixed type of control for the reaction rather than simply phase-boundary control. He agreed with Darken on the point that the ability to write a mechanism in the form of an equation was no assurance of its reality or probability. Although equations are usually molecular in form, this does not assure that the reaction takes place by means of these molecular species. He objected to Philbrook's proposal that concentrations at the surfaces are significant, on the grounds that actually one should be concerned with chemical potentials rather than concentrations.

by C. E. A. Shanahan

The Desulfurization and Dephosphorization of Carbon-Saturated Iron

A considerable amount of time has been devoted to the thermodynamic study of iron and steelmaking reactions generally, and there is no doubt that this type of research is essential to the fundamental understanding of the chemistry of the processes. Moreover, the standard free-energy equations which result from such work are a real practical help in that they reveal the limits to which particular reactions can proceed under given sets of conditions. In cases where the reactions proceed sufficiently rapidly to attain equilibrium under plant conditions, the technician can use thermodynamic data to calculate the amount of refining that he will obtain following a change in operating conditions.

However, a very important practical disadvantage of thermodynamic studies is the complete absence of a time factor; no knowledge is forthcoming on the rates of refining reactions, with the result that an apparently efficient refining scheme (judged from thermodynamic data) may be practically and economically useless because of the slow rates of reactions involved. This fact, although obvious, has been emphasized because past experience has shown that it is often disregarded. Thus, it is quite common to find technicians using Ellingham diagrams to "prove," for example, that the removal of phosphorus from hot metal cannot occur appreciably until most of the carbon has been oxidized; it is tacitly assumed, of course, that the relative reaction rates are so high as to be of no consequence.

Mr. Shanahan is affiliated with the R. T. S. C. Laboratories, Whitchurch, Aylesbury, Buckinghamshire, England.

The author desires to thank Sir Charles Goodeve, Director, and Dr. A. H. Leckie, formerly Head of the Steelmaking Division of the British Iron and Steel Research Association, for many helpful discussions.

From what has already been said, there is obviously a great need for kinetic information on steelmaking reactions generally, and it may be asked why this branch of work has tended to be neglected. In part, this is due to the fact that many of the reactions are largely diffusion controlled and the problem of obtaining high reaction rates is mainly one of rapidly bringing the reactants together and removing the products, that is, of intimately mixing together the reacting phases. A major drawback in the way of conducting useful kinetic experiments arises from the difficulty of drawing general conclusions which might be applied to plant conditions. Unlike thermodynamic studies, the quantitative data concerning heterogeneous reaction rates often cannot be separated from the dimensions, shape, and type of surface of the reaction vessel. Thus the reaction rates of a slag-metal reaction must depend, among other things, on the slag-metal interfacial area and on concentration gradients that are functions of mass-flow patterns within the system. These are functions of the size of the experimental apparatus, and the data cannot be mathematically scaled up to plant conditions with any degree of certainty.

The rapid desulfurization and any dephosphorization of hot metal that can be accomplished is of great technical importance to several British open-hearth plants. It is not uncommon to find such plants forced to use hot metal containing about 0.2% S and 1.5% P, and obviously any cheap and rapid method of removing most of these impurities prior to the open hearth is of great interest. For this reason attention has recently been given to the kinetics of desulfurization and dephosphorization, and some of the work with which the author has been acquainted is now described.

DESULFURIZATION

With the object of using a desulfurized blast-furnace slag to desulfurize hot metal, many kinetic experiments were made at approximately 1450° C in which carbon-saturated molten iron initially containing 0.20 to 0.30% S was desulfurized with a molten slag composed of 40% CaO, 40% SiO$_2$, and 20% Al$_2$O$_3$. Some of these experiments were conducted in a carbon-lined horizontal cylindrical furnace fitted with a carbon wedge longitudinally along the bottom of the melting chamber; radiant heat was supplied by a graphite resistor bar just below the roof of the melting chamber (precise details of this furnace have been given elsewhere[1]). The purpose of the wedge was to ensure intimate mixing of the slag and metal when the furnace was rocked. In each experiment approximately 7 1/2 lb of iron was melted in the furnace, and the sulfur content raised to about 0.25% by the addition of ferrous sulfide. Approximately 2 1/2 lb of the slag, previously melted in a separate furnace, was then added, and the course of desulfurization followed by taking pairs of slag and metal samples periodically. Figure 1 shows typical data obtained from two groups of experiments, namely, one group in which the furnace remained motionless and a group in which the furnace was rocked almost continuously throughout the desulfurization process. It is obvious from Fig. 1 that the desulfurization rate under

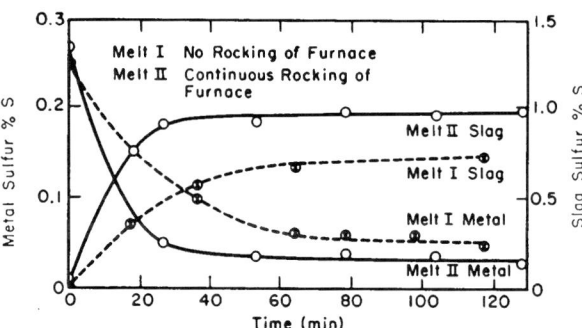

Fig. 1. Slag desulfurization in furnace.

quiescent conditions, that is, no rocking, was diffusion controlled, since a marked increase in desulfurization rate occurred when the furnace was rocked. Surprisingly, however, all efforts to increase still further the desulfurization rate by increasing the intensity of slag-metal agitation were unsuccessful, suggesting that either diffusion had ceased to be the controlling influence or that the increased mixing rates were producing only a negligible increase in intimacy of contact between slag and metal.

Several experiments were made using the same quantities of slag and metal but performing the desulfurization in a carbon crucible and obtaining agitation by means of a nitrogen jet. The nitrogen was introduced to the slag-metal interface through

a carbon nozzle, but although various pressures of nitrogen were employed, the desulfurization rate was in every case very similar to the "rocked" melt shown in Fig. 1.

Attention was now turned to more drastic methods of intimately mixing slag and metal together. These involved atomizing the molten iron and allowing the globules to fall through the molten slag. Although such processes would be impracticable on a large scale and, indeed, the experimental difficulties encountered supported this contention, it was felt that the results obtained would be valuable in assessing the desulfurization rate-controlling mechanisms apparently present in rocking-furnace and nitrogen-blowing experiments. Three methods were used; the first and simplest consisted in pouring the molten iron onto an inclined flat carbon plate projecting from the inside wall of a carbon crucible containing the molten slag. Much of the iron was globularized as a consequence of splashing, and the globules fell into the slag. In a second experiment (see Fig. 2), the molten iron

Fig. 2. Colander experiment.

was passed through a "colander" (consisting of a preheated carbon crucible containing approximately 50 1/8-in. holes drilled in its 3-in. diam base) prior to falling on to the splashing plate. This method was found to produce a more efficient subdivision of the iron. A third experiment involved atomizing the iron by means of a nitrogen jet. The molten iron was held in a carbon crucible which had been provided with a carbon stopper rod in a similar manner to that employed with steel teeming ladles. The stopper rod was hollow so that nitrogen could be passed out through its base. By slowly raising the stopper rod, using a rack and pionion mechanism, the molten iron was effectively atomized by the nitrogen and dropped into the molten slag contained in a crucible below (Fig. 3). In

Fig. 3. Nitrogen atomization.

all three experiments, the molten slag was water quenched as soon as possible after the iron addition. The over-all time between beginning the iron addition and water-quenching of the slag was never more than 60 sec, so that this represented the maximum time available for desulfurization.

Examination of the solidified melts was confined to the globules of iron remaining in the slag, since

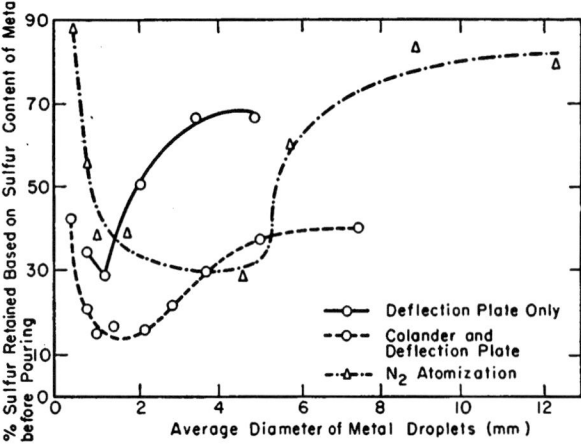

Fig. 4. Influence of mixing methods on desulfurization.

in spite of the atomization techniques much of the iron had passed through the slag as a stream and consequently did not undergo appreciable desulfurization. In each case the slag layer was separated from the metal and carefully crushed, the metal globules being collected magnetically. Residual slag was removed from the iron globule surfaces by "trundling," and any angular globules discarded because these were probably fragments of larger particles broken during the preliminary slag-crushing process. The iron particles were then graded by hand picking and sieving and sulfur determinations conducted on each fraction. Tables 1, 2, and 3 list the results obtained from the three atomizing experiments, and Fig. 4 shows the relationships be-

Table 1. Data from Deflection-Plate Experiment

Average Particle Size (mm)	[S] %
4.85	0.206
3.40	0.204
2.04	0.155
1.21	0.090
0.75	0.106
Molten iron prior to pouring	0.308
Molten slag prior to pouring	0.034 (S)
Slag after experiment	0.046 (S)

Table 2. Data from "Colander" Experiment

Average Particle Size (mm)	[S] %
7.40	0.102
4.95	0.094
3.65	0.074
2.85	0.054
2.14	0.040
1.41	0.041
1.00	0.037
0.75	0.051
0.38	0.106
Molten iron prior to pouring	0.252
Molten slag prior to pouring	0.015 (S)
Slag after experiment	0.054 (S)

Table 3. Data from N_2 Atomizing Experiment

Average Particle Size (mm)	[S] %
12.25	0.240
8.77	0.252
5.66	0.182
4.54	0.087
3.75	0.091
1.75	0.117
1.00	0.116
0.75	0.167
0.38	0.265
Molten iron prior to pouring	0.302
Molten slag prior to pouring	0.014 (S)
Slag after experiment	0.04 (S)

tween the percentage of sulfur retained by the iron and particle size. To check the absence of globule desulfurization due to air oxidation, carbon determinations were performed on most of the particle fractions; the lowest value was approximately 3%.

Interpretation of Data

In view of the many uncontrolled variables present in the atomization experiments, it is impossible to deduce exact kinetic data. The experiments were conducted chiefly to ascertain whether extremely rapid desulfurization rates were possible under favorable conditions, and in this they were successful. However, it is interesting to note that all three curves in Fig. 4 exhibit minima, although the precise reasons for these are unknown. Obviously, the globules were not all in contact with the slag for the same time period, and it is reasonable to suppose that the average time of slag contact decreases with increase in particle size. This fact, coupled with the lower specific areas of the larger particles, probably explains the inefficient desulfurization of the latter. The smallest particles measured also exhibited poor desulfurization, and this could be attributable to their partial or complete solidification prior to entering the molten slag.

The most important conclusion from the atomization experiments is that it is possible to desulfurize carbon-saturated iron with a $CaO—SiO_2—Al_2O_3$ slag at rates greater than 0.0036% S per sec in the sulfur range 0.04 to 0.30%. It must be appreciated that the conditions employed were extremely favorable to desulfurization. Thus:

(a) There was a negligible back reaction, that is, the return of sulfur from slag to metal, since the sulfur content of the slag never rose above 0.054%. In the case of the rocking-furnace experiments, the first pair of desulfurized slag and metal samples could only be taken approximately 20 min after commencement of the experiments. By this time the slag-sulfur content was sufficiently high to produce a significant back reaction and reduce the over-all desulfurization rate.

(b) Because of the low slag-sulfur content, the iron-oxide content of the slag arising from the desulfurization reaction did not exceed that required for the formation of CO bubbles,[2] and hence the well-known retarding effect of the latter was avoided.

(c) The diffusion of sulfur from metal to slag may be approximately described by the equation

$$\frac{\partial S}{\partial t} = -k \frac{A}{V} \frac{\partial S}{\partial x} \qquad (1)$$

where S is the sulfur content of the metal, V the volume of metal, and A the area of contact between metal and slag.

Conditions in the atomizing experiments were such that the A/V values for the iron globules were relatively large. Moreover, the fact that the particles were moving through the slag presumably ensured a reasonably high value for $\partial S/\partial x$, that is, the sulfur-concentration gradient. Both of these effects would promote high desulfurization rates.

It would be of interest to conduct further atomization experiments in which the slag initially contained sufficient iron oxide to be in equilibrium with carbon and CO gas at 1 atm pressure. Any desulfurization would then presumably require the formation of CO gas bubbles, and the effects of this gas reaction on desulfurization rate would be apparent.

DEPHOSPHORIZATION

Recent work on the Continent[3] has shown that by the use of a top-blowing oxygen technique it is possible to remove appreciable quantities of phosphorus from iron containing 2 to 3% of carbon. Such a possibility is of particular interest to British hot-metal open-hearth shops since the removal of part of the phosphorus at the hot-metal stage would considerably lighten the metallurgical load of the furnaces.

Appreciable dephosphorization of hot metal would appear to be impossible if the appropriate Ellingham diagram is examined. This shows that dephosphorization requires an oxygen potential greater than that permitted by the carbon-oxygen reaction, and only when most of the carbon has been eliminated can the oxygen potential reach sufficiently high values to enable dephosphorization to proceed. No doubt this conclusion has been partly responsible for the lack of experimental work on this topic in the past; the Continental findings[3] emphasize the dangers of relying implicitly on thermodynamic data, and it is hoped that greater consideration will, in future, be given to kinetics.

The dephosphorization of hot metal can be explained by two mechanisms. The first, and the simplest, is that the oxygen potential of the system reaches values high enough for dephosphorization in spite of the presence of large carbon concentrations. This implies that the carbon-oxygen-carbon monoxide equilibrium is not operating, that is, that this reaction is sluggish. Evidence in support of this view was obtained during a series of trials in which approximately 10 cwt of hot metal was top blown with oxygen. Periodically, killed bomb samples were taken and analyzed for carbon and oxygen, and the data from two such trials are given in Table 4. It is obvious that the oxygen values corresponding to approximately 2.7% C are considerably in excess of the equilibrium values for the C—O—CO (1 atm) reaction as can be seen from the p_{CO} columns, which give the equivalent equilibrium CO pressures.

A second mechanism for dephosphorization by

Table 4. Analysis of Bomb Samples					
Bomb Sample No.	[C] %	[O] %	Equilibrium p_{CO}, atm		
			1327° C	1427° C	1527° C
Trial 1					
2	2.70	0.0196	16.77	14.35	12.47
3	2.29	0.0151	10.53	9.01	7.84
5	1.22	0.0130	6.35	5.43	4.72
8	0.09	0.050	3.05	2.61	2.27
Trial 2					
2	2.73	0.024	20.68	17.7	15.38
3	2.26	0.026	17.95	15.36	13.35
5	1.73	0.013	7.69	6.58	5.71
8	0.075	0.093	4.53	3.88	3.37

oxygen blowing assumes that the metal surrounding an oxygen gas bubble is not in equilibrium with the remainder of the melt. This is a reasonable hypothesis in view of the large amount of iron oxide fume that attends oxygen-blowing trials.

Thus at positions of oxygen gas, slag, and metal contact, it is possible that small quantities of decarburized and dephosphorized iron exist. These pockets of relatively pure iron can subsequently decrease the proportion of carbon and phosphorus in the bulk of the hot metal by dilution so that the over-all carbon and phosphorus content of the latter decrease together in apparent disagreement with the predictions of the Ellingham diagram.

It is possible to illustrate the above hypothesis using a simple system composed initially of 100 grams of iron containing 3% C and 1.5% P in contact with a dephosphorizing slag. Assume that a bubble of oxygen is responsible for the approximately complete decarburization and dephosphorization of w grams of iron and that this refined iron dilutes the bulk of the hot metal before the introduction of the next oxygen bubble. Then

$$\% \text{ C remaining in hot metal} = 3 (1\text{-}0.01 \, w)$$

$$\% \text{ P remaining in hot metal} = 1.5 (1\text{-}0.01 \, w)$$

If the calculation is repeated for each oxygen bubble, then after n bubbles

$$\% \text{ C remaining in hot metal} = 3 (1\text{-}0.01 \, w)^n$$

$$\% \text{ P remaining in hot metal} = 1.5 (1\text{-}0.01 \, w)^n$$

By giving a hypothetical value to w it is possible to calculate the rate at which the carbon and phosphorus are removed in terms of numbers of oxygen bubbles, that is, time, if the oxygen flow rate is constant. This calculation has been made for $w = 5\%$ of the total metal, and the relationships are shown in Fig. 5. It is obvious that such a mechanism can account for appreciable phosphorus removal in the presence of high carbon-concentration levels.

CONCLUSIONS

(1) Data are presented showing that the desulfurization rate of molten carbon-saturated iron by CaO—SiO$_2$—Al$_2$O$_3$ slags can be exceedingly high, providing the iron has a high area of slag contact/volume ratio, high sulfur-concentration gradients, and that the CO gas reaction and the return of sulfur from slag to metal are avoided.

(2) Two mechanisms are proposed to account for the removal of phosphorus from hot metal in the presence of high carbon-concentration levels by oxygen blowing. One involves the assumption that the C—O—CO reaction is sluggish, and data are provided in support of this. The second mechanism assumes that the metal immediately surrounding a bubble of oxygen is not in equilibrium with the remainder of the melt.

References

1. C. E. A. Shanahan and F. J. Lund, *Iron & Coal Trades Rev.*, 701 (March 27, 1953).
2. G. G. Hatch and J. Chipman, *Trans. A.I.M.E.*, 185, 274 (1949).
3. O. Cuscoleca and K. Rösner (Discussion), *Rev. universelle mines*, 96, 668 (1953).

Discussion

PEARSON discussed briefly the course of dephosphorization and decarburization during top blowing a bath of iron in a basic vessel. With a bath initially containing 1.8% phosphorus, there is a rapid drop in phosphorus with only a slight reduction of carbon. There is also a marked rise in the oxygen content of the bath. Phosphorus is reduced to between 1 and 0.5%, while carbon is still above 2.0% (see Fig. 6). He attributed this

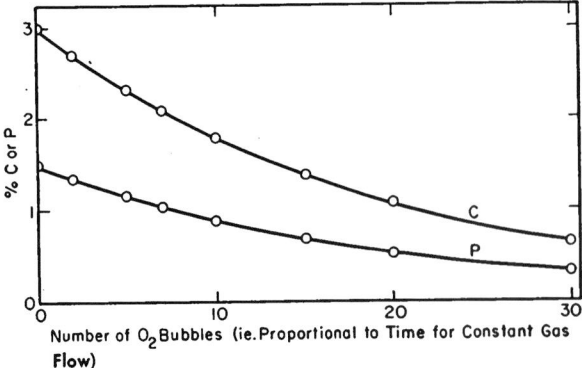

Fig. 5. Calculated rate of decarburization and dephosphorization by oxygen blowing of liquid iron.

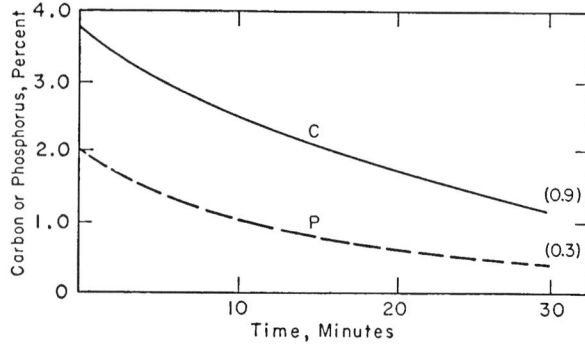

Fig. 6. Course of reduction of carbon and phosphorus during top blowing with oxygen.

result to supersaturation of the bath with CO. The carbon-oxygen product is well above the equilibrium value during the early stages. However, once the carbon reaction gets under way, the oxygen content and the C—O product drop close to equilibrium values. These results were valid for 25-lb and 18-ton heats, provided that a highly oxidizing and basic slag was maintained.

LARSEN told briefly of his experience with a 750-lb pilot-plant oxygen-blown vessel for dephosphorizing and decarburizing high-carbon iron. The metal contained approximately 0.30% phosphorus and 3.5% carbon. The phosphorus dropped to about 0.10% when the carbon was at 2.5%. By slagging off and continuing the blow, it was possible to get below 0.05% phosphorus at 1.5% carbon. The slag was 50% CaO, 15% SiO$_2$, 4 to 8% P$_2$O$_5$, 20% FeO, and 1% MgO.

by G. Trömel

The Effect of Temperature and Silica on the Dephosphorization of Iron with Lime

In the first part of the investigation on the dephosphorization of iron with lime published some time ago,[1] we considered the reaction

$$2[P] + 5(FeO) + n(CaO) \rightleftarrows (nCaO \cdot P_2O_5) + 5[Fe]$$

as occurring in the heterogeneous system Fe—FeO—CaO—P$_2$O$_5$. The distribution of phosphorus between metal and slag in such a system is primarily determined by the equilibria existing in the

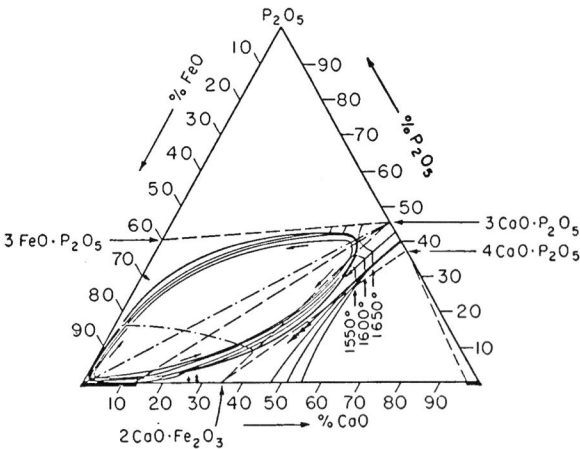

Fig. 1. The system FeO—CaO—P$_2$O$_5$.

slag system FeO—CaO—P$_2$O$_5$. Figure 1 shows the basic features of this system.*

By our experiments with small crucibles, we try

Dr. Trömel is with the Max-Planck-Institut für Eisenforschung, Düsseldorf, Germany.
* The concentrations are given as weight percentages.

to give an answer to the question, what is the lowest phosphorus content in the metal we can get simultaneously with a high P$_2$O$_5$ content in the slag —primarily under conditions similar to those prevailing in the basic converter?

To avoid impurities, it is necessary to use as crucibles the solid phases existing in the system at high temperatures. In the case of the dephosphorization the lime and phosphates rich in lime, 3CaO·P$_2$O$_5$ and 4CaO·P$_2$O$_5$ are still solid at about 1600° C.

By this condition the possible liquid phases are restricted to those saturated with the various solid phases. Furthermore, only such slags can exist which are saturated simultaneously with iron. These conditions largely simplify the equation for the equilibrium constant KP.

On the other hand, these restrictions correspond to the conditions existing in the basic converter or in the basic open-hearth furnace. If we achieve equilibrium in these processes, it must obviously be determined not only by the saturation with iron but also by the saturation with the refractory material of the lining or of the hearth and with the additions which are solid at the working temperature.

In a crucible made of 3CaO·P$_2$O$_5$, two liquid slags are present beside the solid phase and the molten metal. One of the slags is rich in iron oxide and one rich in phosphate of calcium. The relations occurring among these four phases furnish useful indications for the general knowledge of the reaction of dephosphorization. However, they are possible only if the slags are not saturated with lime.

This case is not important for the final state of the reaction, although even in presence of solid lime the slag is probably often not saturated with lime under plant conditions.

The equilibrium of iron with a slag saturated with CaO or with CaO and $4CaO \cdot P_2O_5$ exists in a crucible of CaO or in one consisting of a mixture of CaO and $4CaO \cdot P_2O_5$, respectively.

In this case the equilibrium conditions are very simple. At each temperature only one slag composition is possible at the end of the reaction. With varying temperature, the slag composition must change along the boundary between the fields of primary crystallization of CaO and of $4CaO \cdot P_2O_5$. The lower the temperature, the higher the FeO content and the lower the P_2O_5 content. This explains the fact that, as a rule, the dephosphorization is better at lower temperatures. This effect cannot be very great. The composition of the slag does not vary very much with the temperature under these conditions.

The strong effect of the temperature on dephosphorization as it is known in the technical processes, for example, in the basic Bessemer process, cannot be established in the equilibrium of a pure system. Table 1 and Fig. 2 show the average values

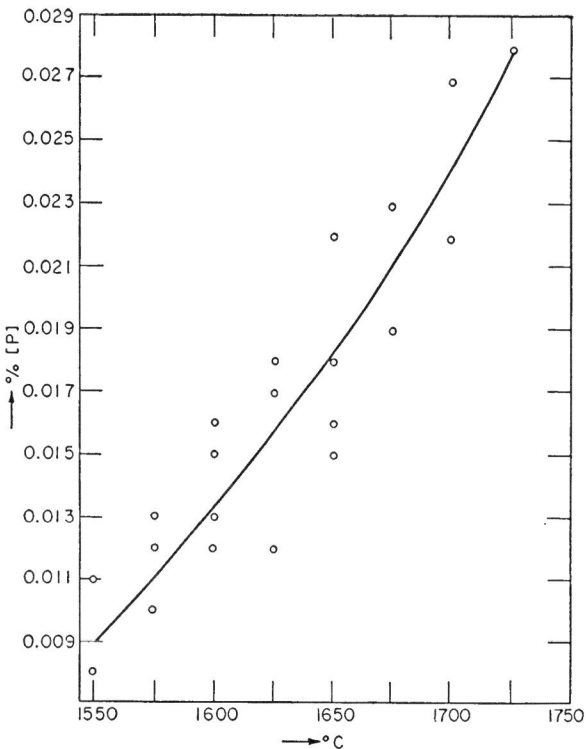

Fig. 2. Effect of temperature on the phosphorus content.

for the temperature range from 1550° C to 1725° C. A number of [O] concentrations are simultaneously listed in Table 1 and shown in Fig. 3.

Figure 4 is a summary of the present values for 1600° C in the system CaO—P_2O_5—FeO—Fe. Part A of Fig. 4 shows the CaO corner of the slag system

Table 1. The Effect of Temperature on the Phosphorus and Oxygen Contents of Iron Melts in Equilibrium with Lime-Saturated Slags*

P_2O_5 Content of Crucible	Temperature, °C	Time, min	Phosphorus in Iron, %	Oxygen in Iron, %	Remarks
25	1550	15	0.011	0.053	
	1550	60	0.011	0.056	
	1575	15	0.012	0.062	
	1575	25	0.013	0.065	
	1575	55	0.010	0.063	Slag flow out
	1600	15	0.016	0.062	
	1600	60	0.012	0.061	
	1625	15	0.012	0.068	Temperature too high for crucible
	1650	10	0.022	0.073	Temperature too high for crucible
12	1550	155	0.008	—	
	1550	60	0.008	—	
			0.008	—	Initial iron with 0.1% P
	1575	15	0.010	—	
	1575	60	0.010	—	
			0.010	—	Initial iron with 0.1% P
	1600	15	0.016	—	
	1600	60	0.013	0.062	
			0.012	—	Initial iron with 0.1% P
	1625	15	0.018	0.072	
	1625	60	0.017	—	
	1650	15	0.018	0.076	
	1650	30	0.015	0.074	
	1650	60	0.016	—	
	1675	15	0.018	0.082	
	1675	30	0.023	—	
	1675	60	0.019	—	
	1700	15	0.022	0.084	
	1700	60	0.027	—	
	1725	15	0.028	0.089	

* Charge: 40 grams Fe + 10 grams slag of initial composition 20% FeO, 55% CaO, 25% P_2O_5.

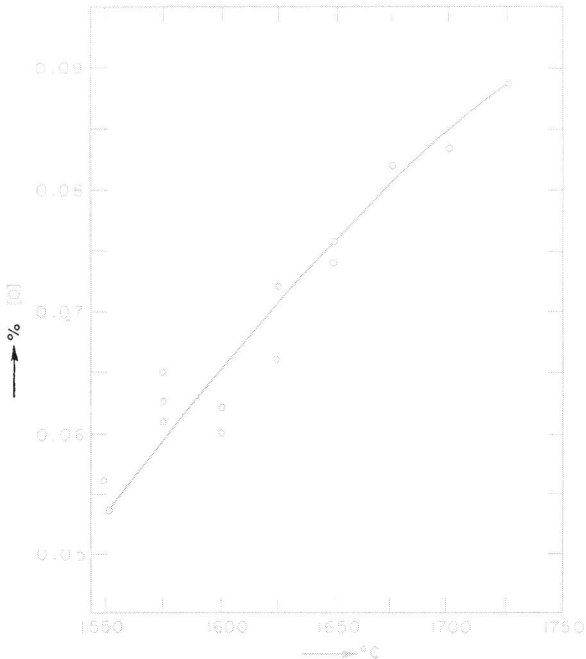

Fig. 3. Effect of temperature on the oxygen content of the steel bath.

CaO—P₂O₅—FeO. Part *B* gives the relation between [P] and (P₂O₅) for the slags in Fig. 4*A*, and Fig. 4*C*, the relation between (FeO) and [O] for the same slags. Finally, Fig. 4*D* shows the relation between [P] and [O] as a part of the system Fe—P—O.

As mentioned before, because of the presence of an excess of solid CaO and 4CaO·P₂O₅, only one composition of the slag is possible in the final state. Consequently we should find only one value for the various relations in each part of Fig. 4. These

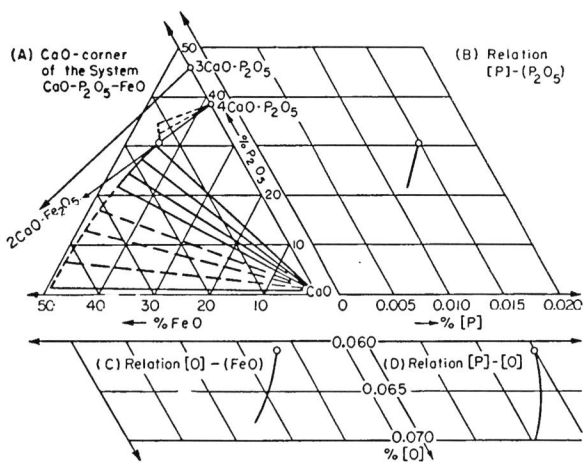

Fig. 4. The effects of slags rich in lime on the [P] and [O] contents at 1600° C.

points are marked by circles. However, as an intermediary state it is possible to get slags saturated with lime alone. The intermediary states are close to the point of the end of the reaction.

The second part of the investigation deals with the effect of silica on the dephosphorization. In this case we must consider the system FeO—CaO—P₂O₅—SiO₂.

As in equilibrium with slags containing FeO, no silicon can be present in iron; only the slag composition can vary. It is known from the ternary system CaO—P₂O₅—SiO₂ that the compound 4CaO·P₂O₅ disappears as a phase if small concentrations of SiO₂ are present. In such slags, the phosphoric acid is combined as solid solutions between 3CaO·P₂O₅ and 2CaO·SiO₂ (designated R$_a$ in this paper).

To determine the residual phosphorus in equilibrium with slags containing SiO₂, melting experiments were again made in special crucibles. Crucibles and initial slag compositions with equal P₂O₅/SiO₂ ratios were used for each series of experiments. The P₂O₅/SiO₂ ratios were chosen in such a way that in one series (P₂O₅/SiO₂ = 5) the liquid phases are saturated with CaO, 4CaO·P₂O₅, and R$_a$, and in another series, with CaO and R$_a$ alone (P₂O₅/SiO₂ = 2). From these conditions, univariant or divariant final states, respectively, had to result.

Figure 5 shows microphotographs of polished sections of some slags. Figure 5*a* is a slag without

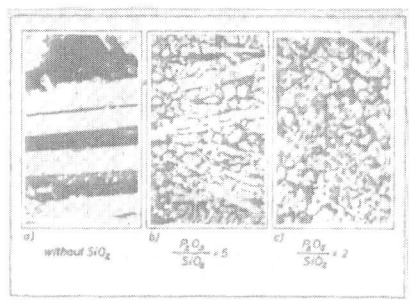

Fig. 5. Microphotographs of polished sections of slags with various SiO₂ contents.

SiO₂. All phosphorus exists as 4CaO·P₂O₅. In Fig. 5*b* a slag is shown with the ratio P₂O₅/SiO₂ = 5. In this case, the small and sometimes hexagonal solid solutions R$_a$ can be recognized together with the long crystals of 4CaO·P₂O₅. Figure 5*c* gives a slag with the ratio P₂O₅/SiO₂ = 2. No 4CaO·P₂O₅ can be seen here. Furthermore, a bright phase with numerous, still brighter precipitations can be seen. It is dicalciumferrite 2CaO·Fe₂O₃ with metallic iron formed by decomposition of the solid solutions of FeO and CaO. The irregular gray-colored structural constituents in Figs. 5*a*, 5*b*, and 5*c* are CaO.

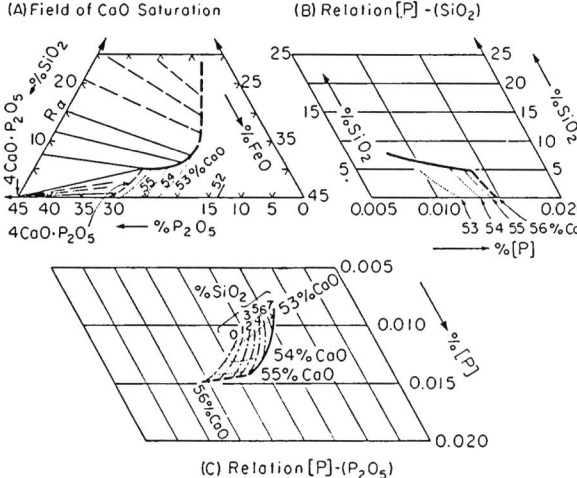

Fig. 6. The effects of slags rich in lime containing SiO₂ on the P content of iron at 1600° C.

In Fig. 6 the observed slag composition and the correlated [P] content of iron for 1600° C are summarized. Part *A* of the figure is a central projection of the field, with lime as primary phase in the concentration tetrahedron CaO—P₂O₅—FeO—SiO₂ from the CaO corner to a plane parallel to the system SiO₂—P₂O₅—FeO at the distance 55% CaO from the CaO corner. Part *B* of Fig. 6 gives the relation between (SiO₂) and [P], and Part *C*, between (P₂O₅) and [P].

In the same way as in the system without SiO₂ (Fig. 4), some experiments gave slags saturated with lime only but very near to the divariant or univariant final states.

The relations in Fig. 6 reveal that the silica in

the slags reduces the residual phosphorus in the iron. By an increase of the silica concentrations up to 8%, the content of P_2O_5 of the slag is decreased and that of FeO is increased. The (CaO) concentration shows only a small decrease. The decrease of (P_2O_5) concentration gives a [P] content in the iron lower than that in equilibrium with slags free of SiO_2.

With slags saturated with lime alone at constant (P_2O_5) concentrations, the [P] content of the metal increases with increasing (SiO_2) concentrations. In other words, the activity of P_2O_5 grows with increasing (SiO_2) content.

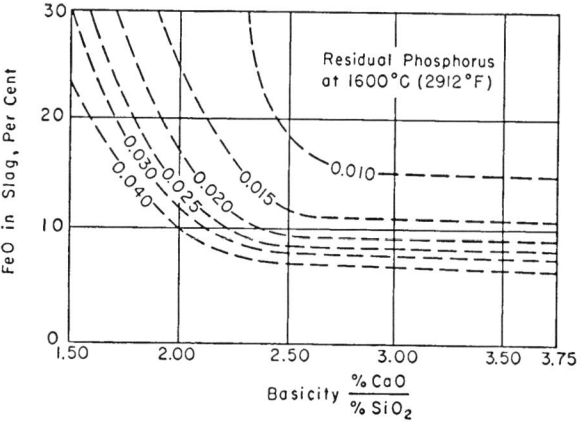

Fig. 8. Effect of lime-silica ratio and FeO on residual phosphorus (Winkler and Chipman).

Table 2. The Effect of Temperature on the Phosphorus Content of Iron Melts in Equilibrium with Lime-Saturated Slags Containing Silica*

Temperature, °C	Time, min	Phosphorus in Iron, %
1550	15	0.006
1575	15	0.006
	60	0.009
1600	15	0.009
	60	0.009
1625	15	0.010
	60	0.014
1650	15	0.016
	60	0.016
1675	15	0.018
	60	0.022
1700	15	0.022
1725	15	0.028

* Charge: 40 grams Fe + 10 grams slag of initial composition 20% FeO, 55% CaO, 16.7% P_2O_5, and 8.3% SiO_2 ($P_2O_5/SiO_2 = 2$).
P_2O_5 content of crucible 8%; SiO_2 content of crucible 4%.

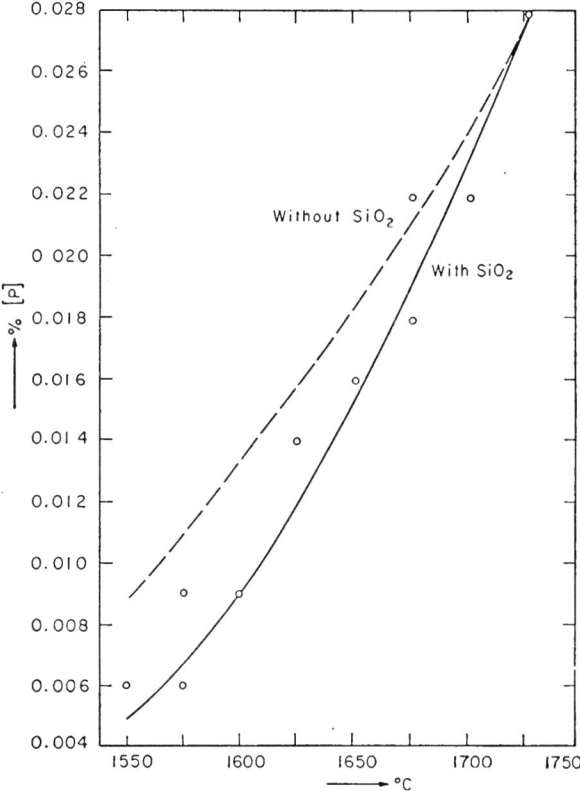

Fig. 7. Effect of temperature on the phosphorus content.

The effect of the temperature was studied for slags with $P_2O_5/SiO_2 = 2$. Table 2 and Fig. 7 present the results for temperatures from 1550 to 1725° C. The [P] concentration increases with increasing temperature. The effect of the temperature on the phosphorus content of iron is again relatively small compared with the observations under plant conditions. The average values lie below those for slags free of SiO_2, as demonstrated by the dotted line taken from Fig. 2.

To compare the results of these laboratory experiments with the conventional open-hearth practice, we have taken the curves showing approximate effect of slag lime-silica ratio (basicity % CaO/ % SiO_2) and slag FeO on residual phosphorus calculated from Winkler and Chipman[2] (Fig. 8). Some of these curves for 0.010% [P] to 0.015% [P] are shown again in Fig. 9A. In Fig. 9B curves for the

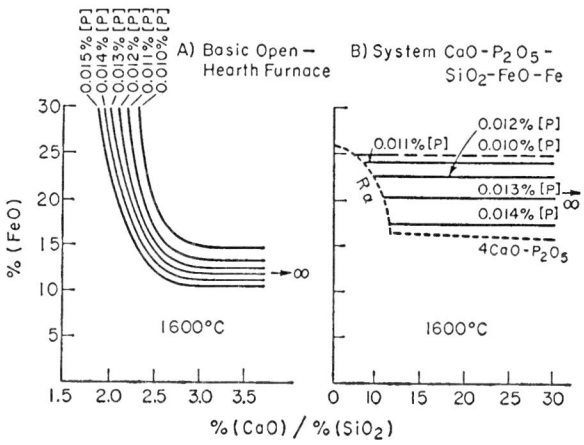

Fig. 9. The effect of (FeO) and basicity on the [P] contents in basic open-hearth furnaces and in the pure system.

same [P] content were obtained by calculating from Fig. 6. Obviously, in a first approximation the effect of (FeO) in slag and of the basicity is of the same type. In both cases at high values for the ratio % CaO/% SiO$_2$, the residual phosphorus becomes independent from basicity. In the pure system, however, [P] concentrations higher than about 0.015%, and values for the basicity lower than about 10 cannot be realized. It will be the task of future investigation to bridge this gap between our results in the pure system and the experiments made under plant conditions.

References

1. G. Trömel and W. Oelsen, *Arch. Eisenhüttenw.*, 26, 497 (1955).
2. T. B. Winkler and J. Chipman, *Trans. A.I.M.E.*, 167, 111 (1946).

Discussion

There was a short discussion among RICHARDSON, DARKEN, and TRÖMEL concerning the agreement of Trömel's data with those of Bookey.* The agreement appeared to be quite good.

*J. Iron Steel Inst. London, 172, 61 (1952).

G. Trömel
B. R. Queneau
Presiding

Section 7

Application of Fundamental Data to Process Development and Metallurgical Problems in the Steel Industry

by B. Kalling

Some Swedish Experiences of the Importance of Physical Chemistry for the Development of Process Metallurgy

The application of chemistry to the metallurgical reactions hardly started to bear fruit until the last few decades. The reason for this is natural. The metallurgical processes had already reached a high degree of perfection, and it did not appear that chemical science could do much more than confirm what experience had already taught. Owing to the high temperatures and complicated conditions under which the reactions take place, moreover, and the severe problems involved in making thermodynamic calculations, metallurgy proved to be an intractable field for research. Above all, there was a lack of the requisite experimental experience, and it was not until this deficiency began to be overcome gradually that physical chemistry acquired the importance for practical metallurgy that it now undoubtedly possesses.

Among the first attempts to apply the laws of thermodynamics to metallurgical reactions, I would mention the studies of Tigerschiöld on the oxides of iron[1] and McCance's notable experiments in the theoretical treatment of the reactions between smelted steel and slag.[2] It was, however, the fundamental research done in the thirties by such pioneering names as Chipman, Körber, Oelsen, and Phragmén—to mention only those whose work has been best known and put to practical use in Sweden—that the results gained from research could

Dr. Kalling is Director of Research, Stora Kopparbergs Berlags AB, Sweden. This paper was presented by U. Kalling.

be applied to any great extent, and with success, to the practical processes. I shall attempt to illustrate by a few examples the significance that the deeper knowledge of the physical chemistry of metallurgical processes has had for their development in Sweden.

REDUCTION OF IRON ORE

There can be hardly any country in which more work has been done in trying to discover technical solutions to the problem of reducing iron ore without smelting—into what we know as sponge iron—than in Sweden. Several processes have been elaborated, among which the Sieurin-Höganäs and Wiberg-Söderfors are the best known. A number of plants in which these processes are applied are now in commercial operation in Sweden. The technical development of the methods has to a high degree been favored by the successive growth in knowledge of the chemistry of the reactions. An illustrative example is the advent and development of the Wiberg-Söderfors process.

Wiberg invented his gas-reduction principle in 1918. At that time Baur and Glaessner[3] had already completed their determinations of the equilibrium ratio CO/CO_2 for the reduction of iron oxides, and Boudouard[4] had studied the course of the reaction $2CO = C + CO_2$ at different temperatures. Although later and more accurate studies in combination with thermodynamic calculations necessitated a not inconsiderable correction of the shape of the curves (Fig. 1), these first results nevertheless enabled the course of the reduction under

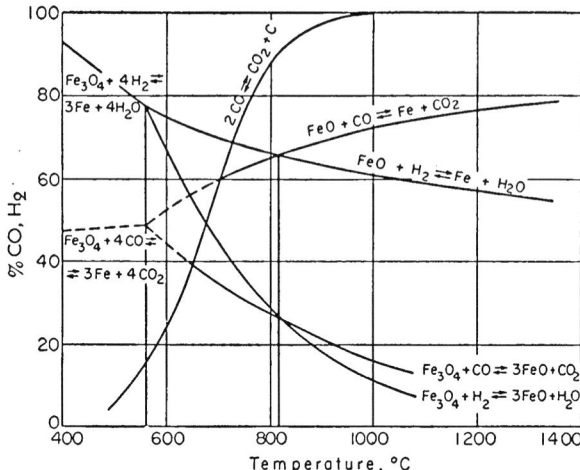

Fig. 1. Equilibrium curves Fe—C—O and Fe—H—O.[6]

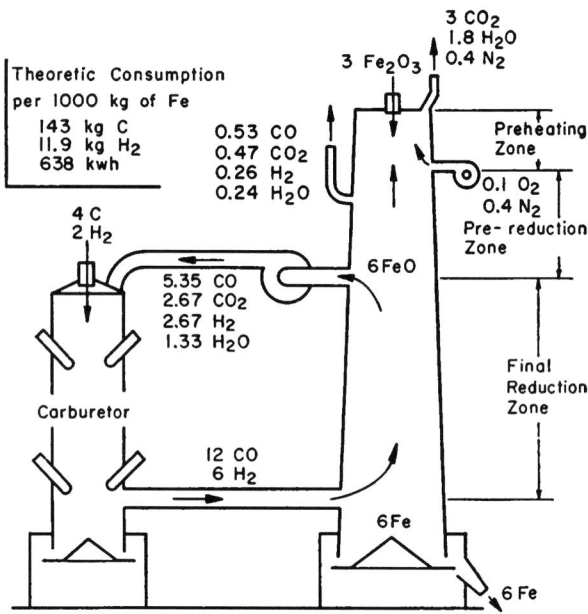

Fig. 2. Schematic view of the Wiberg gas-reduction principle.

different conditions to be anticipated with fair success. Wiberg found that under certain conditions, it should be possible to utilize in this process the entire reducing power of the carbon for iron-ore reduction, that is, to obtain a waste gas entirely free from carbon monoxide by reducing the ore separately from the coke with carbon monoxide generated by allowing part of the gas to circulate through a carburetor in which the carbon-dioxide content of the gas is restored to carbon monoxide. Since the carbon-dioxide content, which theoretically can be obtained during the final reduction stage FeO → Fe as shown in Fig. 1, does not exceed about 30% at normal reduction temperatures (around 1000° C), it is theoretically impossible to attain more than about 50% CO_2 in the waste gas when reducing from Fe_2O_3, in the event that the entire quantity of gas is allowed to pass up through the shaft, despite the fact that the CO_2 content under conditions of equilibrium is about 80% in the reduction of Fe_3O_4 to FeO and practically 100% in the reduction of Fe_2O_3 to Fe_3O_4. If, on the other hand, in accordance with Wiberg's principle, the main part of the gas is led out from the shaft at a lower level, where the reduction to FeO has already largely taken place, and only a relatively small part of the gas is allowed to continue up the shaft, it is possible to have practically all the carbon monoxide in the gas converted to carbon dioxide in the reduction of the higher oxides. The gas led out at the FeO level is carburized in an electrically heated shaft filled with coke and is thereafter returned to the bottom of the shaft.

The Wiberg principle of reduction is illustrated schematically in Fig. 2. It is a typical early example of the importance, when working on metallurgical processes, of knowing the chemistry. Though no commercial application of the method followed un-

til the 1940's at Söderfors, the practical results have been found to agree closely with those calculated by Wiberg in 1918.

Certain iron-ore reduction processes employ hydrogen as reducing agent; and often, as is normal practice in the Wiberg process, the reducing gas consists of a mixture of carbon monoxide and hydrogen. The equilibrium ratio H_2/H_2O in the reduction of iron oxides is now also well known from several investigations. In Fig. 1 the temperature curves for the hydrogen equilibrium are also drawn. In this form the diagram will undoubtedly be found on the desk of every metallurgist who is engaged in the sponge-iron problem. More complete equilibrium diagrams for the system Fe—C—O and Fe—H—O are reproduced in Figs. 3 and 4, mainly as they are summarized by Wiberg[5] and Edström.[6] The curves for the oxygen content of wüstite are, however, taken from some recent investigations by Hovgard and Jensfelt,[7] and their re-

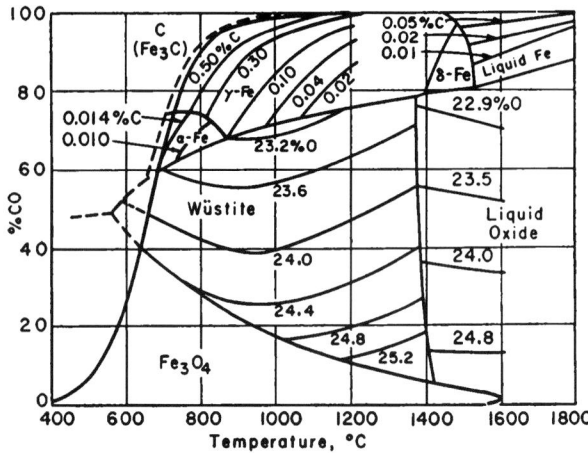

Fig. 3. Complete Fe—C—O diagram.[5, 6, 7]

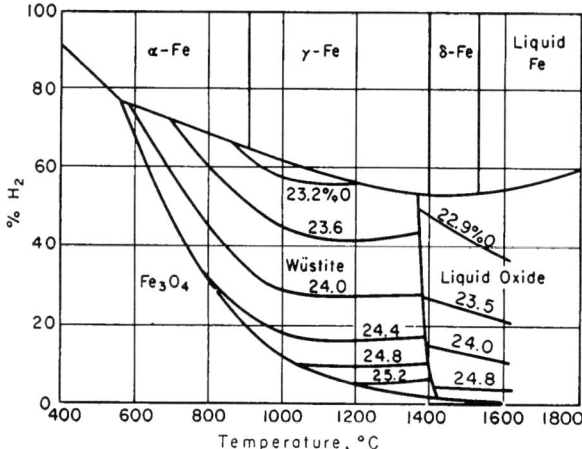

Fig. 4. Complete Fe—H—O diagram.[5, 6, 7]

sults are significantly different from former values, for example, those of Darken and Gurry.[8] These diagrams give a large amount of valuable information for the practical man. In practice, however, the conditions are far more complicated than may appear solely from these equilibrium diagrams. In practical tests in reducing various kinds of iron ores with carbon monoxide or hydrogen, it is in fact found that the speed of reaction varies in a surprising manner under different conditions. It might be expected, for example, that hydrogen with its considerably higher rate of diffusion would penetrate into a piece of ore and complete the reduction very much more rapidly than does carbon monoxide. In reality, at a normal reduction temperature of about 1000° C, it is often only the initial reduction that takes place more quickly with hydrogen, whereas a complete reduction is accomplished more quickly with carbon monoxide. An example of the course the conditions may take is illustrated in Fig. 5.

An explanation of this very important observation was also given by Wiberg.[5] From a microscopic study of partially reduced ore lumps he found that the reason for the greater final speed of reaction of carbon monoxide was that the reducing gas penetrates more easily into the cracks arising between the individual grains of ore during the early phases of the reduction than into the grains themselves. The grains will thereby gradually be entirely surrounded by a compact shell of reduced iron. During reduction with carbon monoxide, carbon diffuses through this film of iron and, as a result of the reaction between this carbon and the enclosed wüstite phase, carbon monoxide is formed under such a high overpressure that the iron shell bursts, permitting the carbon monoxide to penetrate and continue the reduction. If the reduction is performed with hydrogen, on the other hand, no corresponding overpressure can be expected, the shell does not burst, and the reaction is therefore delayed. If a mixture of carbon monoxide and hydrogen is used, however, the reduction can be completed more quickly, as is apparent from Fig. 5.

Some recent Swedish studies that deserve mention in the field of ore reduction are those performed by Edström.[6] He made a thorough study of the reduction process in natural, single crystals of hematite and magnetite with carbon monoxide and hydrogen, investigating especially the causes underlying the long-established experience that hematite ores are generally reduced very much more quickly than magnetites and that the latter, after oxidation to Fe_2O_3, are considerably more easily reducible. The result of one of his experiments with carbon monoxide is given in Fig. 6. By microscopic study

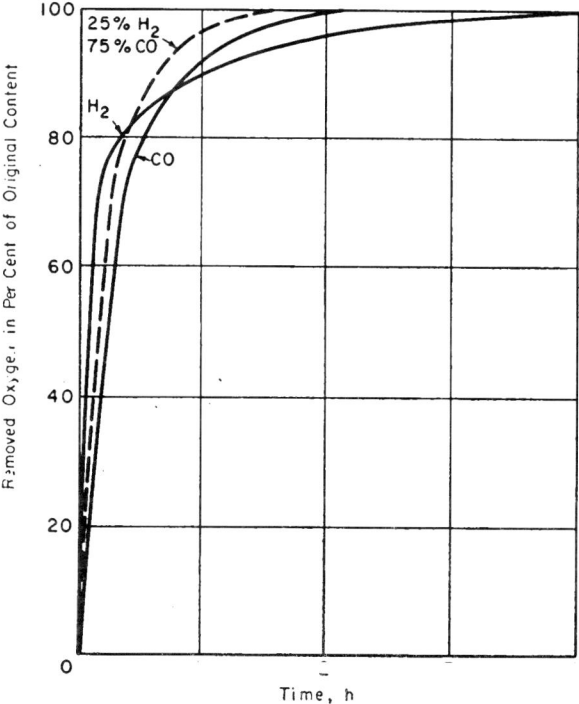

Fig. 5. Reduction of ore lumps with CO, H_2, and CO + H_2.[5]

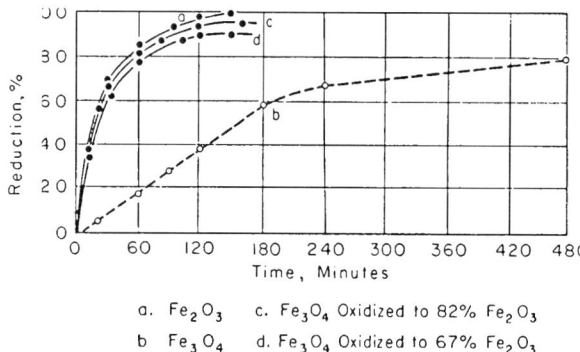

Fig. 6. Reduction of single crystals of Fe_2O_3 and Fe_3O_4.[8]

a. Fe_2O_3 c. Fe_3O_4 Oxidized to 82% Fe_2O_3
b. Fe_3O_4 d. Fe_3O_4 Oxidized to 67% Fe_2O_3

of partially reduced crystals he found, as did Wiberg also, that the reduction of magnetite to wüstite occurs without change of volume under the formation of a completely dense wüstite phase. In order that the reduction may continue under formation of metallic iron, therefore, a diffusion process is required in this solid phase which must have a restrictive effect on the reaction rate. It may be of interest to mention the fact established by Edström and Bitsianes[9] that the transport through the wüstite layer is accomplished by diffusion of iron and not of oxygen. If the starting material instead consists of hematite, a dense magnetite is admittedly formed first, but as the reduction continues, the magnetite —obviously owing to certain defects in the lattice structure—is in this case transformed into a stalky, often greatly swelling, porous wüstite. By this means the transport of gas is promoted to a high degree, and the reduction to metallic iron takes place more rapidly.

These experiences also provide an answer to the question of why hematite ore has proved to have so much greater a tendency than magnetite ores to decompose during reduction, a knowledge of which is, of course, very important in the blast-furnace process. In the reduction to sponge iron, especially when done in a shaft furnace, the properties of the ore in this respect are often determinative of its usability. It must be observed that the iron in all sinter occurs chiefly in the form of Fe_2O_3 and that, for this reason, the sinter may be very prone to decompose in the reduction furnace. This applies especially to sinters of very rich concentrates and, above all, to pellets. Certain additions to the ore prior to sintering, however, may substantially improve the result. Edström studied the effect of a small addition of lime to the strength during reduction.[10] He found that, under the prevailing conditions, a calcium-ferrite phase $CaO \cdot Fe_2O_3$ is formed which is directly reduced to metallic iron and free CaO without passing the FeO stage. The formation of the porous wüstite is thereby avoided, and a fairly compact sponge iron is obtained. The addition of even 1 or 2% of lime greatly improves the strength during reduction. The lime does not appear to have any restrictive influence on the speed of reaction.

BLAST-FURNACE PROCESS

The importance of the research regarding the reaction process in the blast furnace is now generally known and acknowledged. The reactions between iron and slag, in particular, have recently been the subject of extensive studies, among which the many valuable papers by Chipman and Grant and co-workers at M.I.T. should be specially mentioned. Of great interest to practical metallurgy has been, among other things, the elucidation of the condition of manganese and sulfur in the blast furnace, about which I shall say only a few words.

Practice has generally been to so adjust the blast-furnace burden as to obtain a substantial manganese content in the pig iron. The reason for this may be only to obtain the manganese content in the pig iron that is considered necessary for the subsequent steel process. But the opinion is also common that the manganese content has important functions to fulfill in the blast-furnace process as well, primarily perhaps for the purpose of diminishing the sulfur content. It is known that high manganese in the pig iron, even after tapping, may under certain conditions lead to a considerable desulfurization by precipitation of a phase rich in manganese sulfide in the ladle or mixer. The circumstances under which such a desulfurization may occur has been elucidated by Oelsen.[11] He shows, for example, that if the temperature of the pig iron falls to 1300° C, the maximum solubility for sulfur in it is about 0.1% at a manganese content of 1.0%. Generally, however, this saturation limit is seldom reached nowadays, so that the sulfur content obtained will be exclusively dependent upon the composition of the blast-furnace slag and other conditions in the blast furnace. Basic slag and strongly reducing conditions have the effect of lowering the sulfur content, but as to the influence of manganese, investigations by Grant, Kalling, and Chipman[12] and by Grant, Dowding, and Murphy[13] have indicated that the presence of manganese does not tend to a lowering of sulfur under conditions of equilibrium. This is revealed by, among other things, the fact that the remaining manganese content in the slag at the equilibrium is extremely small in the basic slags used nowadays in the blast-furnace process. In addition, the presence of manganese causes a not inconsiderable lowering of the sulfur activity in the pig iron, as shown by Sherman and Chipman.[14]

It is, however, not possible to apply these results directly in practical operation. In Domnarvet we have, for instance, tried to lower the original manganese content in the pig iron from about 1%, and we have found that a marked decrease in manganese could be achieved without increasing the sulfur content. At a manganese content of about 0.5%, however, it was found difficult to keep sulfur down. At this low manganese level the furnace worked more irregularly, and therefore the manganese content had again to be raised, with consequently improved result.

As Chipman and his co-workers have shown, the sulfur distribution ratio in the blast furnace is up to ten times lower than that which corresponds to equilibrium, and therefore it must be of importance not only that the slag has the highest possible basicity but also that it is sufficiently fluid. The beneficial influence of manganese thus must be attributed to the way it affects the melting point and viscosity of the slag.

When operating burdens low in alumina, it has often been found that an increase in the Al_2O_3 content to a great extent has favored desulfurization. The conditions are here presumably analogous to those with manganese. An increased Al_2O_3 content in the slag cannot have any favorable influence on the sulfur equilibrium value itself, but by making the slag more fluid and reactive the reaction may proceed closer to equilibrium. The same can be said about an increase of MgO in burdens low in magnesia.

EXTERNAL DESULFURIZATION OF PIG IRON

I want to say a few words also about the desulfurization of pig iron outside the blast furnace with powdered quicklime by a method elaborated at Domnarvet[15, 16] and now adopted at a number of steel works. This is an illustrative example of a method in which the reaction process has proved capable of reliable calculation from known thermochemical data. The reactions which may occur are the following:

$$CaO + FeS + C = CaS + Fe + CO \qquad (1)$$

or

$$4CaO + 2FeS + Si = 2CaS + 2Fe + Ca_2SiO_4 \qquad (2)$$

The treatment of reaction 1 is simplified by the fact that the activities of both CaO and CaS may be put equal to unity, since the solubility of the two substances in one another has been found to be extremely slight. Making use of the available data[17, 18] for the reactions

$$CaO(s) + H_2S(g) = CaS(s) + H_2O(g)$$

$$H_2(g) + [S] = H_2S(g)$$

$$\tfrac{1}{2}H_2(g) = [H]$$

$$H_2O(g) = [2H] + [O]$$

$$[C] + [O] = CO(g)$$

$$C_{gr} = [C]$$

Eketorp[19] obtained the following value of $\Delta G°$ for reaction 1:

$$CaO(s) + S + C_{gr} = CaS(s) + CO(g)$$

$$\Delta G° = 26,830 - 26.71T$$

When

$$\log K = -\frac{\Delta G°}{4.575T}$$

then

$$K_{1300°} = \frac{a_{CaS} \times p_{CO}}{a_{CaO} \times a_S \times a_{C_{gr}}}$$

or

$$\frac{1}{a_S} = \frac{1}{f_S \times \%} = 128.8$$

According to Chipman,[20] at a carbon content of 4% in the pig iron, $f_S = 3.3$, and the S content at equilibrium will at 1500° C be

$$S = 0.0024\%$$

This value is in good agreement with practical results, which indicate that a sulfur content of this order of magnitude can be attained with carbon as the active reducing agent.

Normally, however, the pig iron contains silicon, and Eketorp has calculated that, in a pig iron con-

taining 0.3% Si, the sulfur content at equilibrium as in reaction 2 amounts only to

$$S = 0.000025\%$$

Thus the S content in siliceous pig iron in the state of equilibrium is extraordinarily low and can, of course, never be attained under practical conditions; furthermore, carbon often takes part in the reaction even in the presence of silicon. At higher sulfur contents, a not inconsiderable development of carbon monoxide is observable at silicon contents up to 0.3–0.4%, but with lower sulfur, below 0.1%, the carbon reaction does not start until the silicon falls below about 0.05%. The reason why the carbon participates in the reaction at all must be that the concentration of carbon in the iron is so much higher than that of silicon, and therefore, an always considerably greater quantity of carbon than of silicon is available at the site of reaction.

The efficiency of solid lime as a desulfurizing agent is to a high degree related to the fact that its activity remains unchanged at unity until all lime has been consumed, that is, transformed to CaS and, during reduction with silicon, of course, partly also to Ca_2SiO_4. To illustrate this fact, it may be mentioned that the sulfur content in a pig iron used in a laboratory experiment was lowered from 0.1 to 0.025% by treating it with a mixture of 90% CaS and 10% CaO in the presence of graphite. After the treatment the powder had a sulfur content of 38.6%. In another case the experiment was started with a sulfur content in the pig iron of 0.01%, which remained unchanged during the experiment when treated with a mixture of CaO and CaS containing 35.3% S.

In order to be able to utilize fully the desulfurizing action of the lime in large-scale production, however, a thorough mixing was found to be necessary. In plants erected earlier the process is performed in a rotary furnace revolving at high speed. Sulfur contents of 0.001–0.002%, and regularly below 0.005%, have been achieved when treated in this way for about 15 min. Powdered lime may find use also for desulfurization of iron in solid form, so long as the iron is sufficiently fine grained. Among other things, this method can be used to advantage for the refining of sulfurous sponge iron in a rotary furnace at about 825° C.

DECARBURIZATION OF SOLID PIG IRON

Before discussion of the ordinary steelmaking processes, a few words about a method where the pig iron is decarburized in the solid state, the RK process[21] may be of interest. Here the granulated pig iron is passed continuously through a rotary furnace, where it is heated to 1000 to 1100° C in an oxidizing atmosphere. The carbon content can in this way be brought down to very low values.

The gas composition has to be regulated so that the carbon is oxidized to the degree wanted, but so that at the same time the iron is protected from oxidation. The circumstances under which both these conditions can be fulfilled can be seen directly in the diagram shown in Fig. 3, which shows that at 1000° C and with a gas containing about 75% CO and 25% CO_2 it will be possible to decarburize a carbon containing iron to about 0.04% C. If $p_{CO + CO_2}$ is lowered from 1 atm to 0.25 atm, for instance, by dilution with nitrogen, the carbon content can be brought down to about 0.01% without danger of oxidizing the iron. In the practical operation of the process it was found relatively easy to get a suitable gas composition in the charge even if the gas on top of the charge was completely burnt by the air supplied. It was only necessary to keep the speed of rotation of the furnace sufficiently high. In this way it was possible to bring about the desired decarburizing effect merely by blowing air toward the surface of the charge, and the heat produced in this way by burning the carbon to carbon dioxide was very nearly sufficient for heating the granulated pig iron to reaction temperature. The process can thus be called a Bessemer process in the solid state.

If the pig iron contains silicon, the equilibrium conditions are changed because silicon under the prevailing conditions has a considerably greater affinity for oxygen than has carbon. At the surface of a grain, silicon will therefore be oxidized in the first place, and not before the silicon content has then dropped to a very low value will the oxidation of carbon begin. The determining factor for the reaction is, however, primarily the diffusion velocity in pig iron of carbon and other elements, which will be oxidized when they reach the surface of the particle. Because carbon diffuses considerably more rapidly than silicon, the net result will be that the carbon will become oxidized almost completely before the silicon content in the center of the particles has decreased appreciably. It is therefore only in the surface layer that a more marked formation of silica can be observed.

It was natural to try to apply a process of this kind, which gives possibilities of oxidizing carbon before silicon, also on pig iron alloyed with other elements having higher affinity for oxygen than carbon has. It was of particular interest to determine if a pig iron containing chromium could also be decarburized in this way. Tests showed that pig irons with chromium contents at least up to 15% Cr could be decarburized to below 0.05% C without any excessive oxidation of chromium. At higher Cr contents the process was made more difficult by a beginning passivation of the surface. Addition of nickel made the conditions definitely worse, apparently because carbon in this case has to diffuse through austenite in which the diffusion velocity *is appreciably lower than in ferrite.* Extensive practical tests have been carried out with this process, but it has not as yet been found possible to use this principle for production of low-carbon chromium

alloys suitable as raw material for stainless steel. Neither did similar tests with pig irons alloyed with manganese give technically satisfying results.

THE ACID OPEN-HEARTH PROCESS

The comprehensive experimental results published in the early 1930's from research in the chemistry of steelmaking found their first application in Sweden in the acid open-hearth process. At that period Körber and Oelsen[22] published their important investigations on the equilibria of the silicon and manganese reactions in slags saturated with silica. A diagram from the work of Körber and Oelsen is reproduced in Fig. 7. Since at least

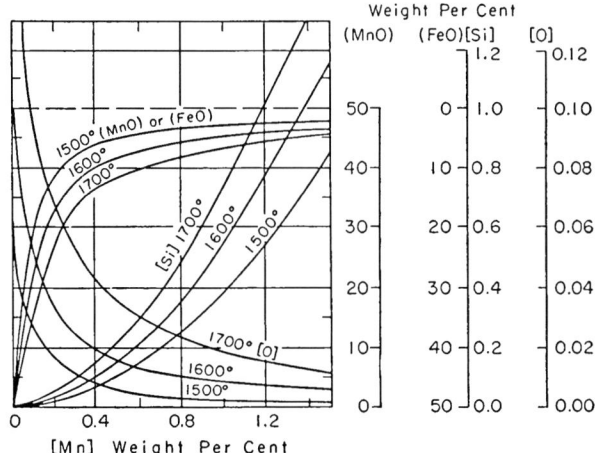

Fig. 7. Mn and Si in equilibrium with silica-saturated slags. According to Körber and Oelsen.[22]

during the later stage of the acid open-hearth process, the slag may be considered as saturated with silica and the furnace lining consists of pure silica, as was the crucible material in Körber's and Oel-

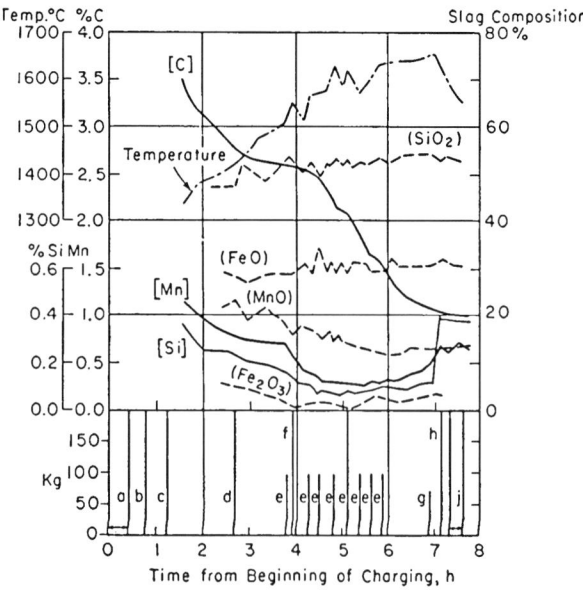

Fig. 8. Swedish acid open-hearth heat.[24]

sen's experiments, a number of factors affecting the course of the reactions could be elucidated simply by comparing the research findings with the conditions obtained in practice. At the same period Vacher and Hamilton's results on the carbon-oxygen equilibrium in molten steel were published,[23] which added greatly to our understanding of the chemistry of the processes. I will briefly touch upon the results that these various investigations at that time led to.[24, 25]

Figure 8 shows the course of a fairly hot-running acid open-hearth charge in accordance with old Swedish practice. The composition of the steel and slag bath immediately prior to deoxidation was as listed, the equilibrium values found by Körber and Oelsen for the same manganese content in the steel being given in the right-hand column for the sake of comparison.

Material	Open-Hearth Heat, %	Equilibrium
Steel: C	1.08	0.03
Si	0.18	0.02
Mn	0.06	0.06
Slag: FeO	31.3	32
MnO	11.7	18
SiO₂	57.0	50

As is seen, both carbon and silicon contents are in this heat considerably higher than the corresponding equilibrium values. First as to carbon, it is obvious that the carbon reaction is very sluggish. According to Phragmén,[26] the reason for this is that the reaction takes place under formation of a gaseous phase, chiefly carbon monoxide. In order that a carbon-monoxide bubble may be newly formed in a homogeneous bath, an immense overpressure is required. It cannot be expected, therefore, that any appreciable evolution of gas will take place inside the melt in the open-hearth furnace, nor in the contact surface between the steel and the molten slag. In the bottom of the furnace, however, the conditions for generation of carbon monoxide are much more favorable. It must be presumed, for one thing, that gas generation is greatly facilitated by the porosities always to be found in the bottom lining, since no new formation of gas bubbles need occur but only a growth from an already existent gaseous phase. Irregularities in the bottom of the furnace, moreover, may conceivably assist in the formation of gas bubbles. The conclusion was reached, therefore, that the normal boil takes place through the generation of gas bubbles at the bottom of the furnace, which thereafter rapidly grow during their passage through the bath up to the surface. That this must be true was apparent, moreover, from some of the experiments carried out by Körber and Oelsen.

The sluggishness of the carbon reaction naturally

affects the course of the other reactions between the steel and slag baths. The only feature that will be dealt with here is the usual Swedish practice of re-reducing silicon into the steel towards the end of the heat. In the heat diagram in Fig. 8, a re-reduction of silicon to 0.18% occurs, as is normal in Swedish practice, despite the fact that the silicon content in equilibrium with the slag would not have amounted to more than about 0.02%. The re-reduction, therefore, cannot possibly have occurred from the slag during the lively stirring caused by the boil that prevailed. At the bottom of the furnace, however, the conditions are entirely different. There a steel bath containing about 1% C is in contact with pure silica. The reaction

$$2C + SiO_2 = Si + 2CO$$

must thus take place at the prevailing temperature. The state of equilibrium for this reaction lies at several percentage of Si in the steel bath. The silicon reduction, therefore, clearly derives from the bottom of the furnace, and the fact that the silicon content does not attain higher values is due to re-oxidation of silicon when it comes into contact with the slag bath. During the acid open-hearth process, consequently, especially in its final phases, there is a continuous migration of Si from the furnace lining to the slag via the steel bath. The silicon content of the steel will accordingly depend on the carbon content, the temperature, and the oxidizing power of the slag.

Through the knowledge of the course of the reactions, moreover, it was found possible to elucidate the factors affecting the composition of the slag in the acid open-hearth process. In the analysis of the slag in the preceding example, the FeO content is about 30%, which is quite normal. The silica content is usually between 50 and 60% and is thus at least close to the saturation limit. Among other slag components, the manganese content predominates, amounting as a rule to about 10 to 15%. Although the high iron content in the slag usually remains practically unchanged throughout the process, as in this case, it is often possible to get the boil virtually to stop toward the end of the heat, even with a fairly high residual carbon content in the steel. The reaction power of the slag ceases. Since the composition of the slag is not otherwise changed than by an insignificant elevation of the silica content, the diminished reaction capacity of the slag is clearly related to an increased viscosity caused by the fact that the saturation limit for silica has been exceeded. The iron content of the slag under the existing conditions appeared to be of subordinate importance. Since, according to Körber's and Oelsen's diagram, a FeO content in the slag of only 10% should be fully sufficient to produce the required oxidizing effect, experiments were made with a view both to raising the manganese content in the heat and also, by the addition of lime to the slag, to attain silica saturation even in slags with lower FeO content. In two such experiments, slags were obtained toward the end of the heat with the analysis shown.

Material		Heat 1	Heat 2	Heat 3
Slag:	FeO	25.1	7.6	13.8
	Fe₂O₃	0.4	0.0	0.3
	MnO	15.1	35.6	15.7
	CaO	0.7	1.8	7.2
	MgO	1.1	2.7	1.5
	Al₂O₃	1.4	1.4	1.6
	SiO₂	56.2	50.9	60.1
Steel:	C	0.31	0.45	0.64
	Si	0.09	0.23	0.136
	Mn	0.09	0.48	0.14

The boiling speed was lowest in heat 1 and highest in heat 3. The result proved, therefore, that slag of a sufficiently oxidizing character is obtainable at FeO contents below 10% by the introduction of other bases such as MnO and CaO. It may be of interest to mention that the slag obtained in heat 2 with its very low FeO content is in close conformity with that resulting from the now abandoned Swedish acid Bessemer process, in which the pig iron normally contained

Si   About 1%

Mn   About 3%

Owing to the higher level of saturation limit for SiO₂ in the presence of lime, a considerable lowering of the iron content in the slag is obtained even for a comparatively low addition of CaO.

THE BASIC OPEN-HEARTH PROCESS

The reactions in the basic open-hearth process are today much more clearly known than before as a result of the comprehensive research work that has been done in this field during the last decades. The difficulties of carrying out reliable experiments are, however, much greater here than in the acid process. As an example of the importance of increased knowledge of the chemistry of the process I want only to touch upon the manganese problem.

Among steel producers it is generally considered that a certain manganese content in the bath is necessary for obtaining a satisfactory working of the heat. A high manganese content, for instance, has been commonly believed to cause a lower oxygen content in the steel. In fact, however, it has been

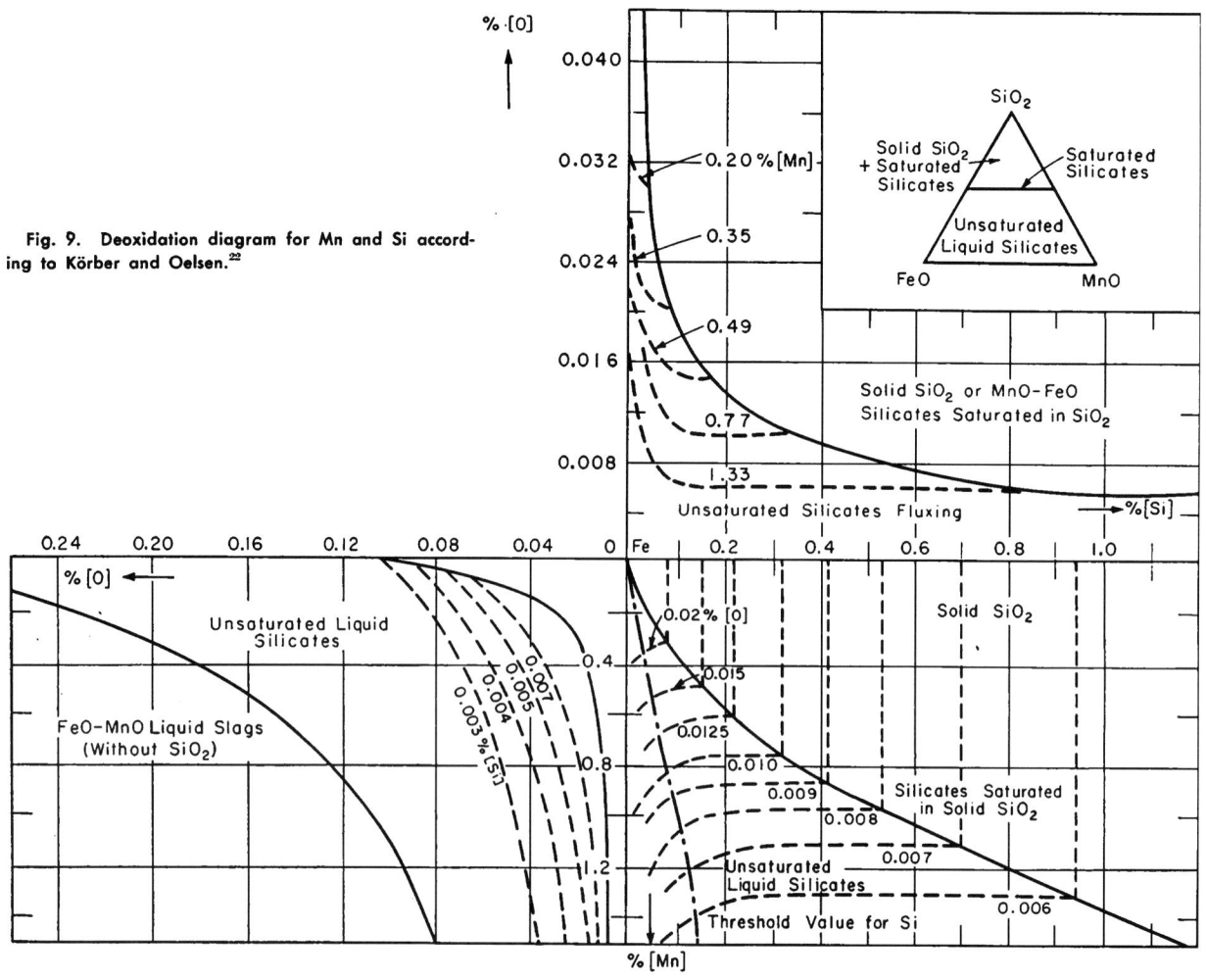

Fig. 9. Deoxidation diagram for Mn and Si according to Körber and Oelsen.[22]

clearly established that the manganese content of the steel cannot influence the oxygen content at the conditions prevailing in the basic open hearth. The oxygen content is primarily fixed by the carbon. A large number of investigations have been carried out in different places in order to make clear in what way the manganese can influence the course of the process. Of the work done I will here only mention a paper by Fornander.[27] By studying a number of heats with varying manganese content and taking special care to obtain reliable samples for determining the oxygen content, Fornander was able to state, as could be expected, that no relationship exists between manganese and oxygen in the steel. Some observations were made, however, indicating that manganese, as in the blast-furnace process, can have a favorable effect on the physical properties of the slag and notably on its viscosity. The sulfur-distribution ratio does not seem to be influenced by the manganese content, according to different investigations.

DEOXIDATION AND TAPPING

Already about 1900 Brinell[28] made comprehensive statistical studies regarding the contents of manganese and silicon necessary for making killed steel. He came to the result that on the average 5.2 weight percentage of manganese gave the same effect as 1% silicon. McCance's first attempts to treat these reactions thermodynamically seemed to confirm the investigations of Brinell, which had attracted much attention. Later investigations, however, based on experimental results, did not show a very good agreement. It was found, for instance, that manganese, in the absence of silicon, was a very weak deoxidizer. According to the data of Körber and Oelsen mentioned before, about 2% manganese is needed for bringing the oxygen content in the steel as far down as with only 0.02% Si.

The reason for this astonishing difference is today generally understood from the work by Körber and Oelsen and from later investigations by, among others, Hilty and Crafts.[29] The manganese reaction, specifically, is influenced to an unexpected degree by the presence of silicon in the system. At first hand this seems to be a result of the fact that MnO is bound relatively strongly in the slag as silicate, whereby its activity is considerably decreased. In the complete deoxidation diagram for silicon and manganese by Körber and Oelsen (Fig. 9), it is possible to get a good idea of how the oxygen content in the steel is influenced by different combinations of silicon and manganese. This diagram may give an explanation of Brinell's results, his conclusions being drawn from the acid open hearth and thus was based on steel never being quite free of silicon. The diagram shows that a silicon content of 0.2% has a corresponding oxygen content at saturation of silica of about 0.013%. At a manganese content of 1.04% (for example, 5.2 × Si content), the presence of not more than about 0.01% Si is necessary to arrive at the same oxygen content; that is, Brinell's data should be correct for an assumption that the Si content in the steel had been about 0.01%.

Among metallurgical problems which have been thoroughly studied in Sweden, I would like in closing to mention the chemical and physical changes during solidification of rimming steels. I want only to refer to the valuable work by Hultgren and Phragmén in this field,[30] which I think is well known also in the United States. A profound study of the chemistry of the reactions was a prerequisite for solving the problems involved.

SUMMARY

With the scattered examples given above, I have tried to stress the importance of enlarged knowledge of process chemistry for the development of metallurgical processes. It may perhaps be rather seldom that an idea for a new process is created by the study of the reactions in the laboratory or by purely thermodynamical calculations, but it is, on the other hand, nowadays inconceivable to try to start or to develop a technical method without access to the results from the extensive research work that has been done in this field during the recent decades.

References

1. M. Tigerschiöld, *Jernkontorets Ann.*, **107**, 67 (1923).
2. A. McCance, *Trans. Faraday Soc.*, **21**, 176 (1925).
3. E. Baur and A. Glaessner, *Z. physik. Chem.*, **43**, 354 (1903).
4. M. O. Boudouard, *Ann. chim. et phys.*, **24** (7th Series), 5 (1901).
5. M. Wiberg, *Jernkontorets Ann.*, **124**, 179 (1940).
6. J. O. Edström, *J. Iron Steel Inst. London*, **175**, 289 (1953).
7. N. A. Hovgard and P. N. Jensfelt, *Jernkontorets Ann.*, **140**, 467 (1956).
8. L. S. Darken and R. W. Gurry, *J. Am. Chem. Soc.*, **67**, 1398 (1945).
9. J. O. Edström and G. Bitsianes, *J. Metals*, **7**, 760 (1955).
10. J. O. Edström (in English), *Jernkontorets Ann.*, **140**, 101 (1956).
11. W. Oelsen, *FIAT Rev. Ger. Sci., 1939–1946*, **27**, Inorganic Chemistry, Washington, (1948). 212
12. N. J. Grant, U. Kalling, and J. Chipman, *Trans. A.I.M.E.*, **191**, 672 (1951).
13. N. J. Grant, J. W. Dowding, and R. J. Murphy, *J. Metals*, **5**, 1451 (1953).
14. C. W. Sherman and J. Chipman, *J. Metals*, **4**, 597 (1952).
15. B. Kalling, C. Danielsson, and O. Dragge, *J. Metals*, **3**, 732 (1951).
16. S. Fornander, *J. Metals*, **3**, 739 (1951).
17. T. Rosenqvist, *J. Metals*, **3**, 535 (1951).
18. *Basic Open Hearth Steelmaking*, p. 638, Second Edition, Physical Chemistry of Steelmaking Committee, A.I.M.E., New York, 1951.
19. S. Eketorp, *Rev. mét.*, **52**, 718 (1955).
20. J. Chipman, *Metal Prog.*, **62**, 97 (1952).
21. B. Kalling and I. Rennerfelt, *J. Iron Steel Inst. London*, **140**, 137 (1939).
22. F. Körber and W. Oelsen, *Mitt. Kaiser-Wilhelm Inst. Eisenforsch. Düsseldorf*, **15**, 271 (1933).
23. H. C. Vacher and E. H. Hamilton, *Trans. A.I.M.E.*, **95**, 124 (1931).

24. B. Kalling, *Jernkontorets Ann.*, **118** (T.D.), 69 (1934).
25. B. Kalling and N. Rudberg, *Jernkontorets Ann.*, **121**, 93 (1937).
26. G. Phragmén, *Thermodynamics*, Stockholm, A.-b. Nordiska bokhandeln, Sweden, 1931.
27. S. Fornander, *Discussions Faraday Soc.*, No. 4, 296 (1948).
28. A. Wahlberg, *Jernkontorets Ann.*, **56**, 210 (1901).
29. D. C. Hilty and W. Crafts, *Trans. A.I.M.E.*, **188**, 425 (1950).
30. A. Hultgren and G. Phragmén, *Trans. A.I.M.E.*, **135**, 133 (1939).

Discussion

RICHARDSON commented that the level to which one could desulfurize iron or steel was governed primarily by the oxygen level in the system. This in turn was controlled by the deoxidizer present, which might be carbon or silicon in the usual cases. In the discussion of whether silicon or carbon would control the oxygen level in the desulfurizing system of Kalling's, CHIPMAN pointed out that only at the relatively low temperatures would silicon be more reducing than carbon in the usual concentrations of carbon encountered. DARKEN proposed the use of the temperature dependence of the silicon reduction so that one could get a better reaction by first reducing silicon to near the equilibrium level at a high temperature. Then by lowering the temperature, one would get a relatively large increase in the reducing capacity of silicon, which would then be higher than that of carbon. The second suggestion was that one first reduces silicon from a fairly acid slag and then uses it with a basic slag for desulfurization, which in turn increases the effectiveness of silicon as a reductant.

FISCHER described an experiment in which the reaction of carbon and sulfur in iron was studied. In a crucible of CaO plus 10% CaF_2 and an "equilibrium atmosphere," the product of [% C] [% S] equaled 0.011 ± 0.002. The carbon range was from 0.15 to 4.5%. Additions of silicon to the equilibrated bath caused a sharp decrease in sulfur content. DARKEN recalled that the strong influence of silicon on desulfurization was also noted by Trentini, Wahl, and Allard, in their paper on desulfurization of iron, as a result of blowing powdered lime through the bath in a Bessemer-like vessel with a blast of nitrogen. This was reported at the 1956 Annual Meeting of the A.I.M.E. ST. PIERRE added that Trentini, Wahl, and Allard's system was favorable to desulfurization because the passage of nitrogen tended to purge the system of CO.

KING raised the question whether the effect of manganese on the slag viscosity was of primary significance. Actually, the ratio of manganese in the metal to MnO in the slag was a good indicator of the oxygen level in the system, and perhaps the effect of manganese as an oxidizer in the slag should be considered. QUENEAU stated that in some production tests, the rate of desulfurization is not particularly influenced by the manganese of either the slag or the metal. With a low slag-to-metal volume, it took 4 to 5 hr for sulfur to reach equilibrium. However, additions of aluminum completed the desulfurization in a very few minutes. Titanium did the same thing. In 15 min they were able to get up to a ratio (S)/[S] of 200 to 1. He also observed that even after careful grinding and magnetic separation of several cycles, their slags still contained some metallic iron; these results raised the question of the true iron-oxide content of a slag. CHIPMAN commented that the work reported by Hatch had shown the same thing.

by G. St. Pierre

Gas Utilization in the Reduction

of Iron Oxides

INTRODUCTION

The purpose of this paper is to summarize calculations of the quantity of reducing gas required to produce a given amount of free iron from its oxide. Gaseous reduction of iron oxide is usually accomplished with mixtures of CO and H_2. The chemical reactions involved have been thoroughly defined to be:

$$FeO \quad + \frac{CO}{H_2} = Fe \quad + \frac{CO_2}{H_2O} \quad (1)$$

$$Fe_3O_4 \quad + \frac{CO}{H_2} = 3FeO \quad + \frac{CO_2}{H_2O} \quad (2)$$

$$3Fe_2O_3 + \frac{CO}{H_2} = 2Fe_3O_4 + \frac{CO_2}{H_2O} \quad (3)$$

The reduction is normally carried out below the melting point of each iron phase. At temperatures below 560° C, wüstite (represented as FeO) is not stable, and the reduction of Fe_3O_4 proceeds directly by reaction 4:

$$Fe_3O_4 + 4\frac{CO}{H_2} = 3Fe + 4\frac{CO_2}{H_2O} \quad (4)$$

The thermodynamics of reactions 1 through 4 are well known. Figures 1 and 2 which are taken from Edstrom[1] summarize the equilibrium data. The equilibrium for reaction 3 lies far to the right at all temperatures, and either CO or H_2 is very efficient in the reduction of Fe_2O_3 to Fe_3O_4; however, reactions 1, 2, and 4 are not so favorably inclined. In each case appreciable quantities of H_2 and CO remain after the reaction has ceased. For Fe_3O_4 to

Dr. St. Pierre was with the Inland Steel Company, East Chicago, Indiana. Now he is with the Department of Metallurgical Engineering, Ohio State University, Columbus, Ohio.

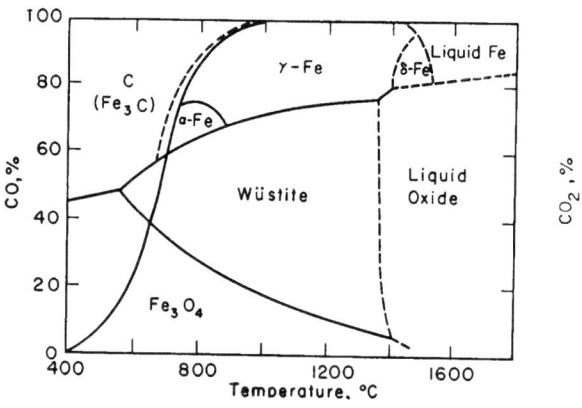

Fig. 1. Iron-oxygen-carbon system.

FeO, the utilization of CO and H_2 is greatly improved by increased reaction temperature. For the case of FeO to Fe, the situation is a little different because increased reaction temperature leads to greater H_2 utilization but lesser CO utilization.

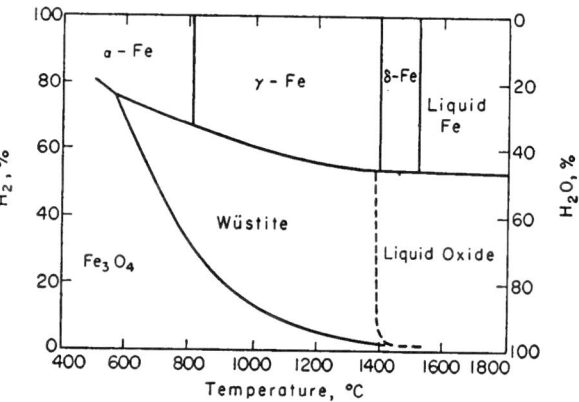

Fig. 2. Iron-oxygen-hydrogen system.

Total pressure has no effect on the equilibrium of each reaction. Consideration of the data of Figs. 1 and 2 indicates that there must exist optimum temperature conditions for the maximum utilization of CO—H₂ mixtures.

CALCULATIONS

Stoichiometric Relations

As previously stated, the reduction of Fe_2O_3 to Fe proceeds in three steps. The reduction of a mass of Fe_2O_3 to Fe is schematically illustrated in Fig. 3.

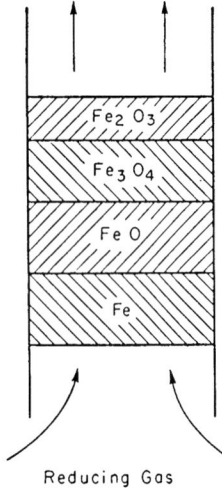

Fig. 3. Schematic representation of gaseous reduction of iron oxide.

If Fe_3O_4 is reduced below 560° C, the FeO field is absent. At each step a particular quantity of oxygen has to be transferred from the solid to the gas phase in order to move iron to the oxide state. For the reduction of 0.5 mole Fe_2O_3 to 1.0 at. wt Fe, oxygen must be removed in each step as follows:

(1) From FeO to Fe, 1.00 at. wt oxygen removed.
(2) From Fe_3O_4 to FeO, 0.333 at. wt oxygen removed.
(3) From Fe_2O_3 to Fe_3O_4, 0.167 at. wt oxygen removed.

If the three reduction steps are completely separated, the fresh reducing gas first contacts Fe and then passes through the successively higher oxides. For the production of one at. wt of iron, the gas must have gained 1 at. wt oxygen from the solid when it reaches the FeO—Fe border, 1.333 at. wt oxygen when it reaches the Fe_3O_4—Fe border, and 1.50 at. wt oxygen when it reaches the Fe_2O_3—Fe_3O_4 border. If Fe_3O_4 is reduced below 560° C, steps 1 and 2 are combined, and in transforming 1 at. wt iron from Fe_3O_4 to Fe, 1.333 at. wt oxygen must be removed.

Equilibrium Relations

The equilibrium relations have already been summarized in Figs. 1 and 2. When the gas reaches

each of the borders indicated on Fig. 3, the gas must contain CO and H₂ in proportion to CO_2 and H₂O to at least meet the equilibrium requirements. There can be an excess of CO and H₂ at any border but never a deficiency.

Combined Relations

For a given set of temperature conditions the minimum amount of a CO—H₂ mixture to produce 1 at. wt Fe from 0.5 mole Fe_2O_3 can be calculated. Several assumptions must first be made:

(1) There is no free oxygen or hydrocarbon in the gas used for reduction.
(2) Essentially pure iron is produced. Carbon and oxygen contents are negligible.
(3) There are no side reactions in the gas phase producing a new component, for example, methane by a reaction like $CO + 3H_2 = CH_4 + H_2O$, or carbon by the reaction $2CO = C + CO_2$.
(4) There is no regeneration of reducing gas or selective elimination of gaseous reduction products, that is, regeneration of CO and H₂ by additions of carbon, $CO_2 + C = 2CO$ and $H_2O + C = H_2 + CO$.
(5) There is no H₂O or CO_2 contained in the iron oxide which will evolve during reduction.

All of the listed assumptions can be removed in more detailed calculations; however, this simplified version of iron reduction serves as a good reference point.

The nomenclature to be used in the calculations is given in Table 1.

Table 1. Nomenclature

Term	Definition
c_0	Fraction CO_2 in the effective* gas entering the system
h_0	Fraction H₂O in the effective gas entering the system
C	Fraction of $(CO + CO_2)$ in the effective gas
H	Fraction of $(H_2 + H_2O)$ in the effective gas
V	Moles of effective gas entering the systems per at. wt Fe
$(ce)_1$	Equilibrium fraction CO for FeO—Fe border. Values given in Fig. 1 divided by 100
$(he)_1$	Equilibrium fraction H₂ for FeO—Fe border; Fig. 2
$(ce)_2$	Equilibrium fraction CO for Fe_3O_4—FeO border; Fig. 1
$(he)_2$	Equilibrium fraction H₂ for Fe_3O_4—FeO border; Fig. 2
$(ce)_3$	Equilibrium fraction CO for Fe_2O_3—Fe_3O_4 border; Fig. 1
$(he)_3$	Equilibrium fraction H₂ for Fe_2O_3—Fe_3O_4 border; Fig. 2
$(ce)_4$	Equilibrium fraction CO for Fe_3O_4—Fe border; Fig. 1
$(he)_4$	Equilibrium fraction H₂ for Fe_3O_4—Fe border; Fig. 2
x_1	Moles CO in the gas mixture per at. wt Fe at FeO—Fe border
y_1	Moles H₂ in the gas mixture per at. wt Fe at FeO—Fe border
x_2	Moles CO in the gas mixture per at. wt Fe at Fe_3O_4—FeO border
y_2	Moles H₂ in the gas mixture per at. wt Fe at Fe_3O_4—FeO border
x_3	Moles CO in the gas mixture per at. wt Fe at Fe_2O_3—Fe_3O_4 border
y_3	Moles H₂ in the gas mixture per at. wt Fe at Fe_2O_3—Fe_3O_4 border
x_4	Moles CO in the gas mixture per at. wt Fe at Fe_3O_4—Fe border
y_4	Moles H₂ in the gas mixture per at. wt Fe at Fe_3O_4—Fe border

* Effective gas is $CO + CO_2 + H_2 + H_2O$.

The stoichiometric and equilibrium relations dictate the validity of three equations at each border.

At the FeO—Fe border:

$$x_1 + y_1 = (C - c_0)V + (H - h_0)V - 1.000 \quad (1a)$$

$$\frac{x_1}{CV} \geq (ce)_1 \quad (1b)$$

$$\frac{y_1}{HV} \geq (he)_1 \quad (1c)$$

Equation 1a simplifies to

$$x_1 + y_1 = \phi V - 1.000 \quad (1d)$$

where

$$\phi = C + H - c_0 - h_0 = 1 - (c_0 + h_0) \quad (1e)$$

because $C + H = 1$.

The minimum value of V to meet the border requirements is found by letting

$$x_1 = (ce)_1 CV$$

$$y_1 = (he)_1 HV$$

and substituting these values into equation 1d. Rearrangement of terms yields the relation

$$V \min 1 = \frac{1.000}{\phi - (ce)_1 C - (he)_1 H} \quad (1f)$$

At the Fe_3O_4—FeO border:

$$x_2 + y_2 = \phi V - 1.333 \quad (2a)$$

$$\frac{x_2}{CV} \geq (ce)_2 \quad (2b)$$

$$\frac{y_2}{CV} \geq (he)_2 \quad (2c)$$

In a similar manner it is found that the minimum value of V to meet the requirements at the Fe_3O_4—FeO border is given by the equation

$$V \min 2 = \frac{1.333}{\phi - (ce)_2 C - (he)_2 H} \quad (2d)$$

At the Fe_2O_3—Fe_3O_4 border:

$$x_3 + y_3 = \phi V - 1.500 \quad (3a)$$

$$\frac{x_3}{CV} \geq (ce)_3 \quad (3b)$$

$$\frac{y_3}{CV} \geq (he)_3 \quad (3c)$$

Similarly the minimum value of V to meet the requirements at this border is given as

$$V \min 3 = \frac{1.500}{\phi - (ce)_3 C - (he)_3 H} \quad (3d)$$

The amount of effective gas which must be supplied to produce 1 at. wt Fe is the largest value of V calculated from the three equations corresponding to each border.

$$V \min 1 = \frac{1.000}{\phi - (ce)_1 C - (he)_1 H} \quad (1f)$$

$$V \min 2 = \frac{1.333}{\phi - (ce)_2 C - (he)_2 H} \quad (2d)$$

$$V \min 3 = \frac{1.500}{\phi - (ce)_3 C - (he)_3 H} \quad (3d)$$

If Fe_3O_4 is reduced below 560° C, then the amount of effective gas required is the larger value of V calculated from the equations

$$V \min 4 = \frac{1.333}{\phi - (ce)_4 C - (he)_4 H} \quad (4a)$$

$$V \min 3 = \frac{1.500}{\phi - (ce)_3 C - (he)_4 H} \quad (3d)$$

The value of $V \min 3$ need never be calculated because the values of $(ce)_3$ and $(he)_3$ are always so low that it is never greater than $V \min 1$, $V \min 2$, and $V \min 4$. This means that the same quantity of reducing gas is required regardless of whether the starting material is Fe_3O_4 or Fe_2O_3.

By means of the derived equations the quantity of a mixture of CO, CO_2, H_2, and H_2O required to reduce 1 at. wt of iron from its oxide can be calculated. The temperature of reduction at each stage must be specified so that the values of $(ce)_1$, $(he)_1$, . . . and so forth may be taken from Figs. 1 and 2.

To illustrate the method, the quantity of a gas of composition 30% CO, 2% CO_2, 15% H_2, 3% H_2O, 50% N_2 required to produce 1 at. wt. Fe will be calculated for reduction temperatures of 1000° C for FeO to Fe and 800° C for Fe_3O_4 to FeO.

% Effective gas $= 30 + 2 + 15 + 3 = 50$

$$c_0 = \frac{2}{50} = 0.04$$

$$h_0 = \frac{3}{50} = 0.06$$

$$C = \frac{30 + 2}{50} = 0.64$$

$$H = \frac{15 + 3}{50} = 0.36$$

$$\phi = 1 - (0.04 + 0.06) = 0.9$$

At 1000° C, $(ce)_1 = 0.72$, $(he)_1 = 0.60$, and

$$V \min 1 = \frac{1.000}{0.9 - (0.72)(0.64) - (0.60)(0.36)} = 4.50$$

At 800° C, $(ce)_2 = 0.28$, $(he)_2 = 0.31$, and

$$V \min 2 = \frac{1.333}{0.9 - (0.18)(0.64) - (0.31)(0.36)} = 1.64$$

Here $V \min 1$ is much greater than $V \min 2$; 4.50 moles of effective gas is required for each at. wt Fe produced. Since the effective gas was half of the total gas mixture, 9.0 moles of gas must be used.

Using the same principles, Austin[2, 3] calculated the quantity of CO necessary to produce Fe from

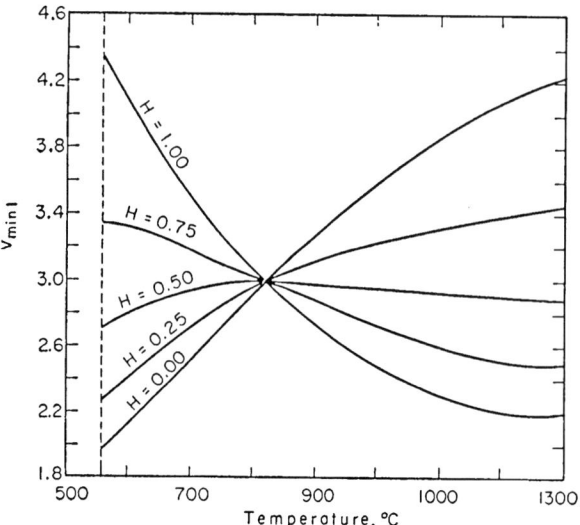

Fig. 4. Gas requirements at FeO—Fe Border; $\phi = 1.0$.

iron oxide at several reduction temperatures. His calculations included only the case of meeting the requirements at the FeO—Fe border. Elliott[4] made more detailed calculations in which he dealt with cases of meeting the requirements at both the Fe_3O_4—FeO and FeO—Fe borders. He considered only CO as a reducing gas and had each step of the reduction process occur at the same temperature. His work clearly showed that there is an optimum temperature for minimum consumption of CO in the reduction of iron oxide.

It will be shown that Elliott's minimum based on the same reduction temperature at each step can be considerably lowered if the temperature at each stage of reduction is an independent variable.

RESULTS

Figures 4 and 5 summarize calculations made in the manner already described. Both curves are for

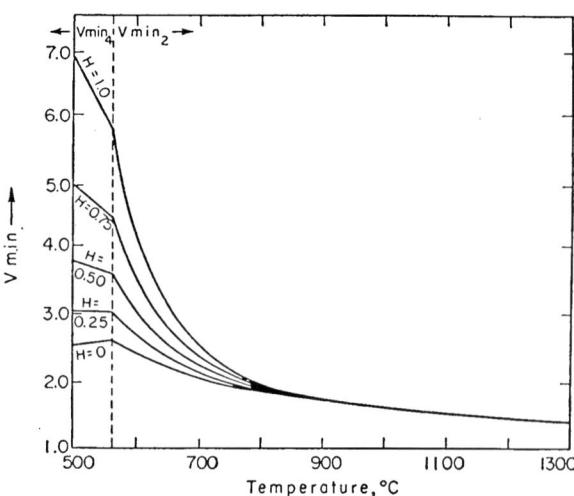

Fig. 5. Gas requirements at Fe_3O_4—FeO border and Fe_3O_4—Fe border; $\phi = 1.0$.

a value of $\phi = 1.00$ which corresponds to no CO_2 or H_2O in the entering gas mixture. From Fig. 4 it can be seen that there is a reversal in the temperature dependence of V min 1 at approximately a half-and-half mixture of CO and H_2. The curves for various mixtures of CO and H_2 all intersect at 825° C, which is the temperature at which $(ce)_1 = (he)_1$. At 825° C the equilibrium constant of the water–gas reaction, $H_2O + CO = H_2 + CO_2$, is unity. Thus, H_2 and CO are equivalent in reducing power at that temperature. Figure 5 shows the temperature dependence of V min 2 and V min 4. The analytical procedure used in the previous example may be replaced by a graphical procedure using Figs. 4 and 5.

Ideal Gas Utilization— Reduction Steps Separated

If the reaction temperature of each reduction step is independently controlled, a temperature program for maximum utilization of reducing gas can be determined from Figs. 4 and 5. The lowest possible value for V min 1 is taken from Fig. 4. Then, from Fig. 5, the temperature required to give a value of V min 2 less than that of V min 1 is determined. In the case of pure CO as the reducing gas, the lowest value of V min 1 is 1.96. Also V min 2 is less than 1.96 when Fe_3O_4 is reduced to FeO at a temperature greater than 750° C. Therefore, the ideal temperature program for maximum utilization of pure CO is to reduce Fe_3O_4 to FeO at greater than 750° C and FeO to Fe at 560° C. In a similar manner the ideal temperature programs for other gas mixtures have been calculated and summarized in Table 2.

Table 2. Ideal Values of Gas Utilization in Reduction of Iron Oxide; $\phi = 1.0$

	Reduction Steps Separated			Reduction Steps Mixed	
Gas Mixture	T_1	T_2	V	T	V
H = 0.00, C = 1.00	560	>750	1.96	635	2.26
H = 0.10, C = 0.90	560	>700	2.08	—	—
H = 0.25, C = 0.75	560	>680	2.28	630	2.51
H = 0.40, C = 0.60	560	>650	2.51	625	2.69
H = 0.50, C = 0.50	560	>635	2.70	620	2.84
H = 0.55, C = 0.45	560	>630	2.81	—	—
H = 0.60, C = 0.40	1300	>650	2.72	1300	2.72
H = 0.65, C = 0.35	1300	>665	2.65	—	—
H = 0.75, C = 0.25	1300	>680	2.50	1300	2.50
H = 0.90, C = 0.10	1300	>725	2.31	—	—
H = 1.00, C = 0.00	1300	>750	2.22	1300	2.22

Maximum Gas Utilization— Reduction Steps Mixed

If the reaction temperature of each reduction step is not independently controlled, it is necessary to consider the reactions occurring at the same temperature. For CO—H₂ mixtures containing less than 55% H₂, the temperature dependence of V min 1 and V min 2 are opposed. For each gas mixture containing less than 55% H₂ (H = 0.55), there is a temperature at which V min 1 is equal to V min 2. At this temperature, gas utilization is a maximum. Figure 6 shows the determination of the optimum temperature for three gas mixtures. For gas mixtures containing more than 55% H₂, both V min 1 and V min 2 decrease with increased temperatures. The optimum temperature for re-

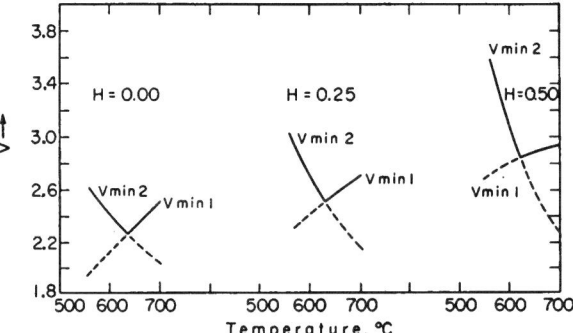

Fig. 6. Determination of optimum temperature for mixed reduction steps; $\phi = 1.0$.

duction with such mixtures is the maximum temperature possible. The figures in Table 2 are for a temperature of 1300° C.

The values given in Table 2 are shown in Fig. 7.

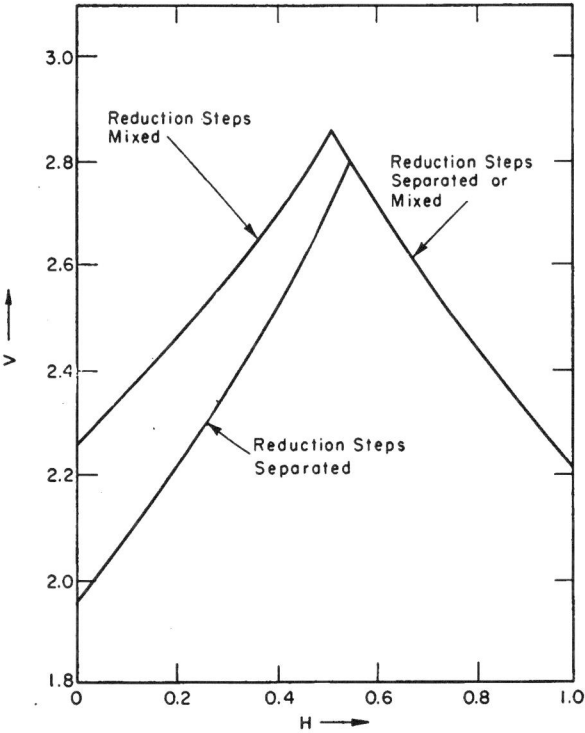

Fig. 7. Ideal gas utilization in reduction of iron oxide; $\phi = 1.0$.

DISCUSSION OF RESULTS

The values of Fig. 7 are the best attainable with each gas mixture. The ideal temperature program for each gas mixture is different, so that Fig. 7 does not correspond to any defined temperature program. Mixtures of H₂ and CO close to 50–50 give the poorest utilization of reducing gas. To achieve the ideal condition for mixtures containing less than 55% H₂, low-temperature reduction (560° C–750° C) is used, but for mixtures containing more than 55% H₂, high-temperature reduction (1300° C) must be used.

The data presented have all been for a starting gas containing no CO₂ or H₂O, $\phi = 1.0$. Figure 8

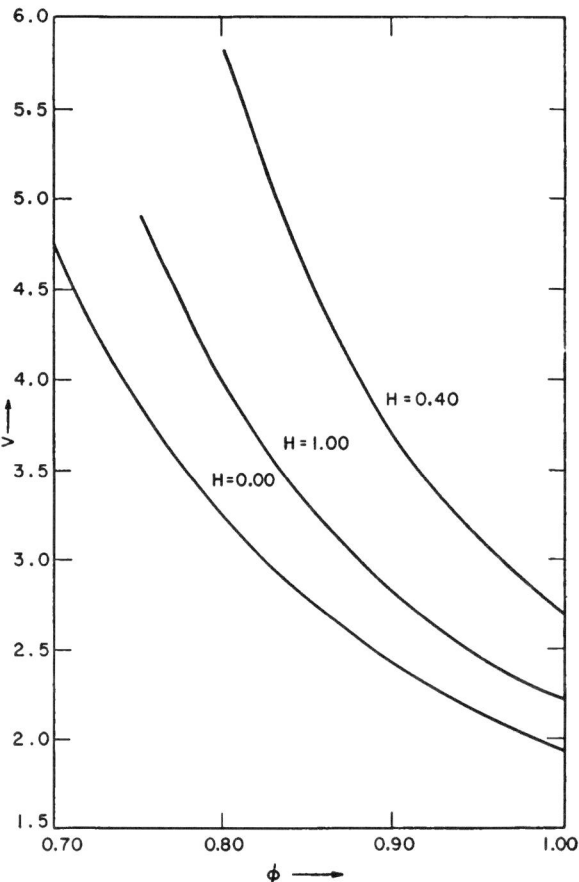

Fig. 8. Effect of ϕ on ideal gas utilization;

$$\phi = \frac{\% \ (CO + H_2)}{\% \ (CO + CO_2 + H_2 + H_2O)}$$

in starting gas.

shows that small amounts of H₂O or CO₂ in the entering gas greatly increase the quantity of effective gas required for reduction.

The assumptions made for this analysis contain one questionable item—carbon deposition. The equilibrium for the carbon-deposition reaction, $2CO = C + CO_2$, is included in Fig. 1. The S curve is for 1 atm total pressure. Although total pressure has no effect on equilibrium for iron-oxide-reduction, it has a pronounced effect on the carbon-deposition reaction. The greater the total pressure, the greater the CO_2 content for equilibrium in the system, $C-CO-CO_2$. The carbon curve falls below the FeO—Fe curve at about 700° C and below the Fe_3O_4—FeO curve at about 650° C. If carbon deposition occurred rapidly, it would be impossible to use CO to reduce iron below 700° C, and the values of V min 1 and V min 2 for CO mixtures below 700° C would be invalid. Fortunately, carbon deposition is a very slow reaction requiring catalysts. At pressures less than 2 atm,[5, 6, 7] it might not interfere with the reported results at suitable conditions for gas-solid contact.

The calculations presented cover only a very simple form of one phase of iron-oxide reduction. It is hoped that more complete pictures will be forthcoming in order that the new processes being proposed for iron-oxide reduction may be more fully examined.

CONCLUSIONS

1. Optimum reaction temperatures exist for which the utilization of $CO-H_2$ mixtures in the reduction of iron oxide is a maximum.

2. Maximum gas utilization is accomplished with low reaction temperature for CO-rich mixtures and with high reaction temperature for H_2-rich mixtures.

3. The poorest gas utilization is with a gas mixture of 55% H_2 and 45% CO. Gas utilization is steadily improved by going from this mixture toward 100% CO or 100% H_2.

References

1. J. O. Edstrom, *J. Iron Steel Inst.*, **175**, 289, (1953).
2. J. B. Austin, *Trans. A.I.M.E.*, **131**, 74 (1938).
3. J. B. Austin, *J. Iron Steel Inst.*, **167**, 358 (1951).
4. J. F. Elliott, Unpublished calculations.
5. L. F. Marek, A. Bogrow, and G. W. King, *Trans. A.I.M.E.*, **172**, 46 (1947).
6. B. Kalling and J. Stalhed, *Jernkontorets Ann.*, **121**, 30 (1937).
7. S. Klemantaski, *J. Iron Steel Inst.*, **171**, 176 (1952).

Discussion

GOKCEN pointed out that the problem of carbon deposition below about 700° C from gases high in CO would be serious in the reduction system as described by St. Pierre, and that it may cause considerable deviation from his idealized system. To this ST. PIERRE agreed. However, he felt that there was great uncertainty in the actual conditions which would be encountered, and that they required experimental evaluation.

by C. E. A. Shanahan

The Study of Slag-Metal Mixing Efficiency by Models

It is generally agreed that slag-metal chemical reaction rates are chiefly limited by the diffusion of reactants and products to and from the slag-metal interface. Evidence for this is provided by certain ladle refining reactions. Thus the desulfurization of hot metal by pouring it onto soda ash placed in the bottom of a ladle is virtually complete within a few minutes; similarly, the almost instantaneous deoxidation and desulfurization of molten steel may be accomplished by tapping into a ladle of a suitable molten slag (Perrin process). On the other hand, the refining of steel within the open-hearth furnace takes several hours, and were it not for slag-metal mixing arising from the carbon "boil," the refining time of this process would be prohibitive.

The problem of decreasing refining times is, therefore, largely one of ensuring more effective slag-metal mixing, and any experiments undertaken to achieve this are of immediate interest to the iron and steel industry. It is not surprising, therefore, that, several years ago, the British Iron and Steel Research Association undertook a series of experiments designed chiefly to develop mechanical techniques suitable for intimately mixing together slag and metal on a large scale. Several schemes, such as the use of a rocking furnace[1] and agitation by gas bubbles were explored using slag and metal on a small scale with some success. However, many difficulties arose from the necessity of using high temperatures, and it was felt that more rapid prog-

Mr. Shanahan is affiliated with the R. T. S. C. Laboratories, Whitchurch, Aylesbury, Buckinghamshire, England.

The author wishes to thank Sir Charles Goodeve, Director, and Dr. A. H. Leckie, former Head of the Steelmaking Division of the British Iron and Steel Research Association, for encouragement and many helpful discussions.

ress might result from experiments with cold slag-metal model systems which could be more easily controlled. A large measure of uncertainty exists in attempting to extrapolate small-scale experimental kinetic data to plant conditions even when linear dimensions are the only differences between experiment and plant. Information deduced from small-scale experiments of use to plant conditions must therefore be approximate only, and the replacement of slag and metal by two other immiscible liquids probably does not seriously affect the usefulness of the experimental results. This paper describes work in which two slag-metal models were employed, namely, the transfer of sodium from mercury to aqueous acid solution and the transfer of methyl red to aqueous zinc-chloride solution from n-hexane. Unfortunately, it became necessary to abandon the work due to circumstances beyond the author's control, but sufficient results (mainly in connection with the amalgam model) were obtained to show the practicability of the technique, and they are given below.

CHOICE OF MODELS

The use of models to study the behavior of large and expensive assemblies such as ships has been particularly successful in the past. This success has resulted from the relatively low cost of model work (as compared with similar experiments on the full-scale prototype) and the ability to extrapolate the data obtained from model performances to full-scale conditions. The relationship between the behavior of a model and its full-scale counterpart is characterized by a set of dimensionless products, the number of which depends on the total variables affecting the process under study (see Appendix to this article). Except in the simplest of models, it is impossible to maintain complete similarity between model and full-scale counterpart, and as a consequence, there is usually some risk involved in predictions based on model experiments. In spite of this fact, it is often found that certain dimensionless products are of only minor importance, so that a very imperfect model still yields valuable data.

It is impossible to obtain a cold model which is exactly similar to a hot slag-metal system, but there are certain pairs of immiscible liquids that emulate slag-metal mixing sufficiently closely to warrant study. The present experiments were designed to study methods for intimately mixing slag and metal together and are not in any way connected with the types of chemical reactions that occur during refining. Intensity of mixing has been measured by the rate of migration of a substance from the "metal" to the "slag" layer, after first establishing that the migration is almost entirely dependent on diffusion. Two model systems have been studied, and the results are given hereafter. It should be emphasized that the work is incomplete; indeed, no relationship has been established between model results and full-scale plant.

MERCURY — DILUTE AQUEOUS ACID MODEL

This system comprised mercury containing 0.2% w/w of sodium as "metal" and $N/10$ H_2SO_4 as "slag." A preliminary experiment confirmed initial thermodynamic calculations that the transfer of the sodium to the aqueous layer takes place almost to completion and also that the reaction is intrinsically very rapid. Thus when 10 ml of the amalgam was shaken with 10 ml of the $N/10H_2SO_4$, neutralization occurred within 2 sec. It was also established that the temperature coefficient of the reaction rate was low; thus the reaction rate constant at 20° C was approximately 34% higher than that at 10° C. By conducting all experiments within the temperature range of 18 to 20° C, the latter never affected the reaction-rate constant by more than about 7%.

A comparison of the following methods of slag-metal mixing has been made since they can easily be used on a large scale:

(1) Mixing by gas blown from a single vertical jet.
  (2) Mixing by pouring the amalgam into the acid.
    (a) From a jet such that the amalgam ran at a constant flow rate.
    (b) By lip-pouring from a second vessel.
  (3) Mixing by the use of a rabble.
  (4) Mixing by blowing gas through three holes drilled in the base plate of the containing vessel, i.e., simulation of a simple form of converter.

These methods of mixing are now discussed separately.

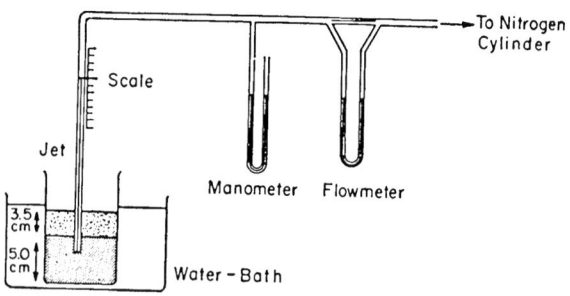

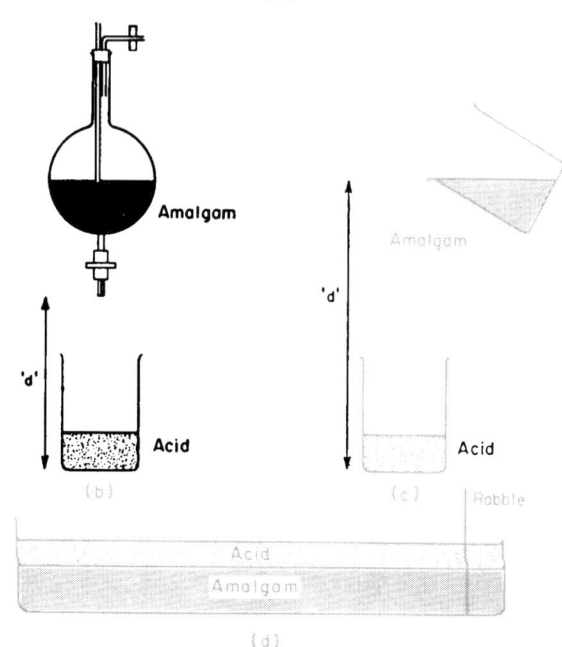

Fig. 1. Apparatus for mixing experiments.

Mixing by Gas Blown from a Single Vertical Jet (Pressure approximately 1 atm abs)

The apparatus used is essentially as shown in Fig. 1a, wherein nitrogen at a pressure of approximately 1 atm is fed by a line containing a flowmeter and mercury manometer to an approximately 0.5-mm jet immersed in the amalgam-acid system. The latter is contained in a 250-cc beaker 11 cm tall and 5.8 cm in diameter, which is surrounded by a water bath. The vertical position of the jet can be varied, the actual position being indicated by a pointer on the jet moving against a fixed scale. A series of experiments was made at varying gas flow rates with the jet in the following three positions:

  (a) In the acid layer.
  (b) 1 cm under the surface of the amalgam.
  (c) 1 cm from the bottom of the amalgam layer, the depth of the acid and amalgam layers being 3.5 cm and 5.0 cm, respectively.

The temperature in all cases was maintained in the range 18 to 20° C.

In an individual determination with the appara-

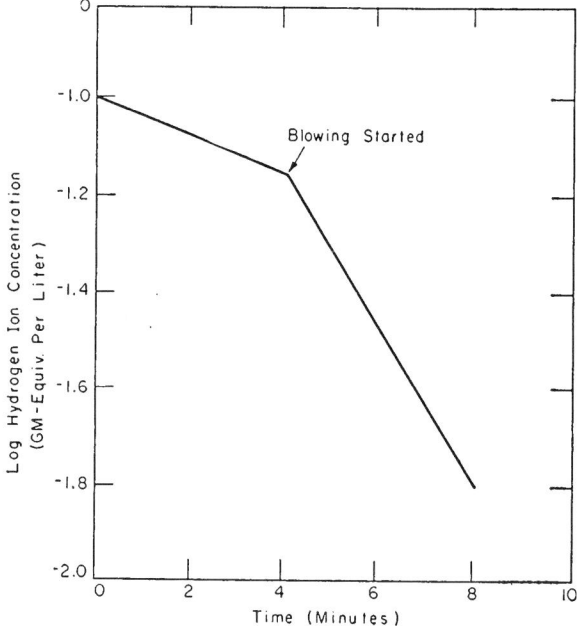

Fig. 2. Single-jet-blowing experiment. Specimen graph for the sodium-dilute acid system.

tus set up as shown, 133 cc of amalgam containing approximately 0.2% of sodium were poured into the beaker and 90 cc of dilute sulfuric acid (approximately 0.1 N) added, a stop watch being started at the time of addition to the acid. Samples of the aqueous layer were removed at approximately 40-sec intervals until the reaction had been proceeding for 4 min. The gas flow, previously set at the required flow rate, was started, and sampling continued for a further 6 to 10 min. The determination of the residual acid in the samples was done volumetrically. From the results so obtained, the reaction rates under quiescent and agitated conditions could be obtained from the slope of the line obtained by plotting the logarithm of the acid concentration

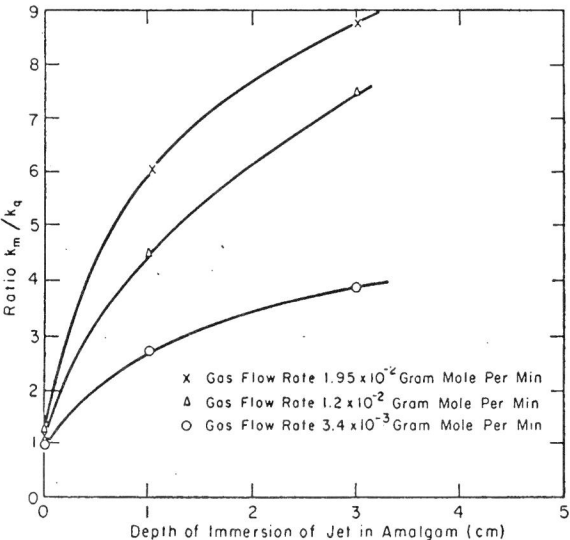

Fig. 3. Single-jet-blowing experiment. Effect of depth of immersion on mixing.

against time, the quiescent rate being designated k_q, the rate when mixed k_m. A typical graph resulting from this procedure is shown in Fig. 2.

The ratio k_m/k_q forms a suitable index for the measurement of the effect of agitation, and the results obtained are shown graphically in this form in Figs. 3 and 4 by means of plots of the index

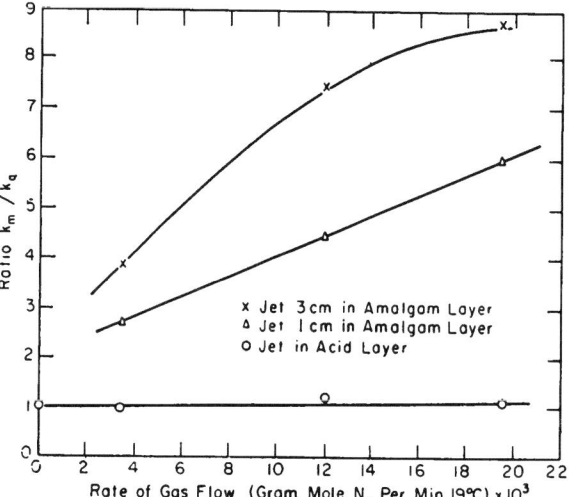

Fig. 4. Single-jet-blowing experiment. Effect of rate of gas flow on mixing.

k_m/k_q against depth of immersion and flow rate, respectively.

Mixing by Pouring the Amalgam into the Acid

From a Jet at a Constant Flow Rate. The apparatus used in shown in Fig. 1b and consists of a simple flow device feeding a jet held at a fixed distance d above the bottom of the beaker containing the acid, a series of experiments being done at various heights using two different jets.

With the height set at the particular distance being used, 133 cc of amalgam were poured into the flask and 90 cc of acid into the beaker. The temperatures of the amalgam and acid were adjusted so as to give a final temperature in the range 18 to 20° C, the same beaker (250 cc) and volumes of amalgam and acid being used in all the mixing experiments. The bottom clip was released and the time of delivery of the amalgam timed by a stop watch, the acid being sampled immediately after delivery had ceased. The initial and final strengths of the acid being known, the rate of reaction could again be calculated from the graph of the logarithm of the acid concentration against time.

Since the containing vessel, the volumes and the concentration of reagents used, and the temperature were very nearly the same in all the various mixing experiments, it is reasonable to assume an average value of k_q (obtained from the nitrogen-blowing experiments) for the remaining experiments and again express the efficiency of mixing as

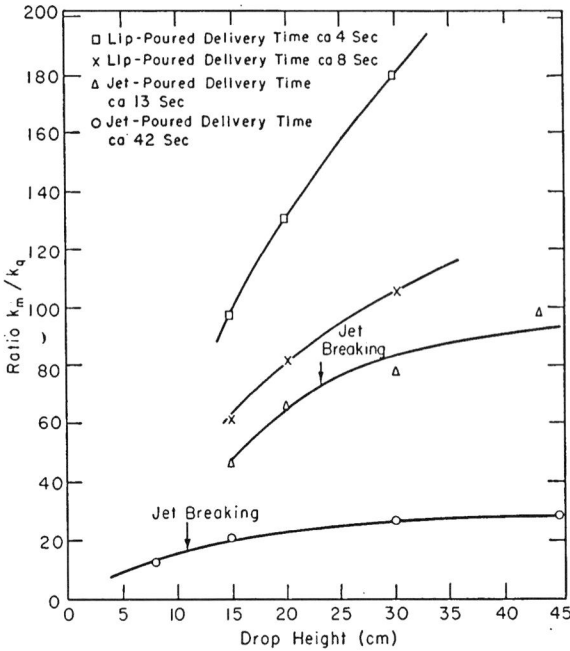

Fig. 5. Pouring experiment. Influence of drop height on mixing. Note, all ratios of k_m/k_q in gas-blowing experiments are below 10.

the ratio k_m/k_q. The results of these experiments are shown in Fig. 5, together with those from the next series on lip pouring, both being graphed as the ratio k_m/k_q against height of drop.

Lip Pouring from a Second Vessel. A total volume of 133 cc of amalgam was lip poured manually from a support at a known height d above a beaker containing acid, d being measured from the lip of the top beaker to the bottom of the second beaker. No attempt was made to obtain a constant flow rate; the delivery time, however, was controlled, and experiments were done at varying heights at two different delivery times. The acid was sampled before and after pouring, the rate of reaction being calculated as above, the results, as previously mentioned, being shown in Fig. 5.

A rough idea of the size of a full-scale plant comparable to the apparatus used is given by taking the beaker as being equivalent to a ladle 6 ft high when the jet used would be equivalent to a pipe of approximately 1/2-in. diam, and a drop height of 30 cm would be equivalent to approximately 18 ft under plant conditions.

Mixing by the Use of a Rabble

The amalgam in this instance was contained in a trough with approximately $5 \times 5 \times 25$ cm dimensions (see Fig. 1), to which the acid was added at the beginning of the run. To obtain a reasonably deep amalgam layer, 266 cc of amalgam and 180 cc of acid were used, that is, double the quantities used in the previous experiments.

After sufficient samples were taken to determine

the quiescent rate, the system was intermittently rabbled by means of a wooden blade 4 cm wide, samples being taken after each rabbling period. The rabbling was done by moving the blade up and down the trough at the fastest rate possible when the blade was touching the bottom of the trough (approximately 12 cm per sec), 10 traverses being done in each rabbling period. Two sets of experiments were done:

(a) With the blade touching the bottom of the trough.

(b) With the blade immersed 0.6 cm in the amalgam layer.

The reaction rate under quiescent and mixed conditions was obtained as before from the plot of the logarithm of the acid concentration against time, a

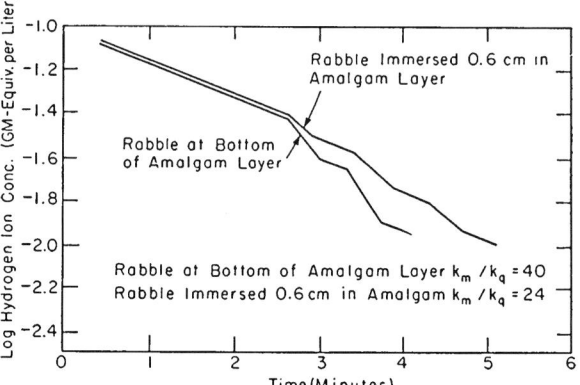

Fig. 6. Rabbling experiment. Specimen graph for the sodium amalgam-dilute acid system.

typical graph being shown in Fig. 6, which also includes the results of the experiments.

Mixing by Bottom Blowing

The mixing vessel consisted of a beaker (of the same dimensions as those used previously) whose glass bottom was replaced by a removable drilled Perspex plate, as shown in Fig. 7. Nitrogen for blowing purposes was fed to the tuyères by a simple "air box" made from a glass funnel.

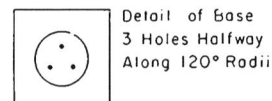

Detail of Base
3 Holes Halfway
Along 120° Radii

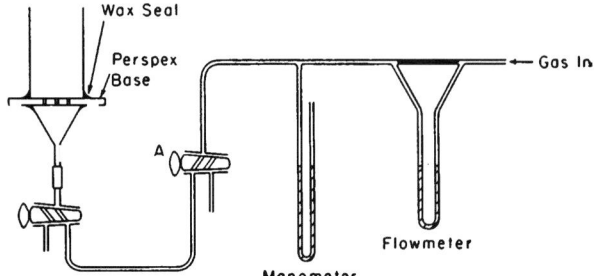

Fig. 7. Model converter.

In an individual experiment a known flow of nitrogen was passed through the system, and 133 cc of amalgam were poured into the container. Tap *A* was then turned to pass the gas to waste, the pressure in the air box being sufficient to support the weight of the amalgam. Ninety cc of acid were poured steadily onto the amalgam surface and samples of acid taken over a period of 4 min, at known time intervals. The gas flow, preset at the required value, was allowed to flow through the contents of the beaker by turning tap *A*, and sampling continued until completion of the reaction. The data obtained from the analyses of the acid samples were sufficient to give the rate of reaction under quiescent (k_q) and agitated conditions (k_m), the ratio k_m/k_q again being used as an index of mixing efficiency.

The results of a series of experiments using a base drilled with 3 holes of approximately 0.80-mm diam (see Fig. 7) are presented as a graph of mixing

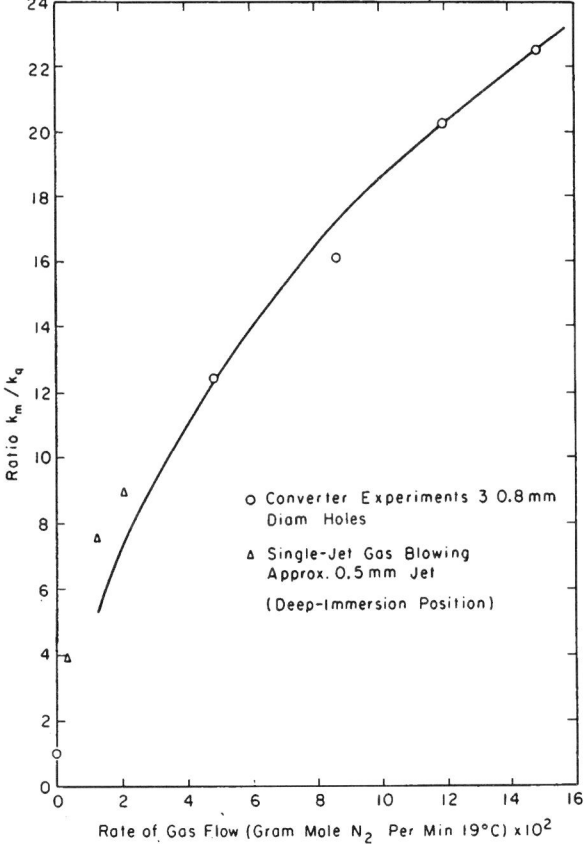

Fig. 8. **Model converter experiment. Variation of ratio k_m/k_q with rate of gas flow.**

index versus rate of gas flow in Fig. 8. The results of the series of experiments using a single vertical jet are also plotted in Fig. 8 for comparative purposes.

DISCUSSION OF RESULTS

The results obtained using the four main types of mixing will be discussed separately and then finally considered as a group.

Mixing by Gas Blowing with a Single Jet

The effect of blowing in the aqueous layer (see Fig. 4) is so small as to be negligible for all practical purposes. Blowing in the amalgam layer, however, causes an appreciable increase in the reaction rate, the increase being greater with increasing depth and gas flow rate. It is nevertheless indicated that, beyond a certain point, increase in the gas flow rate does not effect a commensurate increase in the reaction rate. The practical implications are therefore that

(a) When using gas for mixing purposes, the deeper the jet the better.

(b) Increasing the flow rate beyond a certain limit may be wasteful.

Mixing by Pouring the Amalgam into the Acid

Jet Pouring. The efficiency of mixing increases with increasing drop height, rapidly at the beginning of the curve and finally more slowly. This effect is shown both on the 13- and 42-sec delivery times (4-mm and 2-mm jets). Now with a free-flowing jet, there is a position below the jet where the stream of liquid begins to break into droplets although actual droplet formation may not begin until much lower down. Measurement of the point where the jet began to break (marked with an arrow in Fig. 5) showed it to be associated in both cases with the point in the curve where change occurred from a rapid increase in mixing efficiency to a slower rate of increase (see Fig. 5). It is not suggested, however, that this association is of first-rate importance; there are so many other effects, such as splashing up of the amalgam from the bottom surface, the turbulence caused by the jet, and so on, that the relationship may be fortuitous.

The general drop in the rate of increase of the efficiency, if during mixing the height of pour is increased, may have some bearing on the observations of Yaneske who, when using the Perrin process,[2] stated that there was a critical pour height below which efficient dephosphorization was not obtained, although Perrin states[3] that drop heights of the order used by Yaneske are unnecessarily high.

It will be seen (Fig. 5) that increasing the flow rate increases the efficiency of mixing, an effect which is further exemplified by the results of the following set of experiments on lip pouring.

Lip Pouring from a Second Vessel. These experiments were of necessity of a lower order of accuracy than the preceding ones. To construct a set of apparatus to give a constant and reproducible lip-pouring rate would be time consuming, and since the object was to obtain information of the order of the efficiency of mixing as compared with that shown by the preceding technique, experiments were carried out using only the very simple apparatus shown in Fig. 1c.

The results showed the same trend as those already considered on jet pouring, namely, an increase in the efficiency of mixing with increase of pouring speed. The curve for the 4-sec delivery time shows no signs of bending over such as are exhibited by the jet-pouring curves; the lower curve, however, does show a drop in the rate of increase of mixing efficiency with drop height.

Several experiments were made on the effect of pouring amalgam and acid together. The reaction rates, however, were approximately the same as those exhibited by the direct pouring of amalgam into the acid, a result which could be due to the relative volumes of amalgam and acid (133 : 90) used. Since the majority of the acid flowed into the bottom vessel first, the system then became virtually the same as in that used in the direct pouring of amalgam into acid.

Mixing by the Use of a Rabble

The conditions of the experiment were such as to give the maximum possible rate at the deep-immersion position; that is:

(1) The rabbling blade was relatively wide when compared with the width of the trough.
(2) The fastest possible rate of traverse was used.
(3) The amalgam layer was relatively shallow.

In spite of these facts the maximum reaction rate found was only four times the quiescent rate, this rate being obtained with the rabble at the bottom of the amalgam layer. There was no point in rabbling solely in the acid layer since it was already known both from the literature[4] and from the previous experiments that this has little effect on the rate of reaction.

Mixing by Bottom Blowing

Both techniques of gas blowing, by single jet or by bottom blowing, have similar mixing efficiencies for a given nitrogen flow rate (that is, within the flow rates studied). Unfortunately it was found impracticable to extend the single-vertical-jet data to values of nitrogen flow in excess of about 2×10^{-2} gram mole per min because of excessive "slopping," and consequently, a direct comparison can be made only at relatively low gas flow rates. However, the curves for the two sets of data are probably of the same shape and similar mixing efficiencies are, therefore, to be expected at the higher flow rates. It is interesting to note that the use of the 3-hole base permitted the use of nitrogen flow rates far in excess of those practicable for the single vertical jet. Thus, although the amount of mixing produced with either technique is similar for the same gas flow, a greater gas flow and, therefore, a greater degree of mixing can be tolerated in the 3-hole-base

technique before the advent of "slopping." It is possible that increasing the number of base holes may permit the use of even higher gas flows. The value of the mixing index at the higher flow rates is comparable to that found in slow jet pouring (k_m/k_q values of 12 to 28 at drop height 8 to 45 cm, respectively).

DEVELOPMENT OF SECOND MODEL SYSTEM

The effect of the ratio of densities of the slag and metal phases on the mixing efficiency and the nature of the dimensionless groups (for example, Reynolds number) of importance in relating the model to the prototype are controversial. The amalgam-acid system has a density ratio greater than that of actual slag and metal, and for this reason its use for this type of work may be criticized. Further, the full investigation of the scale factor required to relate model results to the prototype is prohibited by the cost of the mercury which would be required for experiments on a larger scale.

For these reasons a new system was sought which would satisfy the following criteria:

(1) Ability to attain a ratio of the densities of the two liquid phases in the region of 2.5, with some possibility of variation.
(2) Low cost of experiments on a scale involving several gallons of liquid.
(3) Amenability to the use of some material for transfer between the two phases, the concentration of which could be readily determined.
(4) Ability to vary the viscosity of the phases.

Some difficulty was experienced in finding a system which would conform to these four conditions. Finally a model was devised in which the "metal" was represented by aqueous zinc chloride (specific gravity, 1.7 grams/cc) and the "slag" by *n*-hexane. Intimacy of mixing was measured by the rate at which methyl red migrated from the *n*-hexane layer. Throughout any particular experiment, samples of the *n*-hexane layer were obtained periodically and analyzed colorimetrically. Preliminary tests confirmed the high intrinsic rate of transfer of

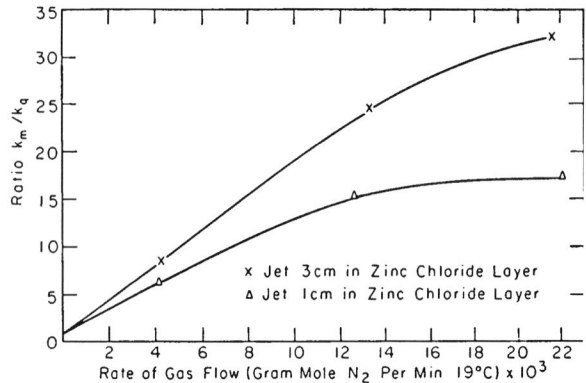

Fig. 9. Single-jet-blowing experiment. Effect of the rate of gas flow on mixing, zinc chloride-methyl red-*N*-hexane system.

methyl red from the "slag" layer; thus 14 ml of n-hexane containing methyl red lost approximately 99% of the latter in 3 sec after shaking in a separating funnel with 20 cc of the zinc-chloride solution. The temperature coefficient of the migration rate was low; thus the rate at 20° C was approximately 25% greater than that at 10° C.

Figure 9 shows results obtained using the above model on the single-jet-blowing experiments already described. No results were obtained from blowing in the "slag" layer only. The results of the two model systems are in general agreement; namely, mixing efficiency increases with gas flow rate and probe depth.

Unfortunately, at this stage experiments were abandoned; it had been intended to use the zinc-chloride system on the remaining mixing techniques already described and then to study the effects of increasing the size of the model while maintaining geometrical similarity.

GENERAL DISCUSSION AND CONCLUSIONS

The extension of the quantitative results obtained in the amalgam-acid system using a particular method of agitation to a slag-metal system would be of dubious value. However, it is reasonable to suppose that the relative efficiencies obtained among the various methods of mixing would be similar to those obtained in a slag-metal system.

In considering the results as a group, it is shown that the efficiency of mixing obtained by pouring the metal into the slag is of much greater order than that obtained either by simple gas blowing or by rabbling; the order of the ratio of k_m/k_q obtained by single-jet gas blowing is marked on Fig. 5 to show this. Further, the faster the amalgam is poured into the acid, the greater the efficiency of mixing, as witnessed by the high ratios obtained in the fastest lip-pouring experiments. It must, however, be borne in mind that although relatively high ratios are found, indicating faster reaction rates, the reaction is proceeding for a much shorter time, and therefore, although a fast reaction rate is obtained, the over-all amount of refining may not be so great as that given by a slower reaction rate operating for a longer time.

To sum up, it has been found that:

(1) Gas blowing in the acid layer has little effect on the reaction rate.

(2) The efficiency of mixing obtained by gas blowing is much less than that obtained by either pouring the amalgam from a jet into the acid or lip pouring the amalgam into the acid.

(3) There appears to be a critical drop height when pouring amalgam into acid beyond which the increased efficiency of mixing is small.

(4) In pouring amalgam into acid from a given height, increasing the flow rate increases the reaction rate.

(5) The efficiency of mixing obtained by rabbling is low and of the same order as that obtained by simple gas blowing.

(6) Bottom blowing from 3 holes produces a mixing efficiency similar to that resulting from the use of a single jet. It appears, however, that a higher total gas flow rate and hence more intense mixing can be tolerated in the former case before "slopping" occurs.

References

1. C. E. A. Shanahan and F. J. Lund, *Iron & Coal Trades Rev.*, 701 (March 27th, 1953).
2. B. Yaneske, *J. Iron Steel Inst. London*, **142**, No. 2, 35P (1940); **143**, No. 1, 425P (1941).
3. R. Perrin, Private communication.
4. W. G. Dunning and M. Kilpatrick, *J. Phys. Chem.*, **42**, 215 (1938).

APPENDIX

Appendix: The Performance of Model Systems in Relation to Their Prototype

To describe completely the course of any process, it is necessary to have a knowledge of the variables concerned. Thus the flow of a liquid through a pipe is characterized by the pressure head, density, viscosity, and possibly surface tension of the liquid as well as the dimensions and surface texture of the pipe. It is possible by dimensional analysis to arrange these variables into groups that are dimensionless: these are called dimensionless products. Any particular system yields a finite number of dimensionless products, and if another system can be devised that is characterized by the same number and values of dimensionless products, the two systems are said to be similar, and one can be employed to study the other. Thus, a process defined by a single dimensionless product, for example the Froude number (v^2/lg) may be studied by any other process characterized by this single number, providing both numbers are numerically equal. This second process could, for example, have linear dimensions a hundredth of the prototype, providing its velocity terms are scaled down ten times. The behavior of the model would then give accurate information concerning the performance of the prototype.

Unfortunately, it is rarely possible to construct a model that is similar to a particular prototype. It is often impossible to define all the variables affecting a process, and even when these are known, the number of dimensionless products is prohibitively large for accurate reproducibility by models. However, theoretical considerations sometimes suggest that certain of the dimensionless products are extremely important to a process, and that providing these are maintained constant (at the expense of large discrepancies among the remaining products) the model behaves in a similar manner to the prototype. Usually it is necessary to conduct parallel experiments with models and prototypes to assess their relationships empirically, all relationships being expressed in dimensionless products.

Numerous variables are involved in the intimate mixing of slag and metal; there are at least eight:

l = linear dimension
v = velocity
ρ_s = slag density
ρ_m = metal density
μ_s = slag viscosity
μ_m = metal viscosity
σ_{sm} = slag-metal interfacial tension
g = acceleration due to gravity

By setting up a dimensional matrix it is possible to show that the slag-metal system is characterized by 5 dimensionless products. These may be written down in several different ways of which the following is one:

(1) Froude number $\quad v^2/lg$
(2) Reynolds number $\quad vl\rho_m/\mu_m$
(3) Reynolds number $\quad vl\rho_s/\mu_s$
(4) Weber number $\quad v^2\rho_s l/\sigma_{sm}$
(5) Weber number $\quad v^2\rho_m l/\sigma_{sm}$

For any particular full-scale slag-metal mixing system, it is possible to calculate the values of the listed dimensionless products. This has been done for several of the systems already described, and, as might be expected, it has been found impossible to devise a model whose dimensionless products are of the same order. However, reasonable agreement can be obtained between Froude and slag Reynolds numbers if water or n-hexane are used as "slags" within the model. Froude and Reynolds number similarity ensure that the results of gravitational forces and flow patterns are similar in model and prototype.

It was hoped to investigate the relative importance of Froude, Reynolds, or Weber numbers by performing a series of mixing experiments on two models of different linear dimensions but geometrically similar. Since the liquids in each would be identical,

For Froude similarity $\quad v_1^2/l_1 = v_2^2/l_2$
For Reynolds similarity $\quad v_1 l_1 = v_2 l_2$
For Weber similarity $\quad v_1^2 l_1 = v_2^2 l_2$

These equations are obviously compatible only when the two models are of the same size. Assessment of mixing efficiency in the two models would show which number is of major importance.

Discussion

CHIPMAN asked if Shanahan's experiments indicated that one should use a high rate of pour or a low rate of pour. Shanahan said that seeking for such an answer was stretching their experimental results a little far, but they had decided that there would be no great difference. DARKEN inquired whether the tailing-off of the reaction as the rate of pouring was increased might not be related to conditions in the open hearth where a steady state was reached with a very good CO bubbling rate. In considering only the kinetics of the system without referring to actual operations, one might expect the reaction to be catastrophic. However, there is a limit where the increase in mechanical motion does not give any greater carbon-monoxide evolution. In response to Darken's question, WAGNER said that the equation of motion gives a steady state where the thickness of the boundary layer is inversely proportional to the fractional power of the rate of motion. Only where the power term is greater than 1 will there be a catastrophe. EMERICK inquired why results did not appear to agree with observed mixing and rates of reaction which were obtained by top blowing the oxygen Bessemer where slag mixing apparently does help in the rate of the reaction. SHANAHAN said that their experimental systems did not duplicate these conditions. He pointed out that a viscous slag would tend to mix more with the metal. TENENBAUM added that, in addition to the effect of the oxygen jet, the turbulence in the metal as a result of CO generation in the top-blown Bessemer would be of considerable significance.

by A. M. Decker

The Chemical Equilibrium in the

Basic Bessemer Process

We have actually no method which allows us a quantitative study of the equilibrium between the slag and the metal at the end of the blow in the basic Bessemer converters. Our paper would suggest a simple solution, based on classical formulas.

At first, we must notice that the isoactivity diagram of FeO of Turkdogan and Pearson[1] cannot be applied in our case, probably because of the high phosphorus-pentoxide content of the slag. Therefore we searched for another method to calculate the activity of the iron oxide. We have supposed that the equilibrium was obtained between the slag and the metallic bath at the end of the blow.

We studied successively the questions of manganese, phosphorus, oxygen, and sulfur.

MANGANESE

J. Chipman, J. B. Gero, and T. B. Winkler[2] have established the free energy of the reaction

$$[Mn] + (FeO) = (MnO) + Fe \qquad (1)$$

for which

$$\log K_{MnO} = \frac{6440}{T} - 2.95$$

This relation must also be verified in the case of our slags provided that we know the respective activities of the oxides.

We have in the pure system

$$a_{MnO} = \frac{n_{MnO}}{N} \qquad a_{FeO} = \frac{n_{FeO}}{N} \qquad n_{MnO} + n_{FeO} = N$$

We have supposed that in our complex system

Dr. Decker is with the C.N.R.M., Liége, Belgium.

the mutual ideality of the two oxides is always realized and that the two relations are still useful. But here n_{FeO} and N are unknown. The FeO content of the cold slag is not necessarily the equilibrium content in the liquid slag.

To explain the relation between the total iron content of the slag and this quantity n_{FeO} which interests us, we have supposed that the MnO content of the slag and the residual Mn content of the metal given by the slag and metal analysis are right and calculated n_{FeO} by

$$n_{FeO} = \frac{n_{MnO}}{[Mn] \, K_1}$$

and examined the relation for several slags between this quantity and various other factors of the slag. We observed a very close relation between n_{FeO} and the total iron content of the slag which can be written as follows:

$$n_{FeO} = 0.0109 Fe_{total} \% + 0.014$$

where n_{FeO} are moles and $Fe_{total} \%$ by weight.

This statistical formula has perhaps no absolute value, but, as the correlation coefficients were of about 0.65 and even of 0.8, we have assumed that the relation would be valuable in all cases of the basic Bessemer slags.

With this assumption, it is possible to calculate the residual manganese content of the metallic bath. Figure 1 shows the analyzed Mn content versus the calculated Mn content for three Belgian steelworks, at the time we established the above relation, and for two German steelworks also. Figure 2 shows the same for 150 blows of another steelworks.

It seems that the formula is valid not only for the basic Bessemer process, but is also valid in some

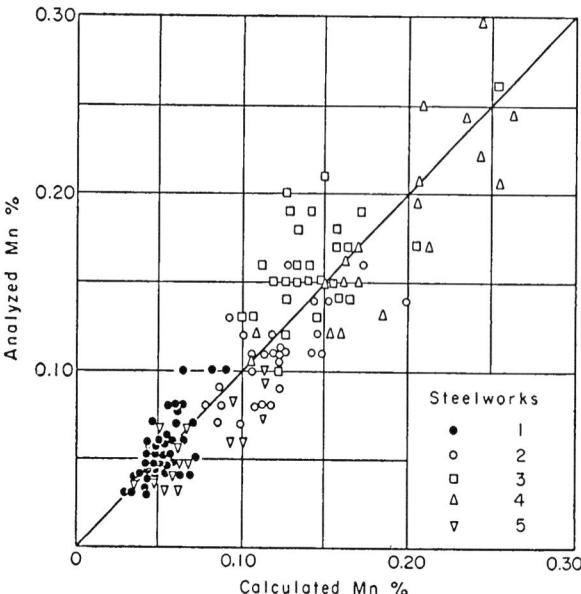

Fig. 1. Relation between analyzed and calculated residual manganese content of the bath.

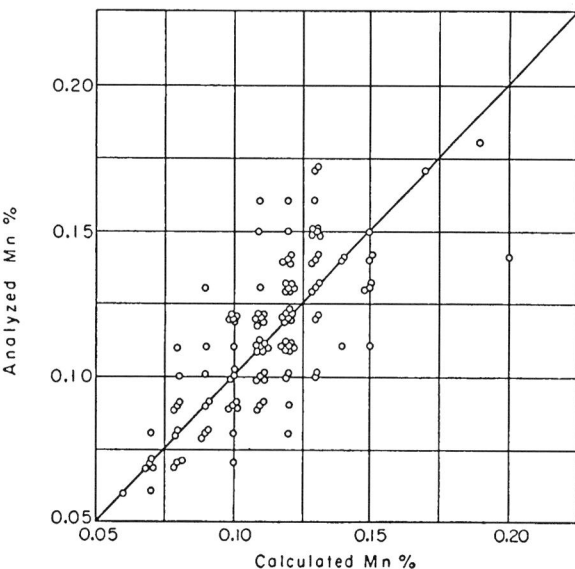

Fig. 2. Relation between analyzed and calculated residual manganese content of the bath for steelworks Number 6.

cases for the open-hearth process. Figure 3 shows the results obtained with the open-hearth heats of Herasymenko and Speight.[3]

PHOSPHORUS AND OXYGEN

We want to recall at the beginning two series of experiments (Bookey, Richardson, and Welch[4]) which seem to us to be very important: first, the dephosphorization experiments in a calcium-oxide crucible at temperatures between 1550 and 1600° C between liquid iron and a solid slag, and, second, similar experiments (Fischer and vom Ende[5]) in lime crucibles at temperatures between 1550 and 1700° C between molten iron and molten slag.

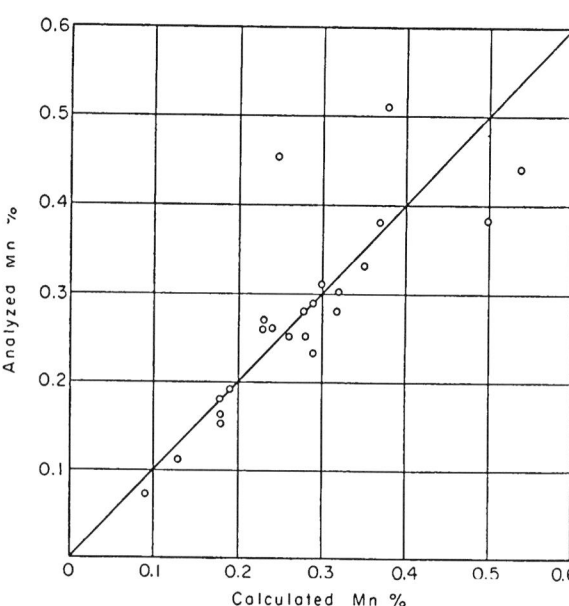

Fig. 3. Relation between analyzed and calculated residual manganese contents of open-hearth heats of Herasymenko and Speight.[3]

Bookey, Richardson, and Welch show that in the conditions of their experiments

$$\log K_P = \frac{1}{[P]^2[O]^5} \qquad (2)$$

Fischer and vom Ende admit the same relation. In other words, we suppose in formulas 2 and 3 that we have saturation in CaO and in phosphate.

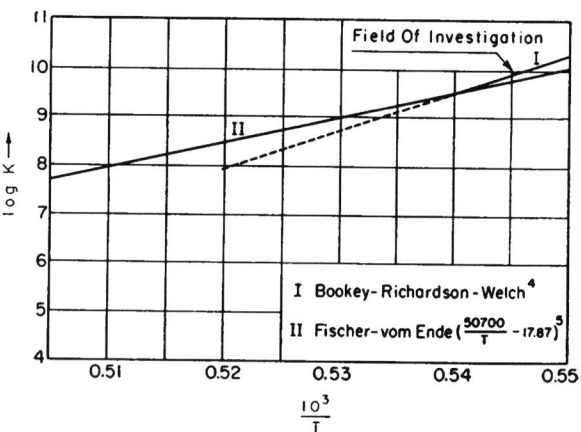

Fig. 4. Comparison of the dephosphorization equilibrium constants from Bookey, Richardson, and Welch,[4] and Fischer and vom Ende.[5]

Figure 4 shows us the two equations, in which Fischer and vom Ende's relation may be written:

$$\log K_P = \frac{50,700}{T} - 17.87 \qquad (3)$$

Speith and vom Ende[6] have taken in the basic Bessemer vessel bombs for O determination and have found that the oxygen contents associated with the phosphorus contents show a good distribution along the above line.

Now the following question arises: how is it possible, from the slag analysis, to confirm or disprove this relationship?

We have transformed the weight analysis of the slag in moles by dividing the CaO, MnO, SiO₂, and P₂O₅ contents of the slag by their respective molecular weights; we took for MgO + Al₂O₃, which generally was not analyzed, 0.05 mole, for FeO the calculated value as explained before, and for Fe₂O₃ the difference between the total iron content and the FeO content. From the total number of moles N, we subtract the CaO combined as $4CaO \cdot P_2O_5$, $2CaO \cdot SiO_2$, $2CaO \cdot Fe_2O_3$, and we found a number N'.

Generally, the lime so combined is lower than the whole CaO content of the system.

We supposed then, as a first approximation, that the ratio n_{FeO}/N' would give us a value of the activity of the iron oxide of the slag. That would demonstrate that FeO behaves ideally in a system of $4CaO \cdot P_2O_5$, $2CaO \cdot SiO_2$, $2CaO \cdot Fe_2O_3$, CaO, MnO, MgO.

With the well-known formula of Taylor and Chipman[7] which gives us

$$\log K_{FeO} = \log \frac{[O]}{a_{FeO}} = -\frac{6320}{T} + 2.734 \quad (4)$$

we may find O'.

We know the phosphorus content by the analysis of the metallic bath, and we may then calculate $\log \dfrac{1}{[P]^2[O]'^5}$ and put the values in a figure versus $1/T$. Figures 5a,b,c, gives the results for three steelworks. We also indicated Fischer's curve. The regression coefficients indicating the slope of the correlation are all very close to the slope of Fischer's curve.

If we look at the position of these curves, we see for steelworks (1) a position lower than the curve, for (2) the same situation, and for (3) a position above the curve.

Looking more closely at these figures, we see that for (1) the heats with a negative CaO content are under the corresponding regression line, and that for (2) and (3) the heats with an uncombined CaO content, lower than the medium value, are situated under the corresponding lines.

In other words, we observe effectively a growing relation of the logarithm as a function of $1/T$, but, for a given abscissa, the position of the different heats depends on the CaO content. Thus our method of calculating [O] or a_{FeO} is only a first approximation.

We said at the beginning that our system was saturated with calcium oxide. In fact, we did not take into consideration the CaO content which we obtained after subtraction of the phosphates, silicates, and ferrites. It is not probable that, whatever the calcium oxide content, which depends on the lime addition at the beginning of the blow, the

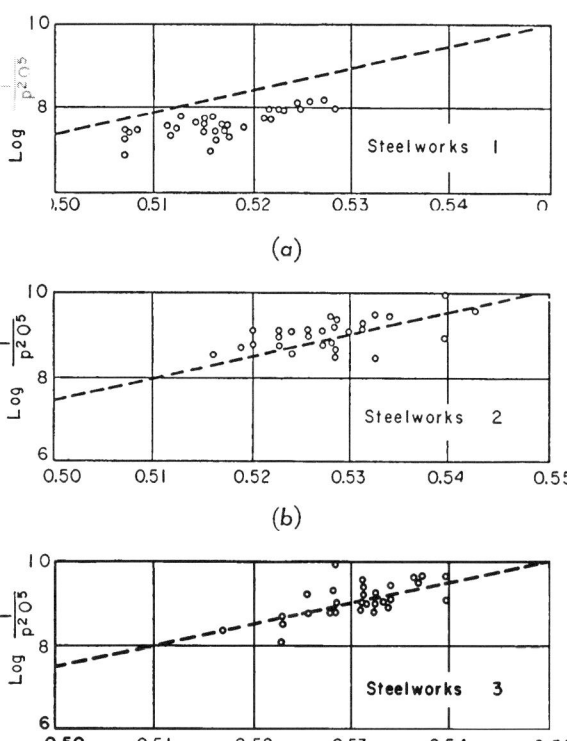

Fig. 5. Dephosphorization constant versus temperature at various steelworks.

finally obtained system would be exactly saturated. It may contain too much CaO or not enough.

In the solid system, which has been studied by Bookey, Richardson, and Welch, the solubility of CaO is approximately 4%, which is very low. For a similar reason, we have supposed that the CaO is not very soluble in our liquid slags, but that this low solubility is proportional to the FeO content of the slag.

We therefore admit that the noncombined CaO is not in the phosphate phase, except for a certain quantity which is dissolved and is a function of the total iron content. We have supposed that above 8% Fe in the slag, we have a solubility of 0.01 mole CaO per 0.8% more Fe, with no CaO solubility under this level. This choice gave us the best results.

In other words, the CaO excess may be calculated by:

$$n_{CaO \; excess} = n_{CaO \; total} - 4n_{P_2O_5} - 2n_{SiO_2}$$
$$- 2n_{Fe_2O_3} - (n_{FeO} - 0.1) \quad (5)$$

This factor must be positive to have saturation. So we obtain a new number N'' of moles by subtracting from N, which we had at the beginning, the whole CaO$_{total}$ minus the dissolves CaO, or what is the same, the combined and the excess dissolved lime. We calculated then, as before, an oxygen content, and made the diagram $\log 1/P_2O_5 = f(1/T)$.

We obtain new figures and following correlation coefficients for heats of five steelworks:

Steelworks	(1)	(2)	(3)	(4)	(5)
Number of heats	40	28	32	125	126
r	0.79	0.56	0.67	0.94	0.80

The correlation coefficients are much higher than before for the same populations, and are even higher in some cases than the correlation coefficients which Fischer found for his laboratory experiments with O analysis by the vacuum method, without slag calculations. Except for the first steelworks the others are now very close to the line of Fischer and

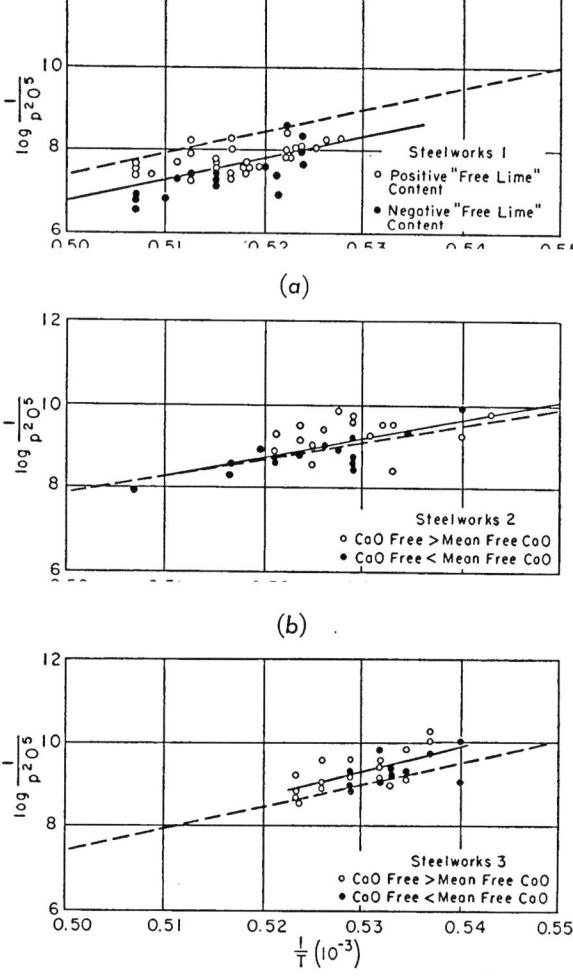

Fig. 6. Influence of method of computing basicity on dephosphorization calculations for various steelworks' data.

vom Ende (Fig. 6a,b,c,), and we have the same regression coefficient.

In Fig. 6a the heats are not saturated with lime, and therefore, we calculated a too high oxygen content.

In conclusion, with simple suppositions about the slag composition, and without oxygen analysis, we have calculated an oxygen content which is in very good agreement with Fischer's formula and which

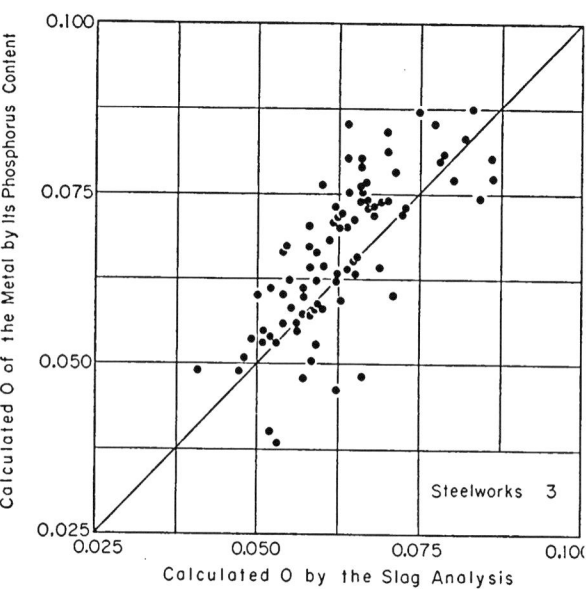

Fig. 7. Correlation of oxygen content of the bath with that calculated from slag analysis for steelworks Number 3.

has been based only on the metal-bath analysis. We therefore think that our method is correct and that our calculations give a good approximation. Figures 7 and 8 give the calculated oxygen by the two methods. Figure 9 gives $\dfrac{\text{O analyzed}}{\text{O saturation}} = a_{FeO}$ of Fischer's experiments versus the a_{FeO} calculated by the slag for 23 of his heats (Fe < 30%, P_2O_5 < 37.5%).

SULFUR

Since we have analyzed the activity of the iron oxide, we may examine the question of sulfur.

Let us admit at first that the whole sulfur of the

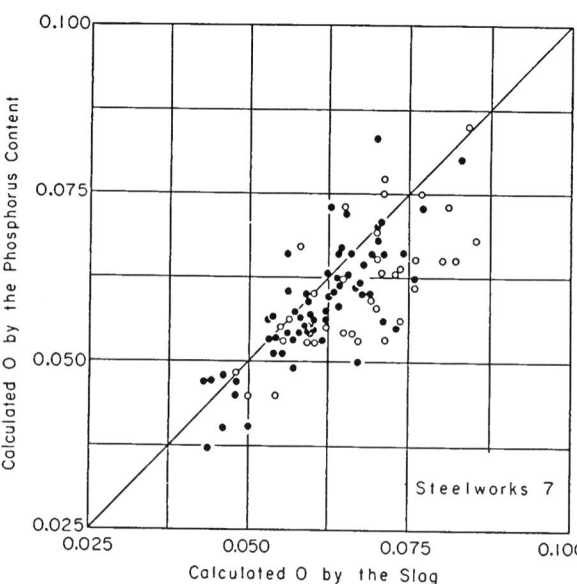

Fig. 8. Correlation of oxygen content of the bath with that calculated from slag analysis for steelworks Number 7.

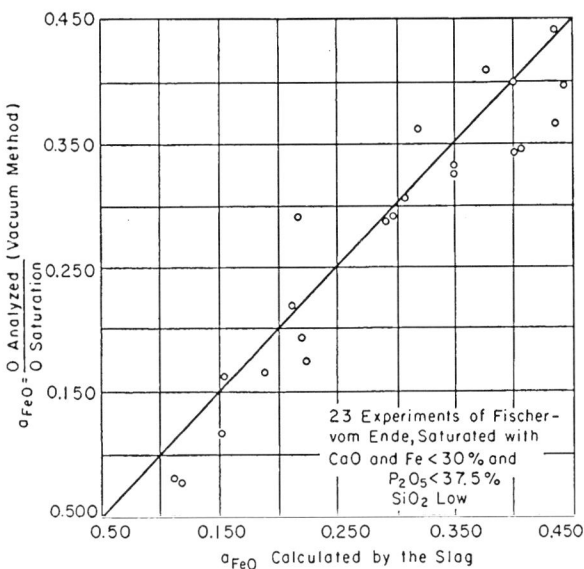

Fig. 9. The activity of FeO from the oxygen analysis of the bath and from the analysis of the slag for the experiments of Fischer and vom Ende.[5]

slag is present only as calcium sulfide. Then, the equation

$$(CaS) + (FeO) = (CaO) + [S] + Fe \qquad (6)$$

is of the greatest importance.

For our lime-saturated slags, we know a_{FeO} and $a_{CaO} = 1$; we must know a_{CaS} and the equilibrium constant.

The calculation of a_{CaS} is naturally difficult, because we have, in addition to our phosphate slag, lime in excess which also has a desulfurizing power. We adopted the following supposition: calcium sulfide distributes itself in the same manner in the phase containing the lime excess and the phosphate phase. In other words:

$$a_S = \frac{(S)}{32(N'' + n_{CaO \text{ excess}})} = \frac{(S)}{32 N'}$$

where N' is the total number of moles minus the lime combined as phosphate, silicate, and ferrite.

We had at our disposal 29 heats of one steel plant, calculated a medium value of the preceding constant, and found 14.7; with this value, we then calculated [S] in the opposite way for this and three other plants. Table 1 shows the results obtained.

Table 1. Statistical Analysis of Sulfur Contents

	Steel Plant	Number of Heats	Mean Analyzed $S_{content}$ (10^{-3})	Mean Absolute Difference (% 10^{-3})	Mean Relative Difference (%)
	3	29	23.2	3.2	13.8
	2	27	34.7	5.3	15.3
	4	10	34.7	5.6	16.1
$S_2 = 14.7\, a_S \cdot a_{FeO}$	5	11	40.6	11.2*	27.6
	Total	77	31.3	5.4	17.2

* All the calculated values are too low.

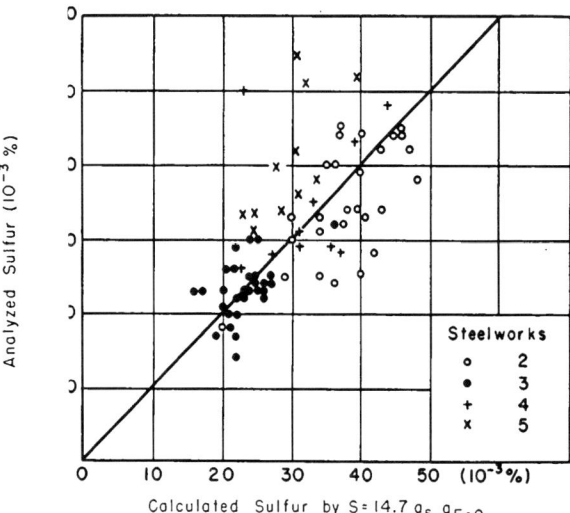

Fig. 10. Relation between the analyzed sulfur contents of the bath.

Figure 10 gives the analyzed sulfur versus the calculated sulfur by this method, and we see that the distribution of the points appears to be good. We started with 3 types of pig irons which were for steelworks 3, 2, 4, and 5 of 1% Mn, 0.5% Mn, and 0.3% Mn. The iron content of the slag varies between 7.11 and 18.25%. There is a big deviation for the points of steelworks 5. As we encounter in this case many heats lower than 1600° C, we explain, as a first idea, the deviation by a still unknown variation of the constant with the temperature. The constant would be valid between 1600 and 1650° C.

It is also possible to calculate the sulfur from the residual manganese content of the steel bath.

The formula [Mn] + [S] = (MnS) keeps its value at steelmaking temperatures and its value is, by

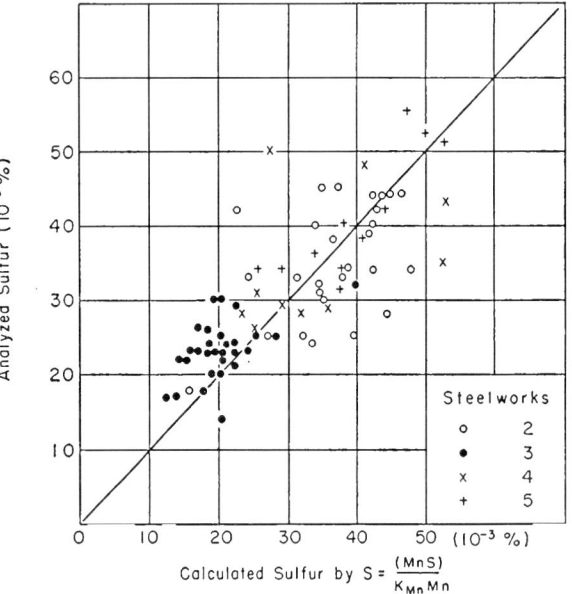

Fig. 11. Relation between the analyzed and computed sulfur of the bath by the use of residual manganese content.

combination of the data of Richardson and Jeffes[8] and Chipman:[9]

$$\Delta G^\circ_{Mn} = -31.250 + 19.23\ T$$

At 1600° C for a steel of 0.1% Mn, 0.03% S, this relation gives

$$a_{MnS} = 0.00083$$

If we assume that $a_{MnS} = (MnS)$ and if we note that $a_S \simeq 0.01$ in the slags, we come to the conclusion that (MnS) is about 10% of the whole sulfur content.

By means of a method similar to those used in the first case, we have found for $(MnO) + (CaS) = (MnS) + (CaO)$ a constant which seems to depend on temperature and can be written

$$\log K_{MnS} = \frac{5430}{T} - 2.72$$

The sulfur content so calculated gives Fig. 11 and seems to be in a good correlation with the analyzed value.

SUMMARY

We have shown that, in the case of the basic Bessemer heats at the end of the blow, there exists a state of equilibrium. It is possible, with the aid of well-known formulas, to calculate the residual manganese, phosphorus, oxygen, and sulfur contents of the steel bath if the slag analysis is known.

References

1. E. T. Turkdogan and J. Pearson, *J. Iron Steel Inst. London*, **173**, 217 (1953).
2. J. Chipman, J. B. Gero, and T. B. Winkler, *Trans. A.I.M.E.*, **188**, 341 (1950).
3. P. Herasymenko and G. E. Speight, *J. Iron Steel Inst. London*, **166**, 289 (1950).
4. J. B. Bookey, F. D. Richardson, and J. E. Welch, *J. Iron Steel Inst. London*, **171**, 404 (1952).
5. W. Fischer and H. vom Ende, *Stahl u. Eisen*, **72**, 1398 (1952).
6. K. G. Speith and H. vom Ende, *Stahl u. Eisen*, **74**, 509 (1954).
7. C. R. Taylor and J. Chipman, *Trans. A.I.M.E.*, **154**, 228 (1943).
8. F. D. Richardson and J. H. E. Jeffes, *J. Iron Steel Inst. London*, **160**, 261 (1948).
9. J. Chipman, *Trans. Am. Soc. Metals,* **30**, 840 (1942).

Discussion

ELLIOTT expressed interest in the point of view brought out in the paper that it was possible in this process to obtain equilibrium when reactions were proceeding so very rapidly. He felt that perhaps some further consideration should be given to the point of view of the theoreticians with regard to kinetic conditions which limit reactions. DARKEN also expressed an interest in this point, but it was generally agreed among the author and two discussers that a contributing factor was the very violent and intimate mixing which was achieved in the basic Bessemer vessel.

TENENBAUM raised the question of how the activity of oxygen in the slag was obtained. DECKER said it was taken from the experimental results of Fischer and vom Ende. Measurement of the oxygen content was not possible. However, they found that generally the oxygen content lies between 0.05 and 0.07% in the bath. The dispersion in the oxygen values is approximately 0.005%. DARKEN asked why one should find equilibrium between the slag and metal rather than the gas and metal. DECKER said their observations showed consistent performance with regard to slag-metal equilibrium, whether they were blowing air or oxygen-enriched air, and they obtained equivalent results for metal analyses. DARKEN also noted that the oxgyen value of the slag was important rather than the oxygen content of the metal. DECKER said that there was no difference in the temperatures of the slag and metal. He closed the discussion by pointing out that there was a big difference between the basic open-hearth furnace and the basic Bessemer with regard to the carbon-oxygen reaction. In the basic open hearth the oxygen level is usually above the equilibrium value for the reaction, whereas in the basic Bessemer it is below. This has been validated by work recorded by Speith and vom Ende at Huckingen.

H. B. Emerick
D. L. Murphy
Presiding

Section 8

Solidification of Castings and Ingots

by M. Tenenbaum

Solidification of Steel Ingots

INTRODUCTION

Any description of the solidification of steel ingots involves considerable speculation regarding the specific mechanisms involved. The interpretation of the results of studies on this subject is by no means universally accepted. In this paper an attempt has been made to present a description of the solidification process which is consistent with most reliable observations and which has proved useful as a basis for controlling the characteristics of the ingot product.

THE TEEMING OPERATION

The teeming period begins when a ladle full of metal has been transferred from the steelmaking furnace to the pouring platform. General practice is to pour directly into the mold through a refractory nozzle located at the bottom of the ladle. In special cases, this phase of the operation is done indirectly either by pouring into a tundish, or basket, or by pouring through a refractory system which leads the metal into the bottom of the ingot mold. This discussion will be limited almost entirely to the direct pour. During passage through the nozzle, there is some erosion of the nozzle refractories. This refractory material is frequently carried along with the molten stream into the molds. In transferring from the ladle to the mold, there is a drop in temperature of the metal which generally ranges between 30 to 60° F. As a result of this temperature change, there is a corresponding change in the concentration of certain deoxidizing elements and deoxidation products.

In top pouring, metal entering the mold strikes the stool surface and, to a greater or less extent, splashes up against the four walls of the mold. As fast as the metal rises in the mold, a thin outer frozen shell forms. If the metal is completely deoxidized, an upper crust or scum may sometimes be seen as soon as the rate of pour slows down. Several minutes after the ingot has been poured, a general shrinkage can be observed in practically all carbon steels, as a result of which an air gap forms between the outer ingot surface and the mold wall.

When considerable amounts of gas form during solidification, bubbles rise to the upper surface, stirring the metal and keeping the ingot top molten. Where the tendency towards gas formation is strong, the gas rises to the surface of an ingot in a rather uniform peripheral rim. This rim proceeds inward as the ingot solidifies. At some later stage this rimming action can be stopped by placing plates over the remaining liquid surface and chilling the upper crust or by deoxidation of the liquid metal remaining in the ingot core. Steels freezing with these characteristics are called rimmed steels.

Intermediate between the steels which solidify quietly and those which rim on solidification, there is the group of semikilled steels. Upon completion of the pour of a semikilled steel, there is a slight tendency toward gas formation. Generally, on these steels, the upper surface crusts over within about a minute. Depending on the extent to which gas forms, the upper surface will either erupt, bulge, remain unchanged, or sink following the crusting over of the top.

GENERAL SOLIDIFICATION PROCESSES

The preceding review of the teeming operation indicates some of the processes which control the behavior and appearance of an ingot during solidification.

The first process is that of heat removal. This is the most obvious process, since the primary purpose of this phase of the operation is to convert liquid steel poured at a relatively high temperature into a solidified ingot at a considerably lower temperature.

Dr. Tenenbaum is Superintendent, Metallurgical Department, Inland Steel Company, East Chicago, Indiana.

The second process is an indirect result of the manner by which heat is transferred from the solidifying ingot to the relatively cold mold. In this process definite temperature gradients are developed between the inside and the outside during ingot solidification. These gradients have a very significant effect on the manner of solidification and the resultant ingot structure.

A third process is that of selective freezing. This is the same phenomenon which is encountered in so many metallurgical and chemical processes wherein a relatively pure constituent freezes from a liquid phase. As a result, solid crystals separate from the metal which are purer than the original liquid, and the remaining liquid is necessarily less pure. Obviously, in a one-component system, there can be no selective freezing.

A fourth process is gas evolution from the steel ingot during solidification. Both carbon and oxygen are dissolved in liquid steel. If their activities in the solidifying ingot are sufficiently high, a reaction occurs, and a carbon-oxide gas forms. This reaction becomes apparent through rimming action, eruptions, or blowholes in the final ingot.

INGOT SOLIDIFICATION

To help describe the manner in which steel ingots solidify, the freezing of several hypothetical ingots will be described. Three conditions will be imposed upon the first such ingot which are not applicable to the solidification of steel. In successive examples, these conditions will be altered to fit more closely the situation existing in the actual freezing of steel ingots. The three conditions imposed on the first hypothetical ingot are:

1. The molten liquid is a single-component phase (such as pure iron).
2. There is no volume change during solidification. (This necessarily means that there is no heat of solidification.)
3. Heat is conducted away rapidly, so that a definite temperature gradient between the liquid and the adjacent solid-liquid interface is always maintained.

In pouring this liquid metal into the mold, an outer zone or shell chills immediately upon contact with a heavy, cold, solid surface. The situation next to the mold wall is shown schematically in Fig. 1. A steep gradient exists between this frozen shell and the adjacent liquid metal. Crystals form ahead of the freezing interface and these appear more or less at random in the outer shell. Under these conditions, freezing proceeds at an extremely rapid rate, and there is insufficient time for any appreciable segregation. As the skin thickens, the mold heats up, and the cooling rate decreases. The temperature gradient is then reduced so that crystals no longer form ahead of the freezing interface.

The iron crystals extending farthest into the liquid metal from the frozen shell lose heat less rapidly than the adjacent liquid, and therefore, the adjacent liquid solidifies faster. Under these conditions, freezing tends to take place along a rather plain surface. There can be no segregation under these conditions since there is only one component being considered in the solidifying system. With the assumed condition of a temperature gradient throughout solidification, freezing then proceeds into the core of the ingot. Some solidification also takes place down from the top surface and up from the stool. The resultant ingot is solid throughout as the result of the assumption that there is no shrinkage.

If, instead, it is assumed that there is a decrease in volume during solidification, the process approaches that actually encountered in freezing of killed-steel ingots. Freezing under these conditions would again start with the formation of a thin shell composed of randomly oriented unsegregated iron crystals. In the first stages of solidification, this shell will tend to shrink as the result of the volume change. This tendency toward shrinkage, however, is opposed by the pressure that the contained liquid metal exerts on the inner faces of the thin shell. In effect, then, this freshly frozen outer skin may actually stretch in response to the force exerted by the contained liquid metal. It is presumed that many of the surface problems associated with the working and utilization of killed-steel ingots arise during this early stage of solidification.

After sufficient metal has frozen to withstand the pressure of the liquid from within, visible shrinkage occurs, and the outer ingot surface separates from the mold wall. (Although the evidence is not direct, it might be expected there would be a change in freezing rate, or at least a reduction in the temperature gradients resulting from the formation of this air gap.)

During this early stage of the solidification process, heat is radiated rapidly from the upper surface, and a solid upper crust forms over the ingot. With further solidification, the liquid-metal level in the mold must drop as a result of shrinkage. A cavity containing a very low pressure gas forms. Thus separated from the balance of the ingot, the upper-crust temperature drops rapidly. Radiation losses

Fig. 1. Schematic drawing illustrating early solidification process. (*Basic Open Hearth Steelmaking*)

between the liquid-metal surface and the cold upper crust must be high. As a result of radiation losses to the cold upper crust, it is possible for the liquid-metal surface in the shrinkage cavity to freeze over and thus form a solid bridge. The metal below the bridge finally separates from this solid crust, and a new cavity begins to form. This process continues until the ingot is completely frozen. This pattern, which was brought out decades ago by studying the freezing of stearine ingots, is shown in Fig. 2.

Fig. 2. Cross section of stearine ingot showing effects of columnar freezing and shrinkage. (*Brearley and Brearley*)

Instead of a pure metal, consider now the solidification of ordinary fully killed steels under conditions where a definite temperature gradient exists *at the freezing interface throughout the process.* Shrinkage depends primarily on the volume change previously described, and this will be essentially unaltered.

In the solidification of fully killed steels, considerable selective freezing occurs. With the metal freezing selectively, continued crystallization depends on (1) the rate at which iron can diffuse toward the crystal surface, (2) the rate at which solute elements diffuse away from the crystal surface, and (3) the rate at which heat can be extracted.

In the formation of the outer skin, freezing is again very rapid. The metal cools very quickly through the freezing range, selective freezing is suppressed, and there is no marked segregation. Nuclei form at a high rate ahead of the freezing interface,

causing random dendrites to appear which interfere with columnar dendrites freezing out from the mold wall. With a reduction in the temperature gradient, the nuclei ahead of the freezing interface also disappear. There is no interference then of dendrites freezing from the mold wall with dendrites forming ahead of this interface. The dendrites are free to grow at right angles to this direction, but in this situation they soon come in contact with adjacent dendrites, and their formation is stopped. Only those perpendicular to the mold wall persist.

Since the protruding points of the dendrites are closer to the mass of liquid with normal iron content, freezing can occur on such protruding points rather than on the surfaces adjacent to the liquid with depleted iron content. This in effect means that freezing progresses in the direction by which iron atoms can be supplied most readily. Eventually, the lower melting liquid between the dendrites freezes. The reaching out of the dendrites at right angles to the mold walls gives a characteristic columnar freezing that can be observed in some degree in practically all steel ingots.

If the thermal gradients persist throughout solidification, the columnar dendrites proceed to the ingot core. Under these conditions, there is considerable microsegregation between adjacent dendrites, but there is no mass rejection of solute elements, and macrosegregation should be practically negligible.

In the solidification of large steel ingots, definite thermal gradients do not persist throughout solidification. In the solidification of most carbon steel ingots, the initial thermal gradients disappear quite rapidly. Even with considerable superheat in the liquid metal being poured into the mold, the temperature of the inner dendrites soon becomes essentially the same as the temperature in the adjacent liquid metal. As the temperature in the freezing zone drops, random dendrites again begin to form ahead of the columnar dendrites. The entire liquid mass eventually enters a temperature range below the liquidus. In the zone just inside of the freezing interface, the liquid then becomes depleted of iron atoms, and the columnar dendrites stop growing. *The random dendrite formation then persists into* the center of the ingot. The solidification pattern resembles that of the ingot shown in Fig. 3.

As the temperature of the liquid metal in the ingot core drops below the liquidus, independent crystals form in the liquid phase. In the transition from liquid to solid, there is a definite increase in density. Consequently, as the crystals gradually grow, they settle to the bottom. The settling takes place over a long period of time, and as the metal freezes slowly in from the side, a cone of settling crystals accumulates at the bottom of the ingot. It would be expected that segregate present ahead of the freezing wall would be trapped in place by this rising cone. The entrapment of segregate between the cone of crystals and the metal freezing in from

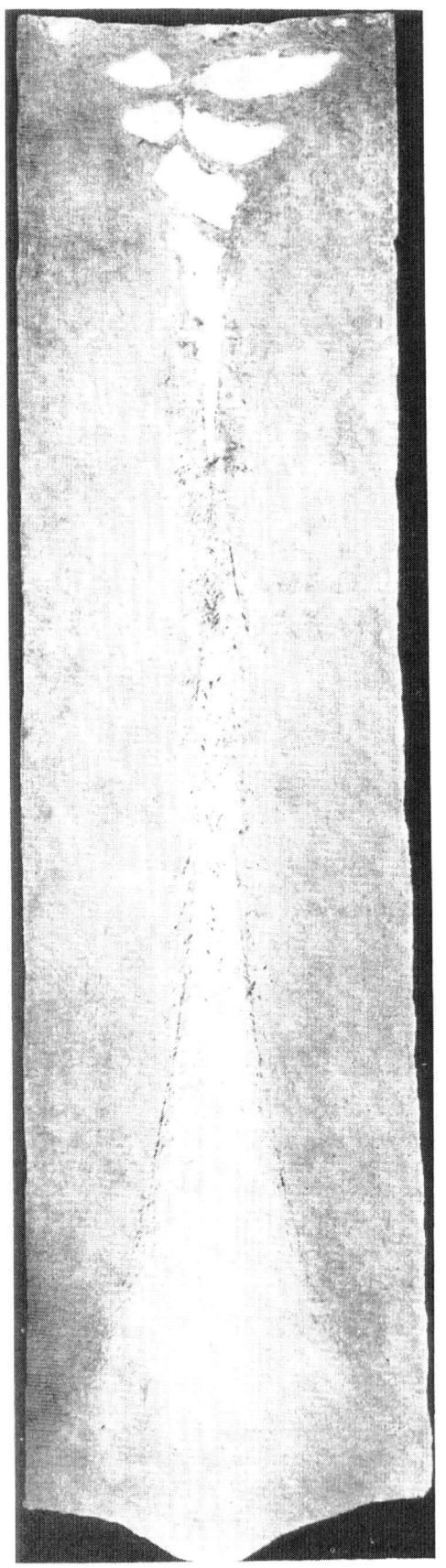

Fig. 3. Sulfur print of low-carbon aluminum-killed open-top ingot.

the mold wall results in an inverted-V segregate pattern.

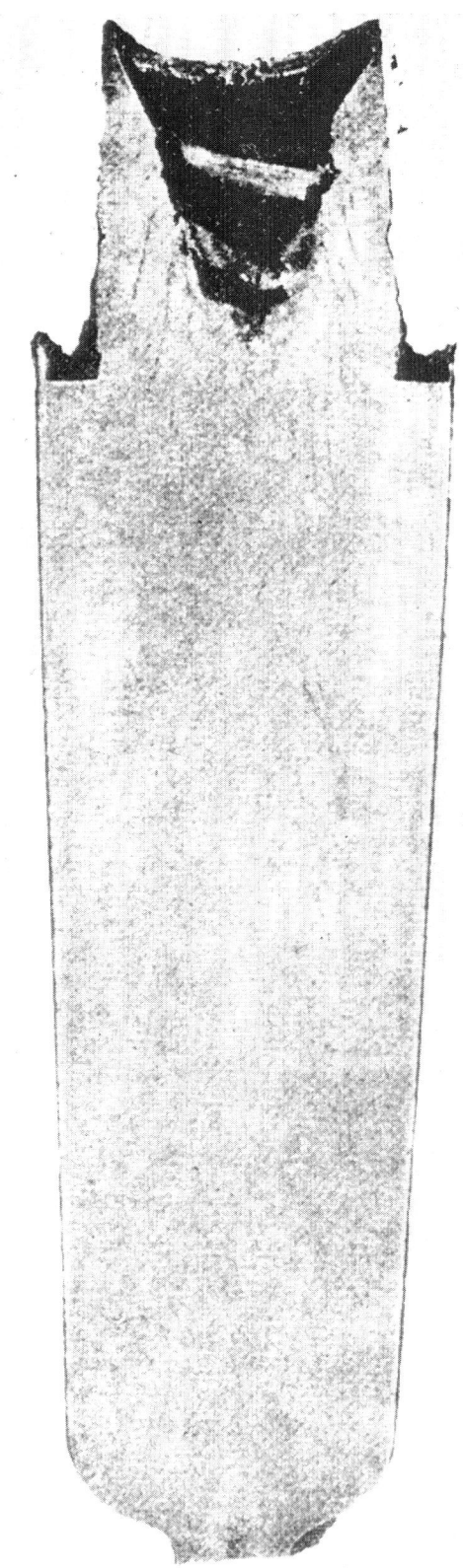

Fig. 4. Structure of split big-end-up hot-topped killed-steel ingot. (*Basic Open Hearth Steelmaking*)

The entire last stage of solidification is one in which the content of iron atoms in the liquid is gradually reduced, leaving a high concentration of solute. The final liquid that can freeze at the base of the pipe cavity is necessarily higher in solute than any other metal. With the final shrinkage, there is some tendency for this segregate to be dragged downward into a mushy core and produce the characteristic central inverted-V segregate.

SOLIDIFICATION IN BIG-END-UP HOT-TOPPED MOLDS

The discussion to this point has been limited to ingots poured into big-end-down molds without hot tops. When killed steels are poured into big-end-up hot-topped molds, the same general solidification processes are involved. The resultant ingot structure however, is somewhat different from that already described. An example is given in Fig. 4. These differences in ingot structure arise from two basic differences in mold construction.

In the first place, metal is poured into the hot top. By intent this metal remains molten during much of the ingot-solidification period. As a result, the drop in the metal level which is caused by shrinkage during solidification may be limited entirely to the sinkhead section. If the metal in the hot top and the upper crust remains molten throughout solidification, the characteristic bridging described in the previous ingot disappears. The last part of the ingot to freeze is located at the base of the shrinkage cavity, and the major segregation is encountered in the same position. If the upper-surface insulation is not complete and an upper crust forms, then the same bridging effect as was mentioned before can be encountered in the hot-top section of the ingot.

The other difference in mold design which causes a difference in the solidification characteristics is the reversed taper of the mold wall. The actual freezing of the zone near the mold wall and the freezing in the columnar zone are quite the same as that described for the big-end-down mold. When freezing has progressed beyond the columnar zone, the thermal gradients are largely dissipated. Dendrites begin to form in the liquid center which settle slowly to the bottom of the ingot. The settling crystals slowly displace liquid metal upward. The segregate which was originally located just inside the columnar zone rises slowly along with the upward movement of the liquid metal. As the metal rises, the segregate tends to leave the solidifying face since there is no interference by a freezing wall in the path of the rise. The result is that streaks appear wherever the solidifying face includes the rising segregate. This solidification sequence is reflected in a different segregate pattern from that described earlier. Instead of a rather continuous segregate line outlining the negatively segregated central cone, a series of short segregate lines appears in the upper part of the body of the reversed-taper ingot. For the most part, these lines appear to start at the end of the columnar zone and rise steeply toward the hot-top section. In general, the segregate lines occur as a series of parallel streaks one within the other, each generally having its base at the end of the columnar zone.

As solidification proceeds, the liquid phase retains a high concentration of solute elements. Again, this high concentration appears as a highly segregated zone at the base of the pipe cavity. As before, V segregate forms in the upper-core zone as a result of the final shrinkage during the transition from a mushy to a completely solid state.

EFFECT OF GAS FORMATION

In the last part of this discussion, the influence of gas formation on the solidification process will be considered. Only the gas-forming reaction $[C] + [O] = CO$ will be considered. It should be recognized that other gases, such as CO_2, H_2, and N_2, can also be released during solidification, but these usually comprise only a small proportion of the total gas volume. It is recognized that the carbon-oxygen reaction is stimulated by increasing concentration of both carbon and oxygen and by reduced carbon-monoxide pressure. In the section dealing with gas formation during solidification, only big-end-down open-top ingots will be considered.

On fully killed steels, the oxygen activity is so low that significant gas is not evolved on freezing. With increasing oxygen activity, some gas formation occurs which is evident in the form of occasional blowholes located at the top central part of the ingot. In an ingot with marginal tendency toward gas formation, the first part of the solidification process is similar to that already described. As solidification proceeds in the upper central part of the ingot, there is a concentration of both carbon and oxygen. Moreover, the tendency to shrink during solidification causes a reduction in pressure at the freezing interface in the upper part of the ingot. As a result, gas bubbles first appear high in the ingot along the freezing interface. Initially, the rate of gas formation is slow, and the bubbles that do form are trapped in place by the solidifying metal. Gas continues to form slowly, and as further solidification occurs, the original bubble is gradually elongated into a channel hole as it follows the freezing interface. In consequence, these channels generally point toward the last part of the ingot to freeze. This situation can be seen in Fig. 5.

If the tendency toward gas formation is very slight, the shrinkage is only partly overcome. This early gas formation will compensate for the initial stages of solidification, while a definite shrinkage cavity forms during the last stage of solidification. Exactly why the gas formation and shrinkage processes occur in this particular sequence is not entirely clear. Tentatively, it might be speculated that earlier in the solidification period, when the liquid-metal temperature is relatively high, oxygen reacts with carbon to form gas bubbles; whereas at the lower temperatures existing toward the end of

Fig. 5. Structure of ingots in which shrinkage is partly compensated by gas formation.

the freezing process, the oxygen reacts with small percentages of deoxidizing elements usually present in these steels.

When the volume of gas formed becomes greater than the amount required to compensate for ingot shrinkage, the top surface can bulge upward. If the gas formation is not so great under these conditions as to result in eruptions through the upper surface, a typically semikilled-ingot structure re-

sults. The central blowhole formation is quite similar to that already described.

With further increase in the tendency toward gas formation, the structure of capped steel ingots is approached. Under these conditions, gas bubbles appear for the first time in observable amounts during the solidification of the subsurface zone. The appearance of subsurface blowholes with increasing tendency toward gas formation is shown in Fig. 6.

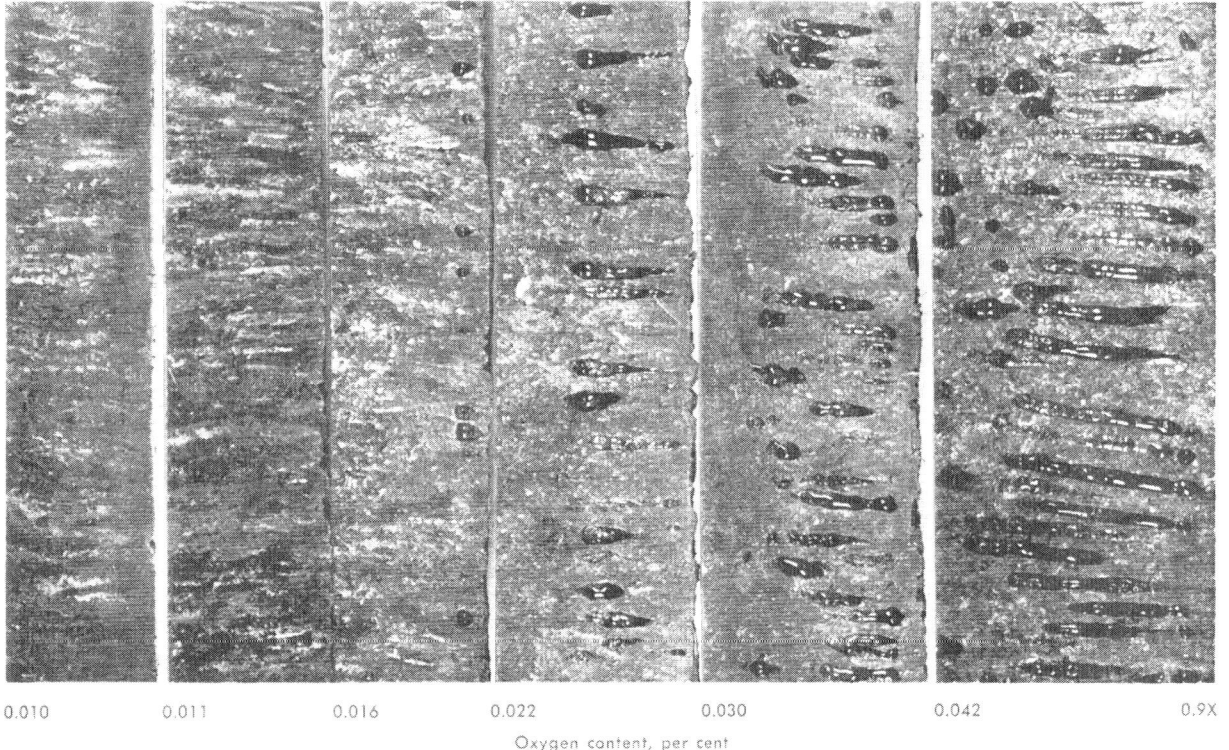

0.010   0.011   0.016   0.022   0.030   0.042   0.9X

Oxygen content, per cent

Fig. 6. Effect of total oxygen content of liquid steel on formation of subsurface blowholes in middle of capped ingots.

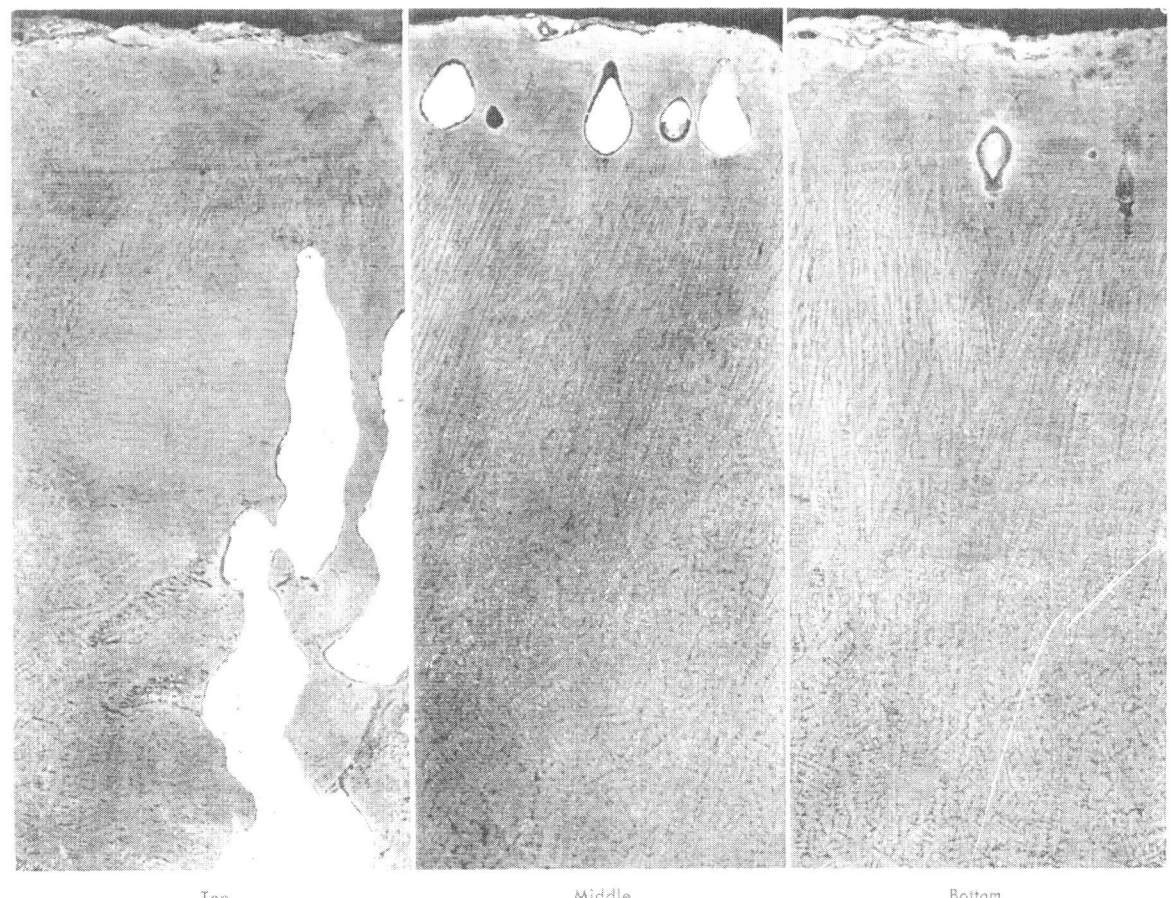

Top   Middle   Bottom

Fig. 7. Subsurface structure of ingot requiring small aluminum-capping addition.

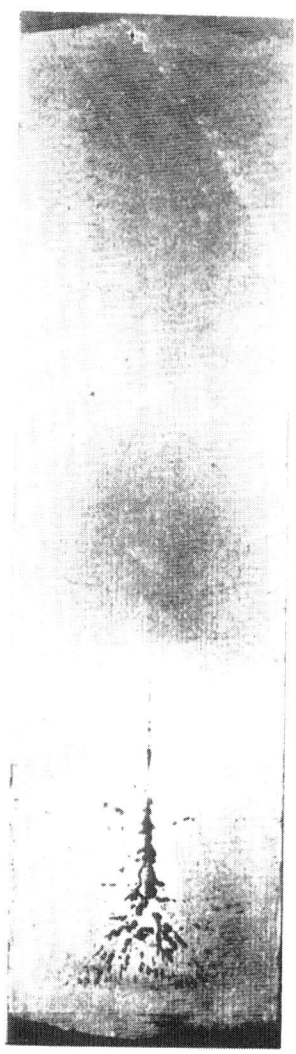

When gases first appear in the subsurface zone, they do not form in sufficiently large volume to evolve spontaneously from the metal. However, during the actual pour there is already a very vigorous motion because of the impact of the metal entering the mold, and it is likely that the early gas evolution is stimulated by this motion. With the completion of the pour, this motion stops, and the bubbles which can no longer rise through the ingots are entrapped. In the lower part of such ingots, it is very likely that gas formation is completely suppressed by the higher pressures that exist in this zone.

The formation of subsurface blowholes interferes slightly with columnar-dendrite freezing of the outer zone. As shown in Fig. 7, when blowhole formation is small, the departures from the columnar-freezing pattern previously described is not very great. If the blowhole formation is rather extensive, the freezing pattern becomes more nearly that of rimming types of steel. This pattern will be described in the next section. The freezing in the upper central zone is very similar to that which has already been described.

The gas bubbles in the subsurface zone probably originate in the higher carbon and oxygen liquid between the growing dendrites. In a quiet ingot these bubbles tend to be entrapped by the dendrite projections. In this way the gas will form continuously between adjacent columnar dendrites and therefore will appear as elongated channels. The extent of visible blowhole-channel formation increases with increasing oxygen content. Some provision must be made on ingots of this type for closing over the top. If such provision were not made, some of the gas that is formed would tend to erupt through the surface. It is common to deoxidize the top surface so as to form a solid upper crust.

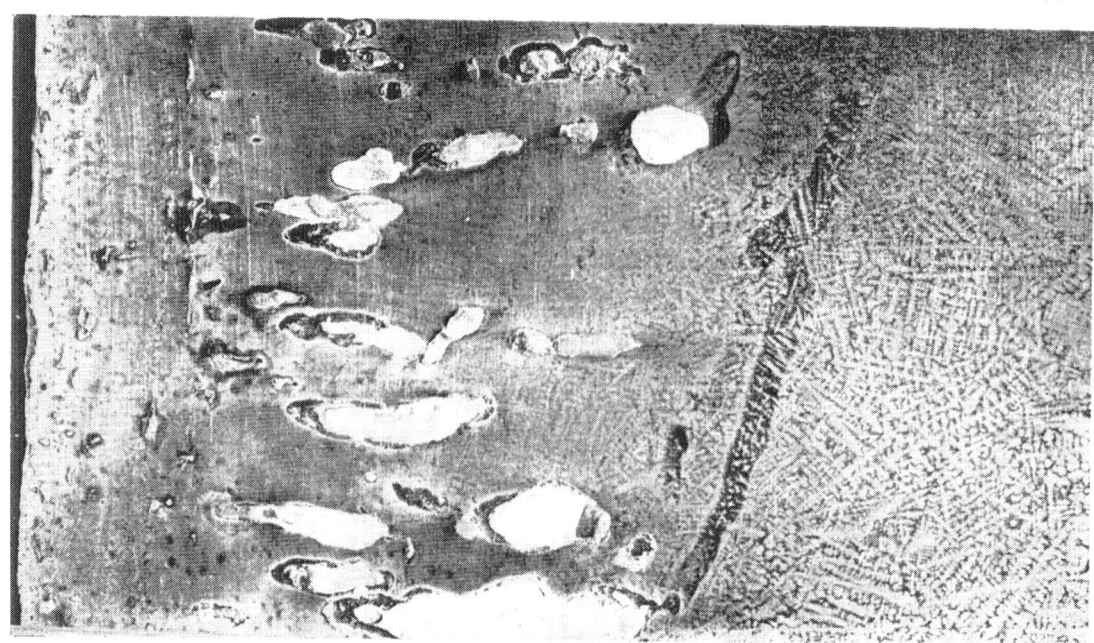

Fig. 9. Rising segregate in aluminum-capped ingot.

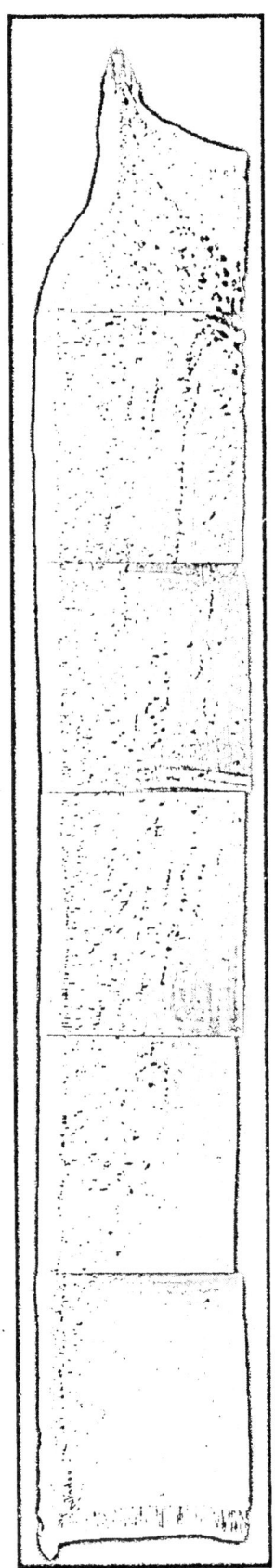

Fig. 10. Structure of split section of mechanically capped ingot.

This is a rather simple procedure when the tendency toward gas formation is moderate. The resulting structure is shown in Fig. 8.

For a short time after capping, bubbles continue to form at the freezing interface and sometimes rise through the liquid. It is this situation which leads to the formation of localized segregate streaks which appear commonly in many capped semikilled steels, an example of which is shown in Fig. 9. Although the exact mechanism by which such streaks are formed is uncertain, it is tentatively assumed that they trace the path along which a gas bubble rose from the solidifying face into the liquid.

At some point with increasing oxygen content, the tendency toward gas evolution becomes so great that the volume occupied by the gases formed in the subsurface zone increases continuously during the initial freezing period. The obvious result of such an increase in gas volume is to raise the upper surface of the ingot. On this type of rising steel a heavy mechanical cap may be placed over the upper surface after the ingot is poured. The function of the cap is to chill the upper crust to increase the pressure in the system and thus to suppress further gas formation. In this type of steel the liquid metal actually rises up to the cap rather than the cap being lowered onto the metal. Unless there is some form of eruption, all shrinkage is necessarily compensated by gas formation. An example of such an ingot is shown in Fig. 10.

RIMMING STEELS

When the volume of gas generated through the carbon-oxygen reaction becomes so great that bubbles rise continuously throughout the solidification period, keeping the top surface open and liquid, the solidification process changes to a marked degree.

The early freezing of rimming steel may be divided into two separate phases. The first phase involves solidification that occurs while an ingot is being teemed. The second phase begins after the mold is filled. During the first phase, a stream of liquid metal from the ladle is entering the mold, causing considerable motion in the body of the ingot. Features of solidification, such as the evolution and formation of gas, could obviously be affected by the physical motion of the metal entering the mold.

During the pouring of steels in which the carbon-oxygen reaction proceeds actively, there may be a very thin chilled outer skin that forms so fast that there is no chance for gas formation. The evidence is that if such a chill zone does form it must be very thin. There is evidence that gas evolution starts very early in steels containing a much lower oxygen content than is found in the normal rimming grades. Consequently, it can only be presumed that gas evolution in the normal rimming ingot begins very close to the outer surface early in the period while the ingot mold is still being filled. Figure 11, which is a photograph of the subsurface section of a mildly rimming steel ingot which was intentionally poured against an effectively water-cooled surface, demonstrates this point. Under these conditions, gas was formed and entrapped in the drastically chilled outer skin.

In the normal rimming ingot, early gas evolution is very likely encouraged by metal movement during pouring. As a result of such metal movement, much of the gas formed during the outer skin formation in a conventional top-poured ingot is washed out, and as a consequence, the outer skin of the rimmed zone is usually sound.

As the metal level builds up in the mold, the pressure in the bottom of the ingot necessarily increases. When pour is completed, the relative pressure in the ingot ranges from 0 to $1\frac{1}{2}$ atm from top to bottom. Gas formation in the bottom of a full ingot is retarded by the higher ferrostatic pressure. In the absence of any motion caused by the metal entering the mold, any gas bubbles generated must develop a rather substantial buoyant force in order to rise through the liquid metal.

It is recognized that the freezing interface offers an ideal surface for the reaction between carbon and oxygen. Gas bubbles form next to the solid surface, and if such gas forms rapidly enough, the bubbles will rise along the freezing interface. After the ingot has been poured, gas continues to form

near the bottom. The rapid motion of the liquid moving past the freezing surface along with the stream of rising gas bubbles serves to sweep away the less pure metal near the solid surface as well as any smaller gas bubbles being generated. This sweeping action also distributes the heat of solidification within the liquid core of the ingot. Freezing under these conditions need not be restricted by the relative diffusion rates of iron and solute atoms but rather is primarily affected by the rate of heat loss. In this situation, any zone closer to the mold wall should freeze more readily than a zone which extends further into the liquid, causing solidification to proceed in a rather smooth layer. Because segregated solute is constantly being swept from the freezing face, microsegregation in the rim of an actively rimming ingot is nearly always rather vague.

With rapid gas evolution, bubbles form rapidly next to the solid surface and extend well into the liquid metal. The rising gases sweep such bubbles upward, and liquid metal flows in behind. As a result, the skin and rim zones can be paradoxically sound, as shown in Fig. 12. In general, the most active sweeping action is encountered in the upper part of rimming ingots, whereas both gas formation and evolution are suppressed somewhat in the lower part of rimming ingots where the higher pressures exist. Resultant rim structures are generally sound in the upper part of the ingot, while the bottom sec-

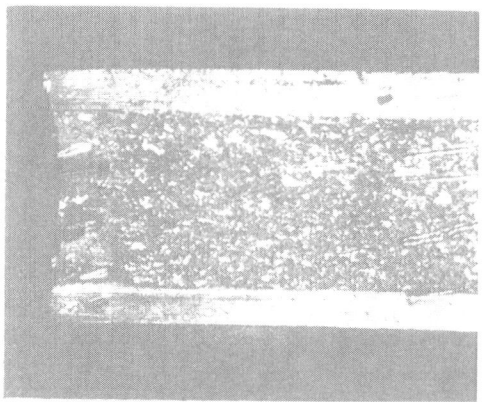

Fig. 11. Freshly broken surface of cast rimming steel section showing chill effect at outer surface.

and rise within the initial chilled skin, thus tending to perpetuate the metal motion that characterized the mold-filling period. With the rapid initial freezing rate, it is very likely that the degree of segregation is relatively small. When the freezing rate has slowed to the point where selective freezing becomes more pronounced, the gas-forming reaction would occur preferably in the zone of higher solute concentration. Freezing must still occur in the direction of thermal gradients, that is, at right angles to the mold wall.

Because of the differences in pressure, the gas will form more readily near the top of a full ingot than

tions characteristically contain channel blowholes in the rim.

As rimming action proceeds, solute elements accumulate and are mixed in the central liquid zone. The longer that rimming is allowed to proceed, the more segregate that is accumulated. To stop this segregation process, an ingot may be capped, thereby suppressing the gas-forming reaction. As in other types of capped steel, the aim in capping is to cause the top surface to crust over and thus entrap the gases formed. Since bubbles cannot be released under these conditions, the gas pressure increases, and the gas reaction is suppressed. At this stage, then,

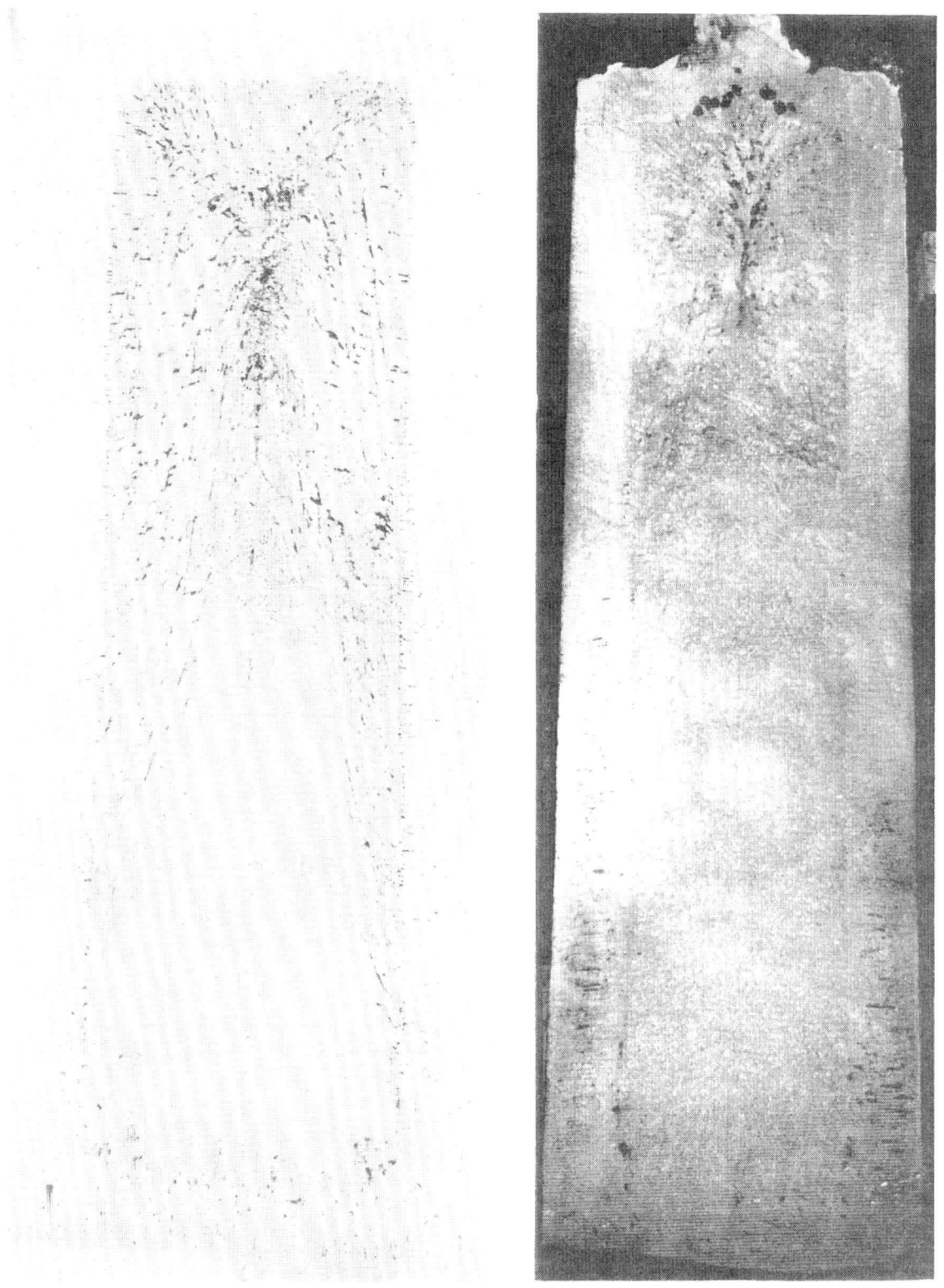

Fig. 12. Sulfur print (left), deep etch (right), rimming steel sections.

Fig. 13. Distribution of blowholes in rimming ingot over which 15 atm pressure was applied. (Hultgren and others.)

gas evolution stops, and the existing gas bubbles remain to outline the inner face of the rim zone. It has been demonstrated that if it were possible to build up sufficient pressure all further gas evolution could be suppressed. On the rimming ingot shown as Fig. 13, it was possible to suppress all visible blowhole formation by applying a pressure of 15 atm on the surface.

Under normal operating conditions, some relatively small amounts of gas form after the cap is crusted over. Most of this gas formation takes place in the upper central part of the ingot where the pressure head on the liquid metal is relatively low and the tendency toward ingot shrinkage encourages the carbon-oxygen reaction.

As soon as the mixing action subsides, the thermal gradient at the freezing interface is reduced. Freezing in from the outside then proceeds in a manner described earlier for semikilled steel, with random dendrites eventually forming ahead of the solidifying interface. Since there is still some gas formation, it would be expected that segregate streaks would again form. With increasing pressures within the ingot body, the blowholes which exist at the interface of the rimmed zone are often compressed, and some segregate appears in the area affected by such gas compression.

With further heat removal there is very likely settling of crystals. As a result, the bottom part of the central core is lower in the analysis of solute elements than is the upper part. The final stage of solidification again is one in which the core zone becomes mushy. In this stage, pressure differences appear in different sections of the ingot, displacing the mushy metal and giving rise to various series of parallel streaks and V-type segregate.

The preceding description of solidification represents a selection of concepts from published reports and from observations made in a single steel plant in which ingots with a relatively narrow range of compositions are made. There have been other concepts reported which are not consistent with the processes described in this presentation. The interpretation of solidification processes given in this paper appears consistent with available information and, more important, has proved useful in establishing procedures for controlling the behavior and performance of the ingot product.

References

Much of the information given in this paper was extracted from the reports and files of Inland Steel Company. In addition, frequent reference was made to the following publications:

1. A. S. Brearley and H. Brearley, *Ingots and Ingot Molds*, Longmans, Green and Co., New York, 1918.
2. Iron and Steel Institute, "Reports on the Heterogeneity of Steel Ingots," *J. Iron Steel Inst. London*, 113, 39 (1926); 117, 401 (1928); 119, 305 (1929). *Special Repts.*, No. 2 (1932); No. 4 (1933); No. 9 (1935); No. 9A (1936) (Discussion); No. 16 (1937); No. 25 (1939); No. 27 (1939).
3. A. Hultgren and G. Phragmén, *Trans. A.I.M.E.*, 135, 133 (1939).
4. *Basic Open Hearth Steelmaking*, Second Edition, Physical Chemistry of Steelmaking Committee, A.I.M.E., New York, 1951.
5. B. M. Larsen, *Trans. A.I.M.E.*, 162, 414 (1945).

6. A. Hultgren, G. Phragmén, S. Wohlfahrt, and J. E. Ostberg, *Trans. A.I.M.E.*, **191**, 101 (1951).

Discussion

MAYO said the inverted-V segregate may not necessarily be a result of the joining of the solidification from the walls with that from the bottom, as presented by Tenenbaum. To confirm this, he showed figures from Marburg's* paper which indicated that the inverted-V segregates started from well within the zone of horizontal solidification. He also described a high-nickel-chrome ingot which had broken in half as a result of mistreatment. The fractured surface revealed details of structure which are not readily apparent on an etched surface. Each segregate streak in the fractured section contained a hollow tube with the ends of the dendrites projecting into the tube. This same ingot showed some subsurface blowholes such as are normally found in semikilled steel ingots. These were also bright and shiny with a very smooth interior surface. It was interesting

* E. Marburg, *Trans. A.I.M.E.*, **197**, 157 (1953).

to note that the segregate associated with the tube was on the side of the tube toward the outside of the ingot. The segregate was the eutectic-type sulfides. The steel analysis was 3.5% nickel, 2% chromium, 0.05% silicon, and 0.25% carbon.

ELLIOTT described an ingot structure which showed the same type of vesicular opening on the outward portion of the segregate streaks. This was a semikilled ingot containing approximately 0.300% sulfur, 1% manganese, and 0.06% carbon. A horizontal cross section through the ingot showed the segregate streaks to be individual rods with the small tubular opening on the outward side. They were distributed over a band approximately 2 to 4 in. wide in a 22 by 22-in. ingot. This was a big-end-down ingot (as was Mayo's). MAYO said that the inverted-V segregate streak with a tube was found also in the fully killed big-end-up and big-end-down ingots which had approximately 2.5 lb of aluminum per ton of steel added as a deoxidizer. TENENBAUM emphasized that one needs to be careful when considering ingot structures to differentiate between those ingots which generate gas and those which are fully killed and do not have a tendency to generate gas.

by P. Vallet

Theoretical Study on the Solidification Progress of Ingots of Finite Dimensions

INTRODUCTION

The main contribution of Lightfoot to the theoretical study of ingot solidification progress, in his fundamental report written in 1929, is to have found the parabolic law expressed by the following equation:

$$x = K\sqrt{t} \qquad (1)$$

This equation expresses the relationship between the thickness x of frozen metal and the duration t of the solidification process, making it possible, from the physical constants of the metal, to calculate constant K with a reasonable approximation.[1]

The many authors who have shown an interest in the solidification progress of commercial ingots of finite dimensions have mainly concerned themselves with checking this law. Now, the latter is valid only during the first third of the solidification time. The concluding stages of the process are in fact much more rapid than implied by this law. This is not due to the more or less approximate nature of the simplifying hypotheses of Lightfoot (contrary to what has been sometimes put foward), but to the fact that usual ingots have finite dimensions.[2, 3, 4]

In the schematic case studied by Lightfoot, the volume v of frozen metal is proportional to $\sqrt{t}$, as well as its thickness, or the quantity of heat dissipated per unit of area of the ingot skin. In the usual case of finite-dimensioned ingots, contrary to a long-credited belief, it is the volume v of frozen metal which is fairly proportional to $\sqrt{t}$, not its thickness x, except at the beginning of the solidification process. In this usual case, v and x are no

Dr. Vallet is affiliated with the Institut de Recherches de la Sidérurgie, Saint-Germain-en-Laye, France.

longer proportional because the surface area where solidification takes place constantly decreases, while it would be invariable in Lightfoot's theoretical case. In the initial stages of the process, the relative decrease of the surface area of solidification is slight; that is why the Lightfoot law is fairly valid.

In usual ingots, the surface along which the metal freezes depends upon the mold shape, and upon its cooling conditions. Hereafter, we shall assume some symmetry in the phenomena, and it will be possible for us to forecast a priori the shape of this solidification surface and to calculate volume v at a given time.

Such are the basic ideas of the theory which will be described hereafter.

CASE A. A RECTANGULAR INGOT

Theory

We shall assume that the metal is poured in a mold whose inside shape is a parallelepiped (right-angled), the edges of which have the following dimensions: $2a$, $2b$, and $2c$. It will be assumed that the thickness of the mold walls (measured normally to the inside surfaces) is uniform, and that the outside surface is at a uniform temperature. Then we shall assume that the thickness x of frozen metal solidified at the time t (thickness measured normally to the inside surface of the mold) is uniform, i.e., that the same quantity of heat is dissipated per unit area of the ingot skin in the neighborhood of the center of the six faces.

It is easy to understand that the ratio of the volume v of solidified metal at the time t to the volume

V solidified at the time T when the solidification process is completed is:

$$\frac{v}{V} = 1 - \left(1 - \frac{x}{a}\right)\left(1 - \frac{x}{b}\right)\left(1 - \frac{x}{c}\right) \quad (2)$$

Assuming, as stated above, that v is proportional to t, we have:

$$v/V = \sqrt{t/T} \quad (3)$$

Eliminating v/V between equations 2 and 3, we have:

$$1 - \left(1 - \frac{x}{a}\right)\left(1 - \frac{x}{b}\right)\left(1 - \frac{x}{c}\right) = \sqrt{\frac{t}{T}} \quad (4)$$

In the case where x is small enough compared with the three dimensions a, b, and c, we can write as a first approximation:

$$x\left(\frac{1}{a} + \frac{1}{b} + \frac{1}{c}\right) = \sqrt{\frac{t}{T}} \quad (5)$$

Here we find again Lightfoot's law, since this equation has the same form as equation 1. A comparison between equations 1 and 5 shows that between the constants therein exists the following relationship:

$$\left(\frac{1}{a} + \frac{1}{b} + \frac{1}{c}\right)^{-1} T^{-1/2} = K \quad (6)$$

In ordinary conditions, the variation of K is slight; it is T which changes simultaneously with a, b, and c.

If a stands for the smallest of the three dimensions a, b, and c, solidification is completed (when $t = T$) as soon as $x = a$. The volume of metal remaining in the liquid state just at the end of the solidification process is then reduced, according to this mechanism, to an infinitely thin plate parallel to the larger faces of the mold. It has the same center as the latter, and the dimensions: $2(b - a)$ and $2(c - a)$.

Experiments

The results obtained by Briggs and Gezelius,[5] relating to three steel ingots poured into sand molds, will be used. One dimension is common to the three ingots: 203.2 mm. Ingot I has a square section with 92.1 mm sides. Both others have rectangular sections: 57.1×158.7 mm for ingot II, and 35.3×258.0 mm for Ingot III.

Table 1 shows for different values of t/T the values for x obtained by experiment and those calculated by means of equation 4. To find T, which was not given by the authors, we have drawn the curve representing the variations of $\log (v/V)$ as a function of $\log t$ and then extrapolated this curve to the ordinate point corresponding to $v/V = 1$ whose abscissa is $\log T$. This method has given for T the following values: 4.5, 3.2, and 1.52 min, respectively, for the three ingots.

Second Theoretical Approach

Experiment shows that equation 4 is not quite true. The curve representing the variations of $\log (v/V)$ as a function of t is not a straight line with an angular coefficient of 0.5, but a curve whose slight concavity is directed downwards. A more satisfactory representation of the phenomena is obtained by replacing equation 3 with the following one, established on an experimental basis:

$$\frac{v}{V} = e^n(1 - \tau^2)\sqrt{\tau} \quad (7)$$

making $t/T = \tau$. By eliminating v/V between equations 2 and 7, an equation is obtained which replaces equation 4:

$$1 - \left(1 - \frac{x}{a}\right)\left(1 - \frac{x}{b}\right)\left(1 - \frac{x}{c}\right) = e^n(1 - \tau^2)\sqrt{\tau} \quad (8)$$

Equations 7 and 8 contain, respectively, equations 3 and 4 as particular cases for $n = 0$.

In the last column of Table 1 are found the values for x obtained from equation 8 for the different values of t/T. The value for n is found experimentally: a graph was drawn representing the variations of $[\log (v/V)]: \sqrt{\tau}$ as a function of τ^2. A decreasing straight line was obtained, with a slope of $-n \log e$ and whose ordinate at the zero point is $n \log e$ or $\log e^n$. Thus have been found the values for e^n 1.20, 1.12, and 1.14 for the three ingots, respectively.

Table 1. Values of x as a Function of t/T for Rectangular Ingots

| t/T | x Measured, mm | | | x Calculated, mm | |
	Ingot I	Ingot II	Ingot III	Equation 4	Equation 8
0.019	3.3	—	—	2.7	3.1
0.026	—	3.6	—	2.9	3.3
0.055	—	—	3.8	3.3	3.8
0.111	8.9	—	—	7.0	8.0
0.156	—	9.6	—	7.8	8.9
0.222	12.4	—	—	10.6	12.2
0.312	—	13.5	—	11.8	13.5
0.329	—	—	10.4	8.8	10.1
0.444	19.3	—	—	16.8	19.4
0.469	—	16.3	—	15.4	17.4
0.493	—	—	12.7	11.2	12.6
0.625	—	20.8	—	19.0	21.0
0.658	—	—	14.5	13.4	14.9
0.667	26.4	—	—	23.6	26.4
0.889	35.6	—	—	33.1	35.2

Table 1 shows that equation 4 gives a fairly satisfactory representation of the solidification process, but the one given by equation 8 is better.*

* It seems that equation 8 no longer gives Lightfoot's law as a first approach for small values of x and t. In fact, the exponential factor of the second member of equation 8 varies slightly as long as τ remains fairly low. Even in the case when $e^n = 1.20$ (for $n = 0.182$), calculation shows that this exponential factor varies from 1.20 for $\tau = 0$ to 1.186 for $\tau = 0.3$. Its relative variation is slightly higher than 1%, which is low compared with other more important disturbing phenomena.

For small values of x and t, equation 8 becomes:

$$x\left(\frac{1}{a} + \frac{1}{b} + \frac{1}{c}\right) = e^n \sqrt{\frac{t}{T}} \qquad (9)$$

This equation replaces equation 5. Similarly, equation 6 would become

$$e^n \left(\frac{1}{a} + \frac{1}{b} + \frac{1}{c}\right)^{-1} T^{-1/2} = K \qquad (10)$$

If K is calculated using the data of Briggs and Gezelius in equation 10, the results are 10.60, 10.90, and 10.43 mm-min$^{-1/4}$, respectively, for Ingots I, II, and III. The agreement among these three values is quite remarkable, considering the number of intermediate calculations.

CASE B. SQUARE, CIRCULAR, OR REGULAR-POLYGON INGOT SECTIONS

If the ingot section is a square, $b = a$ and the two first parentheses in equations 2, 4, and 8 are equal. By reviewing the calculation of v/V, it can be seen that the equations thus obtained (with $b = a$) still apply when the ingot section is a circle whose radius is equal to a, or a regular polygon having a for apothem.

Ingot I of Table 1 has a square section and corresponds to the particular case that will be studied later.

Just at the end of the solidification process in the case where a is smaller than c, the volume of liquid metal remaining is a very thin prism or cylinder of height $2(c - a)$ which is concentric to the ingot. In a short, squat ingot where a is greater than c, the residual liquid volume which is last to solidify would be a thin square, circular, or flat plate. The plate would be concentric with the ingot, parallel to the base, and it would have a radius or apothem equal to $a - c$.

CASE C. INSULATED MOLD BASE

If one of the bases is insulating, the metal does not freeze immediately against this base. The solid metal which appears progressively against this base is produced on the adjacent conducting side walls. The center of the base remains in contact with the liquid metal until completion of the solidification process. The phenomena occurring against this base are almost identical with those observed in the plane going through the center and perpendicular to the ingot axis in the preceding cases. The solidification process of the ingot is then identical to that of one of the halves of the ingot of the preceding cases, the insulating base playing the part of the median section.

The preceding equations can easily be adapted to this new hypothesis. It is as if the new ingot were one-half the height of an ingot of the first type. Therefore, it is enough to replace c with $2c$ in equations 2, 4, 5, 6, 8, 9, and 10. For instance, equation 10 becomes

$$1 - \left(1 - \frac{x}{a}\right)\left(1 - \frac{x}{b}\right)\left(1 - \frac{x}{2c}\right) = e^{n(1-\tau^2)}\sqrt{\tau} \qquad (11)$$

Of course, if the ingot has a square or circular section, or if this section is a regular polygon, b will be made equal to a as in equation 11.

The solidification process of many commercial ingots which solidify in covered molds probably corresponds to this type. Whenever the ingot mold is not widely open to the ambient atmosphere, a gaseous layer builds up over the ingot top. This layer, thermally more insulating than the ingot mold walls, plays about the same role as an insulating upper base for the ingot top.

The solidification of rimming steel ingots studied by Duflot and Richard [6] follows approximately this law. Figure 9 of the original paper, representing the weight variations of the solidified metal as a function of time, agrees with equation 3 in which the ratio v/V is replaced with that of the corresponding weights, e^n corresponds to 1.18, and $T = 67$ min. On the other hand, these ingots have fairly uniform average dimensions: $a = b = 300$ mm and $2c = 1900$ mm. By substituting these values in equation 11, we have:

$$1 - \left(1 - \frac{x}{300}\right)^2 \left(1 - \frac{x}{1900}\right) = 1.18^{(1-\tau^2)}\sqrt{t/67} \qquad (12)$$

In this equation, x is expressed in millimeters and t in minutes. If this equation is solved by trial and error for values of t indicated in Table 1 of the original paper, values are found for x which are of the same order as those of the authors. As shown in Table 2, the calculated values are intermediate to the thicknesses at the foot and top of the ingot.

Table 2. Solidification of Rimming Steel Ingots Studied by Duflot and Richard

	x Measured, mm		x Calculated, mm
t	At the Foot	At the Top	
0 min 25 sec	25	11	13.3
4 min	52	40	43.8
9 min 10 sec	88	57	70.2
9 min 25 sec	85	62	71.3
9 min 35 sec	89	64	72.1
9 min 45 sec	84	62	72.8
15 min 40 sec	117	80	97.0
16 min	115	84	98.3
24 min 45 sec	142	110	130.2
35 min	165	132	165.0
57 min 30 sec	280	175	242.3

In our opinion, it would be illusory to seek a closer numerical agreement between the measured values and the calculated ones, since several hypotheses of the theory are not exactly realized in practice. For instance, filling of the ingot mold is not instantaneous, ingots have the shape of a square-base pyramid, and so on.

CASE D. BOTH INGOT MOLD ENDS INSULATING

Theory

If both bases are insulating, metal freezing occurs only in contact with the side walls of the mold. The process is then identical in any plane parallel to the insulating end. It is as if it were a problem of plane geometry. It can be seen that if $2c$ is the distance between the two insulating ends, this new process can be accounted for by getting rid of ratio x/c in equations 2, 4 to 6, and 8 to 10. Thus, equation 8 becomes:

$$1 - \left(1 - \frac{x}{a}\right)\left(1 - \frac{x}{b}\right) = e^{n(1-\tau^2)}\sqrt{\tau} \qquad (13)$$

Of course, if $b = a$, the equation is simplified, and x can then be easily obtained as a function of τ and n:

$$x = a\left[1 - \sqrt{1 - e^{n(1-\tau^2)}\sqrt{\tau}}\right] \qquad (14)$$

It is also possible to use the dimensionless variable x/a to get:

$$\xi = 1 - \sqrt{1 - e^{n(1-\tau^2)}\sqrt{\tau}} \qquad (15)$$

It is easy to make sure (by calculating directly the ratio v/V) that equations 13, 14, and 15 still apply when the inside contour of the right-angle section of the ingot mold is a circle of radius a, or a regular polygon of apothem a.

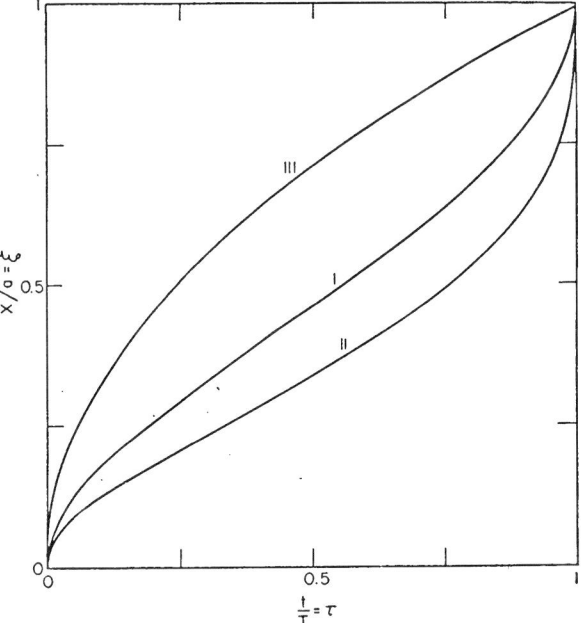

Fig. 1. $\xi = x/a$ as a function of $\tau = t/T$. Curve I is for ingots, the ends of which are insulated, and curve II, for spherical or polyhedral ingots. Lightfoot's curve (curve III) is given for comparison with curves I and II.

On the other hand, in the particular case where $n = 0$, equation 15 would take the simple form:

$$\xi = 1 - \sqrt{1 - \sqrt{\tau}} \qquad (16)$$

Curve I of Fig. 1 shows the variations of ξ as a function of τ in this particular case. This curve has a vertical tangent at each of its ends for $\tau = 0$ and for $\tau = 1$. It has a point of inflexion for $\tau = 4/9 = 0.4444$ whose ordinate is

$$\xi = 1 - 1/\sqrt{3} = 0.422$$

Equation 13 can also be written as follows:

$$x\left(\frac{1}{a} + \frac{1}{b}\right) - \frac{x^2}{ab} = e^{n(1-\tau^2)}\sqrt{\tau} \qquad (17)$$

This equation is simplified if $b = a$. Then, using the preceding notations, we can write:

$$2\xi - \xi^2 = e^{n(1-\tau^2)}\sqrt{\tau} \qquad (18)$$

These two last equations show that if x is represented as a function of $\sqrt{t}$ or ξ as a function of $\sqrt{\tau}$, in the particular case where $n = 0$, we obtain a parabola passing through the zero point of the co-ordinates and having a horizontal axis. In the case where b is different from a, the top of the parabola cannot be reached experimentally, since it would correspond to $x = (a + b)/2$, and the process stops as soon as x reaches the smaller of values a or b. When n is not equal to zero, since the exponential factor varies but slightly, the over-all appearance of the process is unchanged. The curves published by Bishop, Brandt, and Pellini,[7] who have considered the case where $b = a$ and who had fairly low values for e^n, are parabolic curves passing through the origin of co-ordinates and whose vertex is close to $x = a$ and $t = T$, or $\xi = \tau = 1$. Thus, the results of the theoretical calculations are confirmed.

Experimental Comparison

The ingots studied by Bishop, Brandt, and Pellini[7] correspond fairly well to the case where both ends are insulating. The sand base of the ingot mold causes a slower solidification than the side wall and can be considered as comparatively insulating. On the other hand, the hot top keeps a fairly constant temperature on the ingot head and slows down the freezing process in this part of the ingot.

These ingots of 0.06% carbon steel were poured in square-section ingot molds having for side dimension $2a = 2b = 175$ mm, and $2c = 580$ mm in height. The results shown in Table 3 relate to two ingot molds having thicknesses of 105 and 38 mm, respectively.

In Table 3 are given the corresponding t and x/a values measured on the curves given by the authors. These curves can be seen in the graph located in the upper left part of Fig. 7 of the original paper. For each ingot mold will be found the value of x/a calculated by means of the simpler equation 16, as-

Table 3. Values of x/a as a Function of t for Square-Sectioned Ingots with Negligible Solidification at the Ends of the Axis

	105 mm-Ingot Mold			38 mm-Ingot Mold		
	Ratio x/a			Ratio x/a		
t, min	Meas-ured	Calculated		Meas-ured	Calculated	
		Eq. 16	Eq. 15		Eq. 16	Eq. 15
1	0.260	0.175	0.205	0.202	0.168	0.185
2	0.328	0.260	0.305	0.275	0.250	0.274
3	0.405	0.333	0.392	0.347	0.319	0.350
4	0.469	0.392	0.471	0.416	0.383	0.420
5	0.534	0.467	0.547	0.485	0.445	0.487
6	0.611	0.535	0.621	0.561	0.508	0.553
7	0.687	0.609	0.699	0.637	0.574	0.608
8	0.771	0.692	0.776	0.717	0.647	0.692
9	0.885	0.801	0.860	0.801	0.732	0.771
9.5	0.969	0.884	0.933	—	—	—
10	—	—	—	0.924	0.853	0.884

suming $n = 0$, and that given by complete equation 15 with $e^n = 1.15$ for the thick-walled mold, and $e^n = 1.085$ for the thin-walled mold. The values of T determined from the original data are, respectively, 9.70 and 10.45 min. It can be seen that the simpler equation 16 gives an already interesting approximation, but that the approximation given by equation 15 is still better.

Calculation of coefficient K of equation 1 by means of equation 6 gives for solidification in a thick-walled mold $K = 16.2$ mm min$^{-1/2}$ and in a thin-walled mold $K = 14.7$ mm min$^{-1/2}$. The former value is somewhat lower than that which can be calculated by means of Lightfoot's equation when using the recent values for the iron constant, or 17.4 mm min$^{-1/2}$.

CASE E.   REGULAR-POLYHEDRAL OR SPHERICAL INGOTS

Theory

If in equations 2, 4, 5, 6, and 8 to 10 it is assumed that $a = b = c$, the right-angled parallelepiped becomes a cube whose edge is equal to $2a$, the simplest of all regular polyhedrons.

In this hypothesis, equation 8 becomes:

$$1 - \left(1 - \frac{x}{a}\right)^3 = e^{n(1-\tau^2)}\sqrt{\tau} \qquad (19)$$

It is easy to draw $x/a = \xi$ from this equation; we have then

$$\xi = 1 - \sqrt[3]{1 - e^{n(1-\tau^2)}\sqrt{\tau}} \qquad (20)$$

In the particular case where $n = 0$, the exponential

term of the radical is equal to 1, and equation 20 simplifies as follows:

$$\xi = 1 - \sqrt[3]{1 - \sqrt{\tau}} \qquad (21)$$

Curve II of Fig. 1 represents the variations of ξ as a function of τ deduced from equation 21. It shows a vertical tangent at each of its ends (for $\tau = 0$ and $\tau = 1$), a point of inflexion for $\tau = 9/25$, or 0.36 whose ordinate is 0.262. This curve does have the general shape obtained experimentally from solidification of spheres.

In fact, equations 20 or 21 still apply when the ingot has the shape of a regular polyhedron or a sphere. Then, length a stands for the radius of the sphere inscribed into the polyhedral surface or for the radius of the inside of the mold if it is spherical.

Experimental Comparison

We will use the results of Briggs and Gezelius[5] relating to the solidification of steel spheres molded in sand molds. The radii of these spheres I, II, and III were, respectively, 57.1, 76.2, and 114.3 mm. We have found T, which was not given by the authors, by the graphical extrapolation method described above in connection with rectangular ingots. Values of T were found to be 4.0, 7.3, and 17.0 min, respectively, for the three spheres. Table 4 gives for

Table 4.   Values of x/a for Spheres, as a Function of t/T

	x/a Measured for Sphere			x/a Calculated with Equation	
t/T	I	II	III	21	20
0.005	—	—	0.031	0.024	0.028
0.011	—	0.047	—	0.037	0.041
0.059	—	—	0.107	0.088	0.105
0.069	—	0.123	—	0.096	0.110
0.125	0.178	—	—	0.135	0.156
0.137	—	0.167	—	0.143	0.165
0.176	—	—	0.200	0.166	0.202
0.250	0.240	—	—	0.206	0.240
0.274	—	0.253	—	0.219	0.255
0.353	—	—	0.298	0.259	0.317
0.375	0.298	—	—	0.271	0.316
0.411	—	0.333	—	0.289	0.338
0.529	—	—	0.429	0.351	0.430
0.548	—	0.430	—	0.362	0.421
0.685	—	0.513	—	0.443	0.511
0.706	—	—	0.551	0.457	0.550
0.822	—	0.600	—	0.546	0.615
0.875	0.689	—	—	0.599	0.665
0.882	—	—	0.667	0.607	0.693

these three spheres the different values of $t/T = \tau$ (first column), the experimental values of $x/a = \xi$ deduced from the data of the authors (second, third, and fourth columns) and the values of x/a calculated either by means of the simplest equation 21 (fifth column) or by using equation 20 (last column). In the latter case, the values of e^n found empirically, as described in connection with rectangular ingots, and used in the calculation, have been 1.13, 1.13, and 1.17, respectively. These values of e^n fit better for values of τ higher than 0.1. It is felt that this is due to the fairly important variation of thermal conductivity and of specific heat of sand during the first moments following pouring.

On the whole, equation 20 gives a better approach than equation 21, but the latter is still much better than that given by Lightfoot's law. On the other hand, curve II of Fig. 1 shows that the variations of ξ are very rapid when τ tends towards 1: even a slight error in τ then results in a serious error in ξ.

Calculation of coefficient K of equation 1 by means of equation 10, in which $a = b = c$, gives for the three spheres: 10.75, 10.62, and 10.81 mm min$^{-1/2}$, respectively. The agreement of these values among themselves and with those that we have found above for rectangular ingots poured in sand molds is most remarkable, considering the very indirect method by which they have been obtained.

SUMMARY AND CONCLUSION

The parabolic law of Lightfoot applies to commercial ingots of finite dimensions only during about the first third of the solidification process. It is assumed, in agreement with this law, that for finite ingots, the volume of solidified metal (not its thickness) is proportional to the square root of the time elapsed from the end of the pouring operation.

The following equation

$$1 - \left(1 - \frac{x}{a}\right)\left(1 - \frac{x}{b}\right)\left(1 - \frac{x}{c}\right) = \sqrt{\frac{t}{T}} \quad (4)$$

was obtained for right-angled parallelepiped ingots whose edge dimensions are $2a$, $2b$, and $2c$. The derivation of equation 4 assumes symmetrical cooling to provide a uniform thickness x of solidified metal. The thickness x must be measured perpendicular to the mold wall.

This equation can be easily adapted to the following cases:

1. The perpendicular section of the ingot of height $2c$ is a square, a circle, or a regular polygon (case where $b = a$ in the above equation).

2. One of the ends, is insulating (c is replaced with $2c$ in the above equation).

3. Both ends are insulating and the distance between them is $2c$ (x/c is made equal to zero in the above equation).

4. The ingot has the shape of a regular polyhedron or of a sphere ($a = b = c$ in the above equation).

Using the results previously published by different authors, it is possible to verify quantitatively the proposed equation or its different adaptations.

It is also possible to calculate the coefficient of the parabolic law of Lightfoot, and an excellent agreement is found between the values obtained with ingots of different shapes and dimensions (prisms or spheres) when the metal is poured in sand molds.

It is possible to improve the numerical agreement between theory and practice by multiplying the radical of the second member of fundamental equation 4 by the empirical exponential factor $e^{n(1 - \tau^2)}$ in which $\tau = t/T$ and n an experimental constant of the order of 0.12 to 0.19.

References

1. N. H. M. Lightfoot, *J. Iron Steel Inst.*, **119**, 364 (1929).
2. P. Vallet, *Compt. Rend.*, **242**, 2305 (1956).
3. P. Vallet, *Compt. Rend.*, **242**, 2448 (1956).
4. P. Vallet, *Compt. Rend.*, **242**, 2514 (1956).
5. C. W. Briggs and R. A. Gezelius, *Trans. Am. Foundrymen's Assoc.* **43**, 274 (1935).
6. J. Duflot and A. Richard, *Rev. mét.*, **51**, 623 (1953).
7. H. F. Bishop, F. A. Brandt, and W. S. Pellini, *J. Metals*, **4**, 44 (1952).

Discussion

ADAMS* asked if the constants in the various equations changed with different situations. VALLET said they did. The effect of variables was included in the term T which was obtained from experiment.

* Assistant Professor of Metallurgy, Massachusetts Institute of Technology.

by C. W. Sherman

The Effect of Mold Geometry on Vertical Solidification

Several investigators[1,2,3,4,5,6,7,8] have studied the behavior of steel during solidification in the ingot mold both by the "dumping" or "bleeding" technique and by thermometric means. The earlier workers were primarily concerned with the establishment of a relationship for transverse solidification, that is, the freezing from the mold wall to the vertical center line of the ingot. They demonstrated that the parabolic law

$$D = k\sqrt{t} \tag{1}$$

where D is distance frozen in inches and is equal to a constant times the square root of time in minutes, describes the earlier stages of freezing relatively well and that it is a good guide in estimating times of final solidification. There are certain factors, however, which affect this equation, such as mold-wall thickness, rate of filling mold, amount of superheat in the metal, and the formation of an air gap between the mold wall and ingot, although the degree of the effect of each is not universally agreed upon by the various authors.

More recent work has been concerned with the rate of solidification in the vertical direction, that is, the advance of the freezing front from the stool to the hot top. The implication in several papers[5,7,8] has been that transverse and vertical solidification are two separate phenomena and may be influenced independently. The lack of agreement here is rather substantial. The remarks of Feild, Philbrook, Ekholm, and Taylor during the discussion of a paper by Nelson[8] pointed this out. They indicated that they felt that the geometry of the mold had a pronounced effect on the apparent course of vertical solidification. This idea loosened my own curiosity somewhat, and the present paper is an attempt to predict how the vertical solidification might behave if estimated by mathematical means, using solid geometry alone.

In order that the rate of apparent vertical solidification might be calculated easily, several simplifying assumptions were made. These were:

1. That the parabolic law (equation 1) was obeyed in the transverse direction.
2. That the advancing solidification front might be treated as a plane and not a zone.
3. That the volume of liquid freezing was proportional to the area of the interface between the solid and liquid.

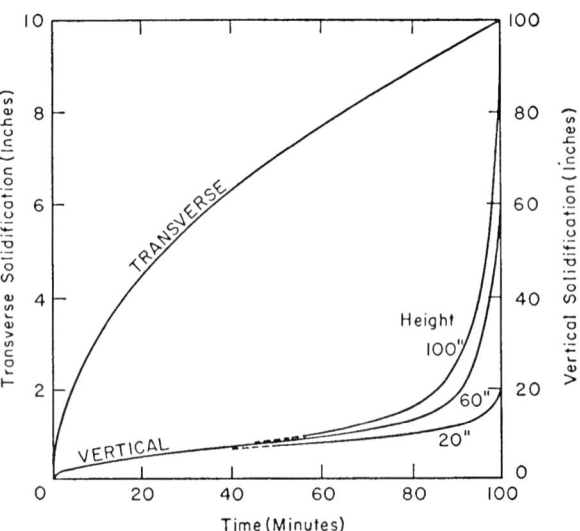

Fig. 1. Calculated vertical and transverse freezing times for three square, straight-walled ingots.

Dr. Sherman is Senior Research Associate, Jones and Laughlin Steel Corporation, Pittsburgh, Pennsylvania.

4. That no shrinkage took place in going from liquid to solid.

5. That transverse and vertical solidification end simultaneously.

The old-fashioned "onionskin" analogy was employed in computing the distance frozen in the vertical direction. For each increment of ingot radius, going from the outside of the ingot to the center, the height of the unfrozen liquid was determined and the distance frozen from the bottom found by difference. The volume of liquid that froze during the second increment was considered proportional to the area of the inside of the first increment, and so on.

Calculations on three different ingot sizes are recorded in Table 1 and plotted in Fig. 1. The gen-

Table 1. Calculated Freezing Times for Straight 20 in. × 20 in. Square Ingots

Transverse Solidification, in.	Time, min (K = 1)	Vertical Solidification, in.		
		Ingot 20 in. Tall	Ingot 60 in. Tall	Ingot 100 in. Tall
1.0	1.00	1.00	1.01	1.01
2.0	4.00	2.00	2.02	2.04
3.0	9.00	3.01	3.06	3.12
4.0	16.00	4.03	4.13	4.24
5.0	25.00	5.05	5.25	5.45
6.0	36.00	6.09	6.46	6.82
7.0	49.00	7.17	7.83	8.49
7.5	56.25	7.73	8.64	9.56
8.0	64.00	8.33	9.63	10.93
8.5	72.25	8.99	10.95	12.92
9.0	81.00	9.83	13.13	16.43
9.1	82.81	10.04	13.79	17.54
9.2	84.64	10.28	14.65	18.91
9.3	86.49	10.56	15.60	20.64
9.4	88.36	10.90	16.91	22.92
9.5	90.25	11.34	18.70	26.07
9.6	92.16	11.95	21.35	30.75
9.7	94.09	12.90	25.70	38.49
9.8	96.04	14.70	34.29	53.89
9.9	98.01	19.90	59.90	99.90

eral shape of the curves conforms to observations. It is interesting to note that the height of the ingot does not appreciably affect the time at which "accelerated" vertical solidification takes place. The three ingots differ, of course, in the distance that must be frozen.

The three ingots in Fig. 1 were straight-sided ingots. In an effort to determine the effect of ingot taper, the calculations were repeated for various tapers on an ingot of constant height where

% Taper (on each side)

$$= \frac{\frac{1}{2} \text{ width (top)} - \frac{1}{2} \text{ width (bottom)}}{\text{height}} \times 100$$

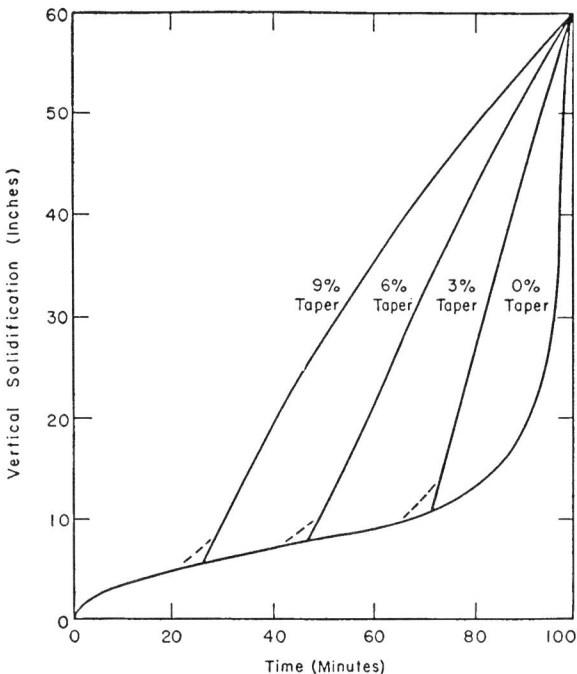

Fig. 2. Calculated vertical freezing times for square, tapered-walled ingots.

It soon became apparent that the distance frozen from the bottom was affected by the closing of the pipe from the sides. This "pinching" by the freezing in the transverse direction caused the vertical solidification to accelerate earlier and in a linear manner. Each increment freezing had the same shape as the outside of an inverted pyramid. The results are shown in Table 2 and Fig. 2.

Table 2. Calculated Freezing Times for Tapered 20 in. × 20 in. Square Ingots

Transverse Solidification, in.	Time, min (K = 1)	Vertical Solidification, in. Ingot 60 in. Tall			
		0% Taper	3% Taper	6% Taper	9% Taper
1.0	1.00	1.01	—	—	—
2.0	4.00	2.02	—	—	—
3.0	9.00	3.06	—	—	—
4.0	16.00	4.13	—	—	—
5.0	25.00	5.25	—	—	—
6.0	36.00	6.46	—	—	15.56
7.0	49.00	7.83	—	10.00	26.67
7.5	56.25	8.64		18.33	32.22
8.0	64.00	9.63		26.67	37.78
8.5	72.25	10.95	10.95	35.00	43.33
9.0	81.00	13.13	26.67	43.33	48.89
9.1	82.81	13.79	30.00	45.00	50.00
9.2	84.64	14.65	33.33	46.67	51.11
9.3	86.49	15.60	36.67	48.33	52.22
9.4	88.36	16.91	40.00	50.00	53.33
9.5	90.25	18.70	43.33	51.67	54.44
9.6	92.16	21.35	46.67	53.33	55.56
9.7	94.09	25.70	50.00	55.00	56.67
9.8	96.04	34.29	53.33	56.67	57.78
9.9	98.01	59.90	56.67	58.33	58.89

The effect of ingot taper appears to be very great. Some of the investigators that have dumped ingots have found this sudden change of slope in the vertical solidification curve, but the interpretation has been varied. Marburg[7] found it but said that taper had no effect.

One of the subtle inferences that came to mind in constructing these curves was that the copper and ferroalloy people should be concerned with "accelerated" transverse solidification since they pour their ingots in flat cakes and shallow slabs. The treatment used in this analysis would suggest that "speeded-up" horizontal freezing might take place.

Unfortunately it is not possible to check the results with much of the published data since ingot tapers are seldom given for molds. Also, some molds have rounded bottoms and multiple tapers which make the calculations infinitely more complex. The results, however, are being used as a guide in our own work on solidification, and it is felt that others might find the treatment useful.

References

1. A. L. Feild, *Trans. Am. Soc. Steel Treating*, **11**, 264, 338 (1927).
2. L. H. Nelson, *Trans. Am. Soc. Metals*, **22**, 193 (1934).
3. J. Chipman and C. R. FonDersmith, *Trans. A.I.M.E.*, **125**, 370 (1937).
4. J. W. Spretnak, *Trans. Am. Soc. Metals*, **39**, 569 (1947).
5. J. W. Spretnak, *Iron Age*, **167**, 107 (1951).
6. H. F. Bishop, F. A. Brandt, and W. S. Pellini, *Trans. A.I.M.E.*, **194**, 44 (1952).
7. E. Marburg, *Trans. A.I.M.E.*, **197**, 157, 1553 (1953).
8. L. H. Nelson, *Proceedings Electric Furnace Conference*, *A.I.M.E.*, **11**, 226 (1953).

by C. E. A. Shanahan

The Importance of the Hydrogen Content of Steel Slabs and Its Determination in Sheet Steel

Although it has been shown on numerous occasions[1] that hydrogen dissolved in steel has an adverse effect on its mechanical properties, new data have recently been obtained in the author's laboratory which, it is thought, will be of interest. The investigation began with a general chemical examination of two plain carbon-steel slabs; one of these was found to contain up to about 15 ml H_2/100 grams and it was, therefore, compared mechanically and metallographically with a similar slab containing much less hydrogen (up to 1.7 ml/100 grams). Although it was expected that approximately complete removal of the hydrogen would take place in processing the slabs to sheet steel, experimental verification of this was desirable in view of the deleterious effects of hydrogen on the enameling properties of sheet steel.[1] This resulted in an investigation into the determination of hydrogen in steel sheet which produced some interesting data showing that most of the hydrogen found in sheet arises from atmospheric contamination during preparation of the sample.

EXAMINATION OF STEEL SLABS

Table 1 shows the average chemical analysis of two steel slabs designated R.201 and R.209. The

Mr. Shanahan is affiliated with the R.T.S.C. Laboratories, Whitchurch, Aylesbury, Buckinghamshire, England.

The author desires to acknowledge the encouragement given by Mr. R. A. Hacking, O.B.E., M.Sc., F.I.M., Director of Research of the R.T.S.C. Laboratories, and his permission to publish this paper.

Table 1. Analyses of Slab Section Samples

Sample No.	% C	% S	% Mn	% Si	% P	% N	% Al (total)
R. 201	0.05	0.022	0.33	<0.005	0.025	0.005	0.040
R. 209	0.04	0.021	0.37	<0.005	0.029	0.005	0.038

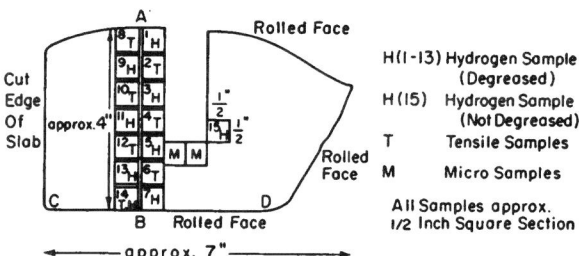

Fig. 1. Position of hydrogen, mechanical, and micro test pieces taken from the transverse section of slab section R.209.

hydrogen content of R.201, as judged by the vacuum-heating technique described in detail elsewhere,[2] did not exceed 1.7 ml/100 grams in any position of the slab. Preliminary tests on R.209, however, showed it to contain approximately 15 ml H_2/100 grams at its center, and a more detailed examination was therefore made. Figure 1 shows a transverse section of slab R.209 and also the positions from which test pieces were taken for hydrogen determination, mechanical tests, and micro-examination. Figure 2 shows the results of the tests and clearly illustrates the adverse effect of hydrogen on the ductility of the steel as measured by the

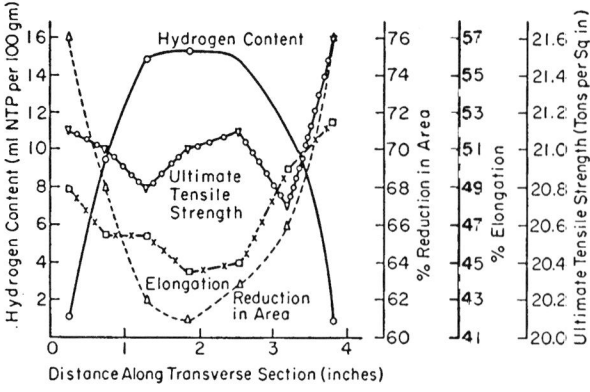

Fig. 2. Comparison of hydrogen content and physical properties with position in transverse section of slab section R.209.

elongation percentage and reduction in area of the tensile test pieces.

In order to develop a theory to explain the effect of hydrogen content on ductility, autographic load-

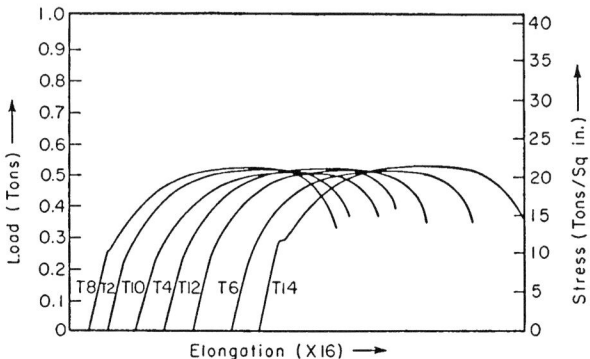

Fig. 3. Load elongation curves from transverse section of slab section R.209.

elongation curves were obtained from the tensile test pieces mentioned above. The curves are reproduced in Fig. 3, and it is evident that high-hydrogen

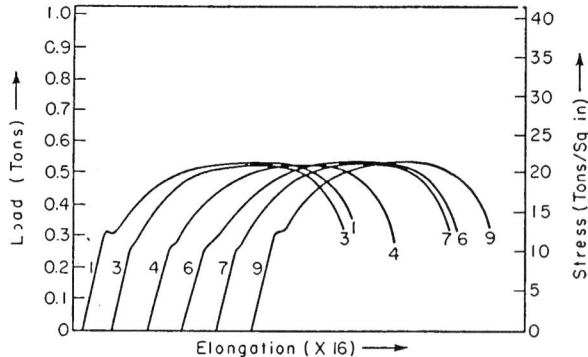

Fig. 4. Load-elongation curves from transverse section of slab section R.201.

positions have little or no yield point. It would appear, therefore, that the high-hydrogen contents have caused considerable straining* of the steel

* 10 ml H₂/100 grams corresponds to a pressure of approximately 14,500 atm.[1]

structure beyond the yield point so that the latter is not detected by mechanical testing. Figure 4 shows a similar set of load/elongation curves taken from a traverse of slab section R.201, in which the maximum hydrogen content was 1.7 ml/100 grams. In this case the yield point has persisted throughout the slab section, and moreover, the "reduction-in-area" values (Table 2) are independent of position.

Table 2. Mechanical Tests on Test Pieces Taken from Transverse Section of Slab Section R.201

Test Piece	Ultimate Tensile Strength, tons/sq in.	% Elongation, on 2 in. gage length	% Reduction of Area
1	21.4	50.5	75.5
2	22.8	33	75.5
3	21.2	46.5	76.5
4	21.1	48.5	76
5	Broke near grip	—	—
6	21.4	52.5	74.5
7	21.3	46.5	75.5
8	Broke near grip	—	—
9	21.3	57	76

Should the foregoing theory of the influence of hydrogen be correct, it would seem reasonable that the structure of samples possessing high-hydrogen contents should show visual evidence of strain. This indeed proved to be the case, as can be seen from Fig. 5 taken from the center of slab section R.209. The voids between the ferrite crystallites are easily distinguishable, arising presumably from the enormous bursting pressure of the dissolved hydrogen. Figures 6 and 7 show similar micrographs taken from the edge of slab section R.209 and the center of slab section R.201, respectively; in these cases there is a complete absence of grain boundary

Fig. 5. R.209. Center of slab showing voids in the ferrite grain boundaries.

Fig. 6. R.209. Sample T14 from edge of slab—no intergranular voids present.

voids, presumably because of their relatively low hydrogen contents.

THE HYDROGEN CONTENT OF SHEET STEEL

In view of the possibility of steel slabs containing high concentrations of hydrogen, it was important to verify theoretical predictions which suggested that negligible amounts of hydrogen remained within the steel after the latter had been processed from the slab to the sheet stage. As a consequence, several samples of sheet steel were prepared for hydrogen determination by cleaning with emery paper and degreasing. The hydrogen values obtained increased with the surface area of the sample and were, therefore, mainly attributable to hydrogen pickup from the atmosphere and not to hydrogen present at the slab stage. In view of this complication when using samples of relatively large

Fig. 7. R.201. Center of slab showing hot-rolled structure of ferrite and pearlite—no voids.

surface area, the following detailed investigation was made on a plain carbon and a stainless steel whose analyses are given in Table 3.

Table 3. Chemical Analysis of Mild Steel and Stainless Steel Samples

Material	% C	% Mn	% P	% S	% Cr	% Ni
Mild steel	0.17	0.58	0.024	0.03	—	—
Stainless steel	—	0.57	—	—	16.8	8.9

The most suitable sample design was found to be that of a set of 25 disks each of approximately 2.5 cm in diameter and 1 mm in thickness, having a total surface area of approximately 250 sq cm. Two sets of these disks were therefore prepared, one set from 1-in. diam mild steel bar and one set from 1-in. diam bar stainless steel. After cutting, the disks were ground on a 3 M Company 120-grade silicon-carbide paper to give a surface finish roughly comparable to that obtained by drawfiling.

In order to obtain an accurate estimate of the amount of hydrogen absorbed as a consequence of surface abrasion, it is preferable to employ samples which initially contain no hydrogen; any hydrogen subsequently found in the specimens can then be ascribed to surface treatment. All the disk samples were, therefore, thoroughly degassed by vacuum heating before each experiment.

The following series of experiments was performed.

Series (a). 25 mild steel disks were completely degassed at 750° C and were then treated as follows:

1. 25 disks were abraded on all faces and edges using the 120-grade silicon-carbide paper. (The disks were placed in a holder with a circular recess for abrading to prevent contamination by handling.)

2. All the disks were stood separately and vertically in a metal rack in the laboratory atmosphere until a period of 1 hr had elapsed from the commencement of abrasion.

3. The disks were degreased in petroleum ether (bp 40° C–60° C) and the hydrogen content determined by analysis of the gases evolved during vacuum extraction at 750° C.

4. To ensure that all the hydrogen was removed before beginning the next experiment, the disks were degassed for 1 hr and allowed to cool before they were taken from the vacuum-extraction apparatus.

This procedure was repeated several times, but on each occasion the number of disks abraded was varied; thus in different experiments, 19, 12, 5, and 0 disks were treated with the 120-grade silicon-carbide paper. In every case, the abraded disks were placed together with the remainder of the set

Table 4. Experimental Data in Present Investigation

Series No.	Sample			Hydrogen Resulting from Abrasion, ml, NTP	Remarks
	No. of Abraded Disks	No. of Unabraded Disks	Apparent Area of Abraded Disks, sq cm		
(a)	25	0	278	1.6	*Mild steel disks*
	19	6	211	1.3	Variable number of disks abraded—
	12	13	133	1.1	25 disks exposed and analyzed.
	5	20	56	0.5	Degreased with petroleum ether.
	0	25	0	0.2	
(b)	25	0	278	1.4	*Mild steel disks*
	19	6	211	1.0	As above. Degreased with carbon
	12	13	133	0.8	tetrachloride.
	5	20	56	0.4	
	0	25	0	0.2	
(c)	25	—	278	1.6	*Mild steel disks*
	19		211	1.4	Variable number of disks abraded
	12		133	0.9	and these disks only exposed and
	5		56	0.4	analyzed. Degreased with petroleum ether.
(d)	25	0	280	1.0	*Stainless steel disks*
	19	6	213	0.7	Variable number of disks abraded—
	12	13	134	0.5	25 disks exposed and analyzed.
	5	20	56	0.3	Degreased with carbon tetrachloride.
(e)	25	0	278	1.6	*Mild steel disks* Exposed for 15 hr. Degreased with petroleum ether.
(f)	0	25	278	0.25⎫ mean 0.15⎬ 0.2 0.20⎭	*Mild steel disks* 25 unabraded disks exposed 1 hr. Not degreased.

of 25 disks in the metal rack, and after the allotted time period, they were degreased and the hydrogen content determined on all 25 disks together. In this way it was possible to estimate the hydrogen content of a sample which had a constant mass and total area but a variable abraded area.

Series (b). The whole series as in series (a) was repeated, but in this instance carbon tetrachloride was used for degreasing.

Series (c). The next set of determinations aimed at measuring the contributions of the unabraded disks to the total hydrogen contents determined in

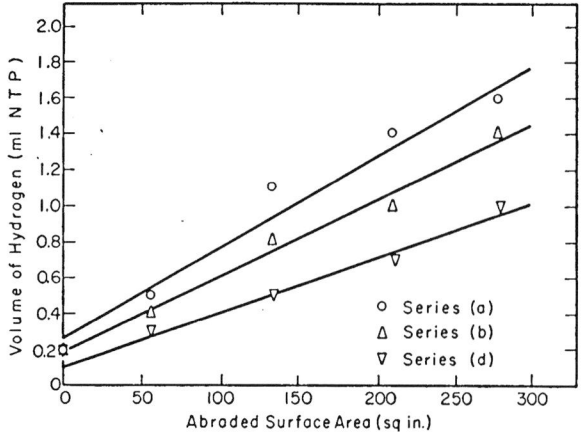

Fig. 8. Volume of hydrogen formed by the abrasion and exposure of steel disks.

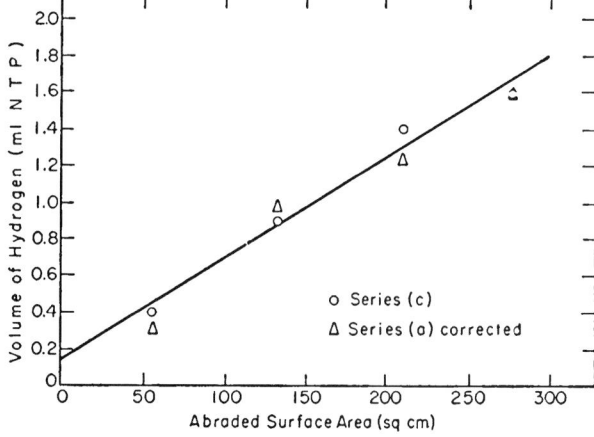

Fig. 9. Volume of hydrogen formed by the abrasion and exposure of steel disks with a constant total area and a variable abraded area.

series (a). Various numbers of mild steel disks were abraded, stood for 1 hr in the laboratory atmosphere, and the hydrogen contents determined after degreasing with petroleum ether. On different occasions, 25, 19, 12, and 5 disks were treated. No unabraded disks were involved.

Series (d). The experiment proceeded as in series (b) above, using 25 stainless steel disks degreased in carbon tetrachloride.

Series (e). The abraded disks in all the previous experiments were exposed for a period of 1 hr. In order to obtain some estimate of the effect of exposure time, 25 mild steel disks were abraded, exposed for a period of 15 hr, and degreased in petroleum ether before determination of the hydrogen.

Series (f). The influence of degreasing was investigated by standing 25 degassed mild steel disks in the atmosphere for 1 hr and determining the hydrogen content without any degreasing process.

The results of all of these experiments are presented in Table 4, and dimensional details of the disks are given in Table 5. Graphs of the variation in hydrogen content with apparent abraded area* for the series (a), (b), and (d) are shown in Fig. 8 and for (c) in Fig. 9.

Table 5.   Dimensional Details of Sample Disks

Material	Remarks	Average Thickness, mm	Average Diameter, mm
Mild steel disks	Before experiments	1.23	25.38
	After experiments	1.20	25.38
Stainless steel disks	Before experiments	1.32	25.41
	After experiments	1.32	25.41

Abrasion had little effect on the dimensions of the disks; the only apparent change was a decrease of 2.5% in the thickness of the mild steel disks. Since this is equivalent to a maximum error of only 0.5 sq cm in the area of the disks, it has been neglected. Values given for the hydrogen content are estimated to be accurate within ±0.1 ml, while the amount of gas given off during the degassing period prior to exposure averaged 0.09 ml.

Analysis of Data. Two major facts are apparent from the results presented, namely, that there is an increase of hydrogen with increase in abraded area and also that there is a possible slight increase in hydrogen content when unabraded disks are exposed to the atmosphere. Consider an experiment in which 25 disks made up of N abraded and $(25 - N)$ unabraded disks are exposed to the atmosphere. If A is the average disk area, then

$$V = NAK + AK_1(25 - N)$$

or

$$V = NA(K - K_1) + 25AK_1 \qquad (1)$$

where

V = volume of hydrogen in ml resulting from exposure and abrasion

K = volume of hydrogen pickup per unit area resulting from abrasion

K_1 = volume of hydrogen pickup per unit area resulting from exposure of unabraded disks

Similarly, for the set of experiments using N abraded mild steel disks, only, as in series (c),

$$V = NAK \qquad (2)$$

In both cases, it is assumed that hydrogen pickup occurs reasonably quickly and is almost complete at the end of the 1-hr period. This assumption is borne out by work reported by Winterbottom on the oxidation of iron in air and by Sloman on the hydrogen content of steel millings after various periods of exposure, and in addition, it has been shown experimentally that there is no increase in the hydrogen content after 15-hr exposure, as in series (e).

Equations for the lines of best fit have been calculated from the experimental data by the method of least squares and are as follows:

Series (a) $V = 0.00505\ NA + 0.255$ (3)

Series (b) $V = 0.00422\ NA + 0.189$ (4)

Series (c) $V = 0.00554\ NA + 0.137$ (5)

Series (d) $V = 0.00305\ NA + 0.184$ (6)

For the mild steel disks, equations 1 and 3 give the gradient of the line $(K - K_1)$ as 0.00505 ml hydrogen per sq cm while the intercept

$$(25AK_1) = 0.255$$

or

$$K_1 = \frac{0.255}{278} = 0.00092 \text{ ml hydrogen per sq cm}$$

Therefore $K = (0.00505 + 0.00092) = 0.00597$ ml hydrogen per sq cm.

Further values of K and K_1 may be calculated as above, using equations 1 and 4, and equations 2 and 5. All values of K and K_1 are given in Table 6.

Table 6.   Hydrogen Pickup Due to Abrasion and Exposure to the Laboratory Atmosphere

Experimental Series	K, ml hydrogen per sq cm apparent area	K_1, ml hydrogen per sq cm apparent area	$K - K_1$, ml hydrogen per sq cm apparent area
(a)	0.00597	0.00092	0.00505
(b)	0.00490	0.00068	0.00422
(c)	0.00554	—	—
(d)	0.00342	0.00037	0.00305

* All areas have been calculated assuming the disks to be perfect cylinders, and no allowances have been made for area increases arising from surface irregularities.

The data of series (a), after correcting for the hydrogen increment of the unabraded disks, have been plotted on Fig. 9 for comparison with the results obtained in series (c) and may be seen in general to be consistent within the estimated experimental error of ±0.1 ml hydrogen. Both values of K obtained from these two series using petroleum ether as a degreasing agent have been shown by a statistical analysis to be significantly greater than the value of K derived from series (b) in which carbon tetrachloride was used for degreasing. As a consequence of that, the lower estimate of K, namely, 0.0049 ml hydrogen per sq cm of apparent abraded area has been taken as the value for the amount of hydrogen formed as a result of abrasion. The corresponding value for the stainless steel sample is 0.0035 ml of hydrogen per sq cm of apparent abraded area.

Subsequent experience has shown that the correction of 0.0049 ml/sq cm of apparent abraded area appears to be approximately independent of atmospheric temperature and humidity, providing the conditions of abrasion and standing time of the specimen prior to the hydrogen determination are maintained as in the above experiments. Table 7 shows

Table 7. Hydrogen Content of Steel Sheets

Sample No.	Weight of Sample, grams	Hydrogen Evolved, ml, NTP	Correction for Surface Hydrogen, ml, NTP	Hydrogen Evolved from within Sheet, ml, NTP	Hydrogen Content ml, NTP/ 100 gram
R.207	39	0.4	0.56	0	0
	39.5	0.4	0.57	0	0
R.211	34.4	0.5	0.53	0	0
	34.5	0.6	0.51	0.1	0.3
R.213	60.9	0.6	0.60	0	0
	64.8	0.7	0.59	0.1	0.2
R.209	40.6	0.6	0.55	0.1	0.2
	41.6	0.4	0.55	0	0

the application of this procedure to the determination of hydrogen in steel sheets made from high-hydrogen slabs similar to R.209. The amount of hydrogen remaining in the sheets from the slabs is negligible.

CONCLUSIONS

1. Evidence is provided to show the deleterious effects of hydrogen on the ductility of steel. This appears to be associated with a straining of the steel structure, as judged by metallographic examination and load-elongation curves.

2. Much of the hydrogen found in steel samples of large surface area arises from atmospheric contamination. It is possible to assess this amount for standardized conditions of abrasion, and so forth, and therefore to calculate the hydrogen remaining in a sheet sample from the slab stage.

3. Using the technique referred to in the second conclusion, it is found that the hydrogen present in a slab makes only a negligible contribution to the apparent hydrogen content of the final sheets.

References

1. C. E. Sims, *Gases in Metals*, p. 119, *American Society of Metals*, Cleveland, Ohio, 1953.
2. G. E. Speight and R. M. Cooke, *J. Iron Steel Inst. London*, 160, 397 (1948).

Discussion

CHIPMAN inquired what special precautions were taken during the drilling in order to avoid the loss of hydrogen. SHANAHAN replied that after extensive tests they found that very little was lost during the drilling operation and no special precautions were taken to chill the samples. Drillings were held in a glass tube under mercury over an extended period of time, and only a very small quantity of hydrogen was observed to diffuse out to be collected in the vial. He also stated that a cold cut face showed only a very slight loss of hydrogen even after having been exposed to room temperature over a number of months. It appeared that diffusion from this cold face was very slow.

During a general discussion, SHANAHAN expressed the view that by some undefined means hydrogen could be adsorbed on the surface. Operations such as grinding which roughened the surface also could contribute to the hydrogen contained on the surface of the steel. DARKEN stated that they had found a similar phenomenon several years ago in which an abraded steel specimen had on its surface an appreciable amount of hydrogen which would be released when the specimen was put into a high vacuum. Whether the specimen remained in the open air for a few minutes, after being abraded, or for several hours did not seem to affect appreciably the amount of hydrogen on the surface. He asked for further history on the treatment of samples by Shanahan. SHANAHAN stated that first the samples were made and then abraded. Subsequently, they were degassed in a vacuum-heating apparatus. It was on this surface that the experiments were started. An unabraded surface which had been degassed and re-exposed to the atmosphere did not pick up an appreciable amount of hydrogen. Degassing occurred at about 750° C. If the sample was degassed and then abraded again in the air, hydrogen was picked up.

by D. J. Carney
and B. R. Queneau

Solidification of Stainless Steel Ingots

The mechanism of solidification of steel in ingot form is very important, but in stainless steels it is especially critical since ingot structure not only influences steel quality but also affects hot workability. Comparatively little has been done to control ingot structure in stainless ingots in spite of the extensive theoretical and experimental work carried out on the freezing of metals. An ideal nucleation and growth process is depicted in Fig. 1 [1]

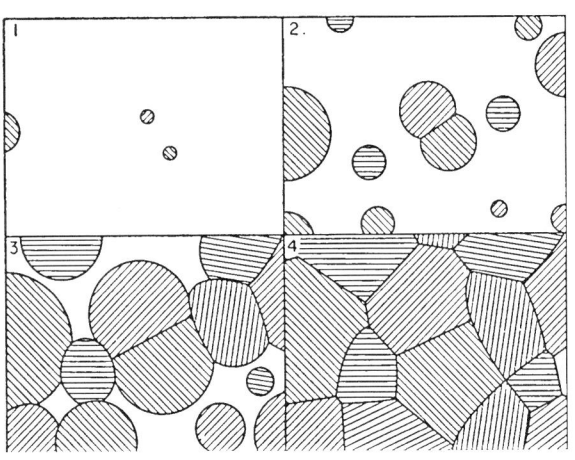

Fig. 1. Depicting an ideal nucleation and growth process in four successive time periods. (Mehl[1])

for four successive time periods. The crystal size developed in the solid metal will depend on the relative rate of formation of nuclei in the melt and on the rate of growth of the individual crystals. Figure 2, after Tammann, shows that at temperatures close to the freezing point few nuclei will form, and rapid growth of the crystals will result in

a coarse-grained structure. At lower temperatures, many nuclei form before freezing is complete, and the resultant structure will be fine grained.

When liquid metal is poured into a mold, the initial chill will result in rapid nucleation and a fine-grained chill layer. Further freezing can occur only by the extraction of heat through the chill layer, and the rate of growth relative to the rate of nucleation is usually sufficiently rapid to form col-

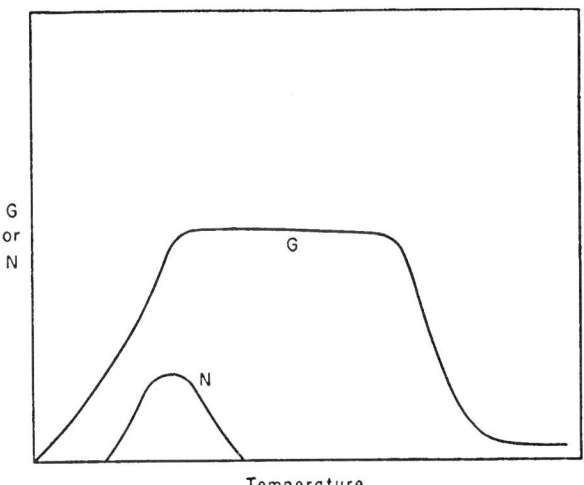

Fig. 2. Rate of nucleation, N, and rate of growth, G, on freezing, as a function of temperature. (Tammann)

umnar grains perpendicular to the mold wall. Further decrease in rate of freezing with increase in thickness to the solid layer will permit nucleation in the liquid and growth of crystals to form an equiaxed zone in the center. A good example of the resultant structure is shown in Fig. 3. When metals are cast from just above their melting points, the ingot structure is usually fine grained; whereas, if the melt has considerable superheat, the ingot will be coarse grained.

Dr. Carney and Dr. Queneau are Division Superintendent, Steel Producing, and Chief Metallurgist, respectively, Duquesne Works, United States Steel Corporation, Duquesne, Pennsylvania.

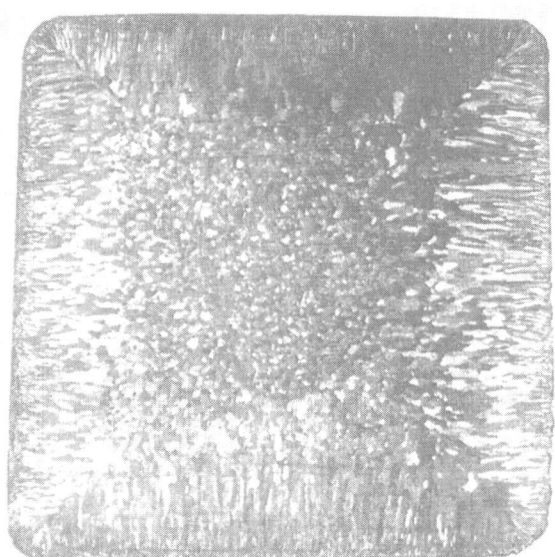

Fig. 3. As-cast crystal structure of 17% Cr stainless steel.

The present authors discovered that the solidification of steels in large molds could be altered by the addition of artificial nuclei. It was found that these artificial nuclei should possess the same crystallographic structure and approximately the same lattice dimensions as the solid at the solidification temperature. The increased number of nuclei promotes a finer as-cast grain size, would result in a more uniform distribution of inclusions, and would decrease both porosity and segregation. It was to be expected that most metallic additions used as nucleating agents would dissolve rather quickly in liquid steel. However, since nuclei are effective at extremely small particle sizes, there would be a finite time before the metallic addition agent would completely lose its identity in the liquid.

EXPERIMENTAL PROCEDURE AND RESULTS

Various grades of stainless steels were melted in an induction furnace having a capacity of approximately 26 lb. Temperatures were obtained with Pt-Pt/Rh thermocouples, with care being taken to record both the maximum temperature attained during melting and the temperature at tap. The melts were poured into 3-in. diam graphite molds with refractory hot tops, and the ingots were split in half, polished, and etched in hot 50% hydrochloric acid. The average length of columnar grains and the size of the equiaxed grains in the center of the ingots were determined, and the alteration in grain size was taken as a measure of the effectiveness of nucleating agents.

27 and 17% Chromium Stainless Steels

Initial experiments were conducted with 27% chromium steels (AISI 446) since these steels solidify over a narrow temperature range (similar to a pure

Table 1. 27% Chromium Stainless Steels

Max Temp, °F	Tap Temp, °F	ASTM Grain Size*	Length of Columnar Crystals, in.†
3060	2840	5–6	1
3020	2855	8	3/4
2895	2840	7–8	5/8
3060	2820	5–6	1
2900	2810	6	3/4
3010	2790	7–8	1/2
2950	2745	9	1/8
2895	2750	8	3/16
2850	2700	9	1/4

* Based on visual examination at 1X using ASTM grain size chart for 100X.

† Since diameter of ingot is approximately 3 in., a columnar length of 1 1/2 in. indicates absence of equiaxed grains.

metal) and are known to develop large, easily identified, columnar grains. Results of experiments with 27% chromium heats were gratifying and remarkably consistent, as shown by Table 1. All 27% chromium heats with and without added nuclei were fine grained and had little columnar crystal formation when tapped at temperatures below 2790° F. At 2790° F, a well-marked columnar zone occurred, and the equiaxed grains in the center were slightly larger. At tap temperatures of 2820° F, and with higher maximum temperatures attained during melting, the structure of the ingots consisted primarily of columnar grains with a few coarse equiaxed grains at the center.

Since chromium stainless steels have a body-centered cubic lattice structure, initial experiments to nucleate heats of 27% chromium steels were made with metallic powders having similar structure, such as ferrochromium and ferromolybdenum. The ferrochromium and ferromolybdenum were crushed to 40- to 60-mesh screen size and were whipped into the melt just prior to tap. As seen in Table 2, these additions apparently were effective

Table 2. 27% Chromium Stainless Steels

Addition	Max Temp. °F	Tap Temp, °F	Length of Columnar Crystals, in.	Equiaxed Grain Size
None	2895	2840	5/8	7–8
FeCr	2915	2860	5/8	8
FeMo	2890	2890	1/2	8
None	3060	2820	1	5–6
FeCr	3060	2825	0	9
FeMo	3100	2835	0	9

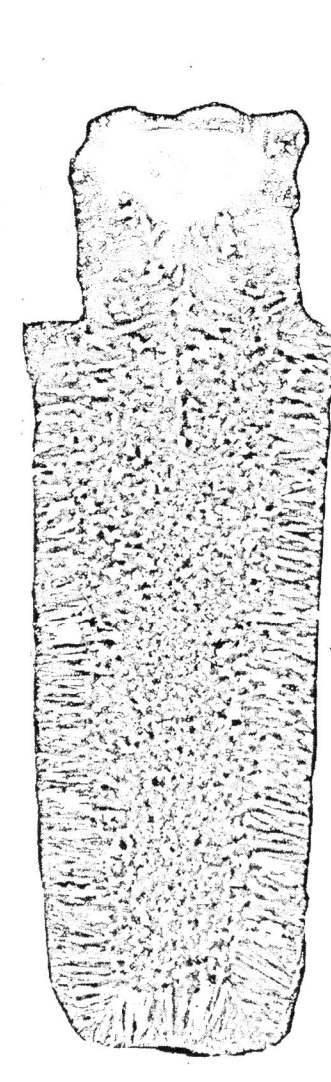
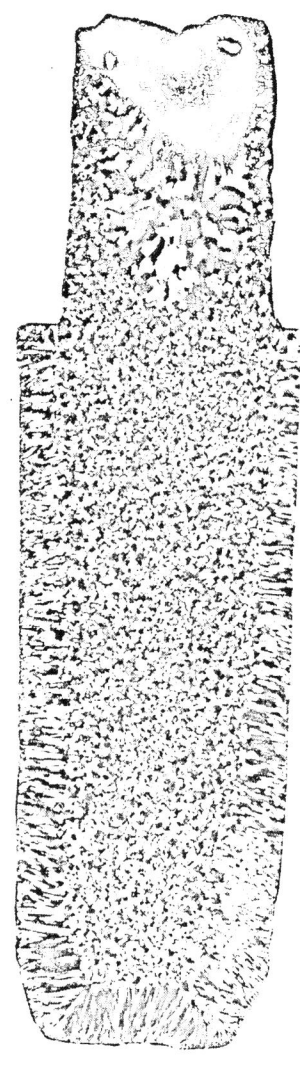

Tap temperature—2850° F.
Addition—none

Tap temperature—2850° F.
Addition—2 lb/ton FeMo

Fig. 4. Effect of addition of metallic powders on as-cast structure of 27% Cr ingots.

in refining the ingot structure of the 27% chromium ingots. Typical ingots of 27 and 17% chromium are shown in Figs. 4 and 5. Experiments with powders having different crystal lattices, such as ferrotitanium, nickel, and alumina powders, did not affect the cast-ingot structure of similar melts. Experiments with ferromolybdenum powders to determine the amount and size best suited to develop a fine-grained ingot structure indicated that the 2 lb per ton originally selected and 60-mesh powder were optimum for the conditions under which these small ingots were made.

18–8 Stainless Steels

The austenitic stainless steels containing nickel solidify over a wide temperature range and have a greater tendency to form columnar grains than the straight chromium grades. Thus, the 18–8 stainless steels poured into the same mold used on the straight chromium grades had entirely columnar

Table 3. 18–8 Stainless Steels

Addition	Max Temp, °F	Tap Temp, °F	Length of Columnar Crystals, in.	Equiaxed Grain Size
None	3040	2800	1 1/2	—
Ni	3090	2835	1/2	7–8
18–8 Powder	3035	2850	1/4	9
FeMn	3060	2825	3/4	8
FeCr	3035	2810	1 1/2	—
FeTi	3040	2860	1 1/2	—

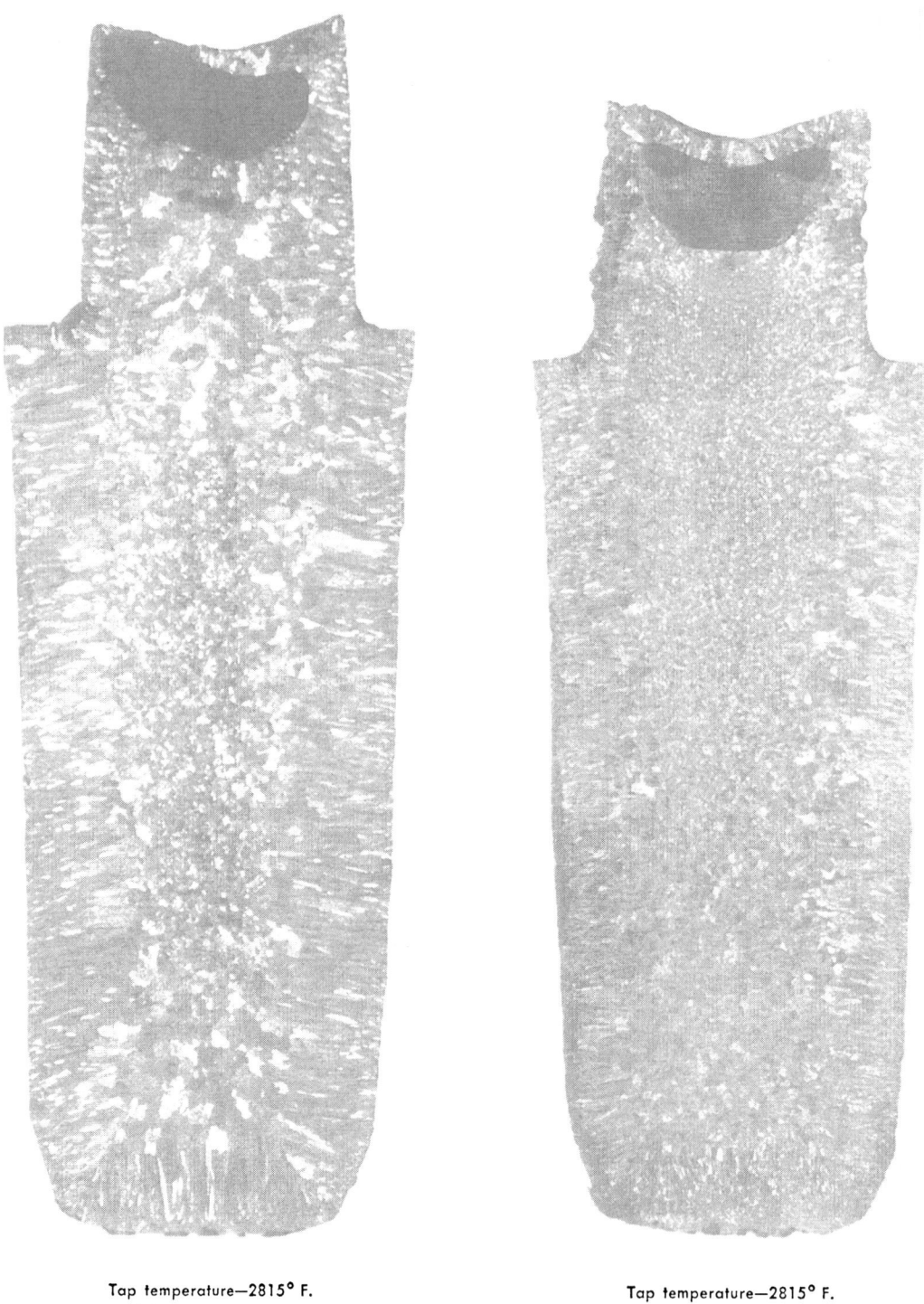

Tap temperature—2815° F.
Addition—none

Tap temperature—2815° F.
Addition—2 lb/ton FeMo

Fig. 5. Effect of addition of metallic powders on as-cast structure of 17% Cr ingots.

structures regardless of the tapping or maximum melting temperatures attained. Powders having face-centered-cubic lattices similar to the austenitic steels, such as nickel, 18–8, and ferromanganese, were all somewhat effective in refining the ingot structure. However, the results were not always consistent, and the data in Table 3 should be considered as optimum results since some heats made with apparently identical practices were entirely columnar. With powders having dissimilar lattice structures, such as ferrochromium, it was not possible to alter the as-cast structure of austenitic stainless steel ingots. Typical ingot structures of 18–8 ingots are shown in Fig. 6.

Trials with Commercial Ingots

In view of the effectiveness of nucleating agents

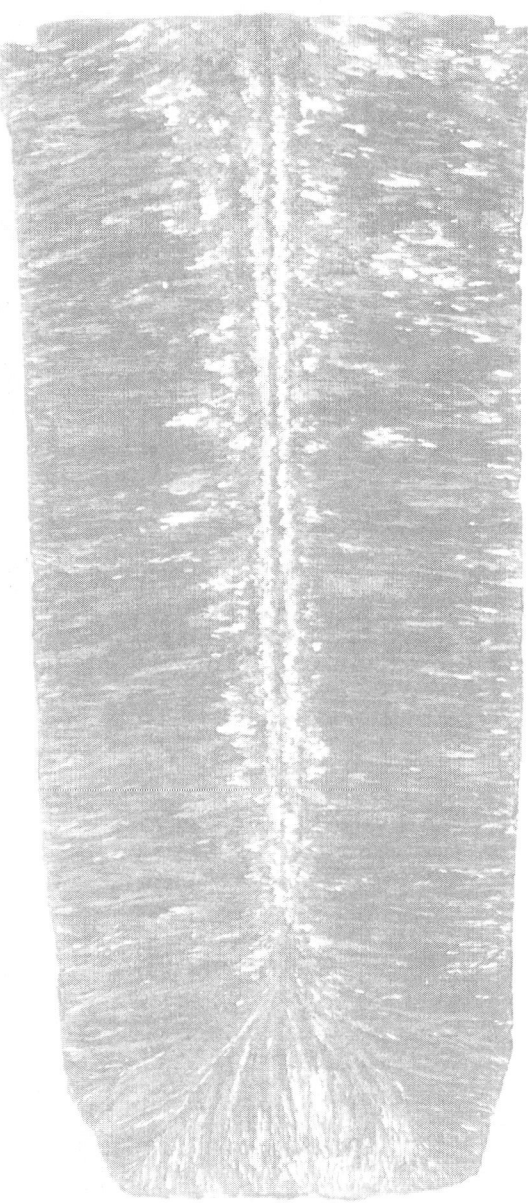

Tap temperature—2800° F.
Addition—none

Tap temperature—2805° F.
Addition—2 lb/ton 18-8

Fig. 6. Effect of addition agent on as-cast structure of 18-8 ingots.

in refining the as-cast grain size of small induction-furnace ingots, a trial was made on a commercial heat of 17% chromium steel. Two 16 × 16 × 60 in. hot-top ingots were poured. One was poured in a normal manner, whereas the adjacent ingot had 2 lb per ton of ferrochromium powder added uniformly to the stream during the pouring operation. The ingots were split and etched, and the structures are shown in Figs. 7 and 8. Although the refinement in ingot structure is not quite as marked as obtained with the small 27% chromium stainless ingots, it is clear that the addition of ferrochromium powder has reduced the length of the columnar grains and resulted in finer equiaxed grains in

the center. The presence of this refined ingot structure has also reduced center porosity and improved the quality of the ingot.

SUMMARY

Based on the premise that artificial nucleation of solidifying metals is possible, it was observed that the addition of certain types of metallic powders to ferritic types of stainless steels produced a finer as-cast grain size. This was demonstrated on both large and small ingots and various grades. Some success was also obtained with austenitic steels, but results were not consistent. This fact was believed

SOLIDIFICATION OF STAINLESS STEEL INGOTS—213

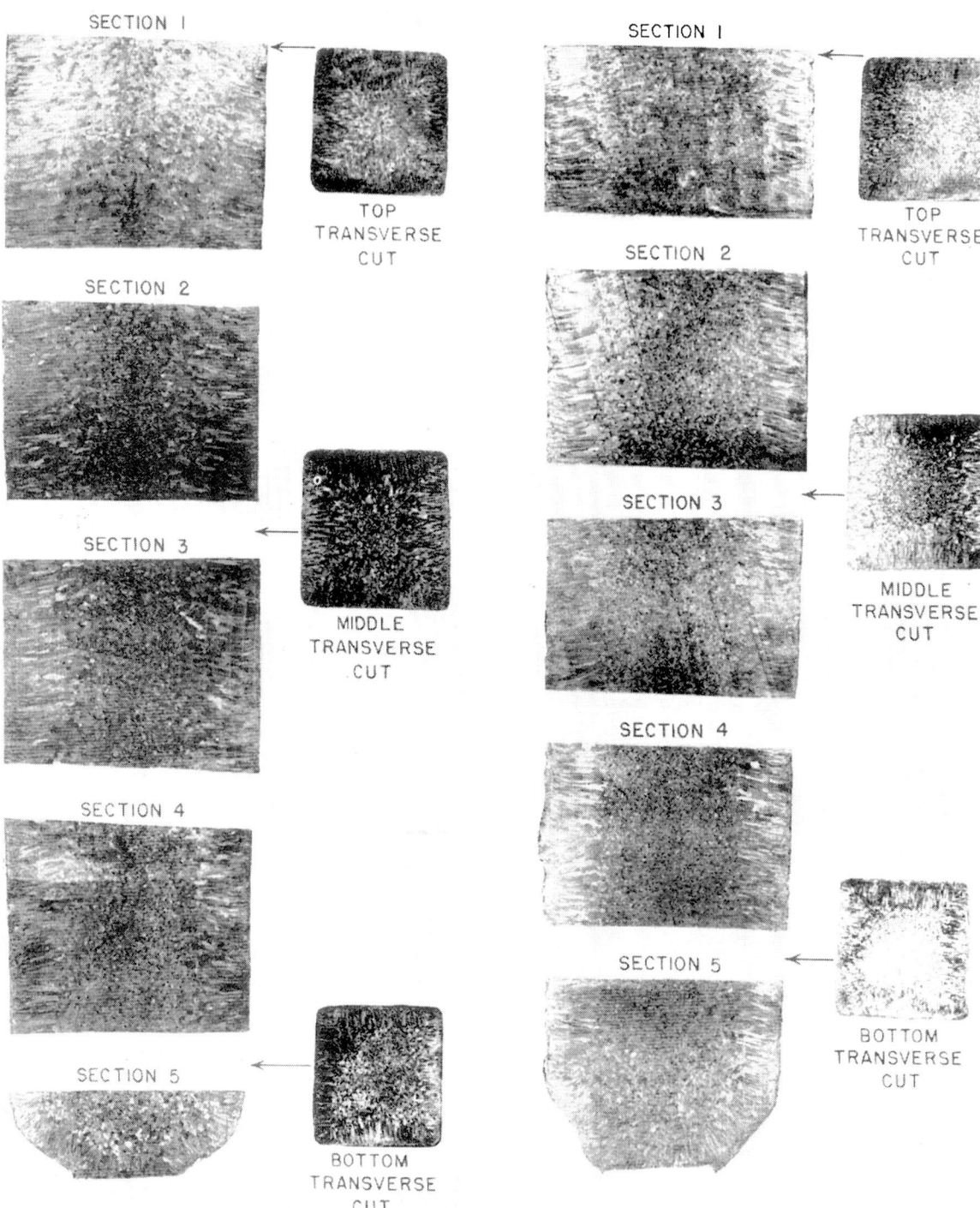

Fig. 7. Untreated commercial 16 × 16 × 60 in. ingot of 17% Cr steel.

to be associated with the wide freezing range of these alloys.

References

1. R. F. Mehl, *Solidification of Metals and Alloys*, 24, A.I.M.E., New York, 1951.

Discussion

SHERMAN suggested that it may be possible that non-

Fig. 8. Nucleated commercial 16 × 16 × 60 in. ingot of 17% Cr steel showing finer grain than ingot in Fig. 7.

metallic inclusions could also cause nucleation just above the melting point so that a fine-grained ingot structure would result. QUENEAU was not certain that this was so. Another possible explanation could be that a metal brought to just above its melting point could have a "memory" of the solid structure which would provide numerous sites for nucleation. DARKEN proposed that highly refractory, nonmetallic precipitates in the powders might carry the lattice structure of the parent metallic powder. They would persist, whereas the metal of the powder itself would melt, and consequently, they could act as nuclei instead. He found it hard to visualize how the influence of the metal of the powder could be maintained for 5 min at 100° C above the melting point of the steel. ELLIOTT suggested that the small, solid, metal-lic-powder particle would leave behind as it melted a cold spot primarily due to its latent heat of fusion. Over very short periods, this spot might provide a point for nucleation.

by C. M. Adams, Jr.

On Segregation in Steel Castings

Segregation in solidifying alloys, such as steel, may occur in widely variant patterns, although the fundamental driving force for segregation remains unchanged. A solid solution in equilibrium with a liquid solution will, in general, have a composition different from that of the liquid solution. Consequently, the first crystal to solidify in an ingot or casting of a dilute solution will tend to have a different, usually lower, concentration of solute than the liquid from which it solidifies. Thus, the classic picture of positive segregation in a steel ingot has the surface of the ingot low in solute, since crystallization initiates at the surface. The central portion of the ingot is rich in solute, and the degree of separation is dependent upon the segregation coefficient of the solute.

However, several mechanical, chemical, and thermal factors, in addition to the segregation coefficient, can influence the degree, the direction, and the average distance of solute migration during solidification: (1) rate of solidification, (2) slope of the liquidus, thermal conductivity, and liquid mass diffusivity of the alloy in question, (3) solidification shrinkage, (4) liquid convection, (5) solid thermal contraction, and (6) amount of superheat with which the metal is poured. These factors affect segregation either by modifying mass transfer through the liquid or by causing local flow of solute-rich liquid.

In the absence of pronounced dendritic solidification or wide separation between the liquidus and the solidus, the liquid-solid interface is essentially planar, and positive segregation occurs in a direction perpendicular to the mold wall. The degree of separation increases with: (1) increasing segregation coefficient, (2) decreasing solidification rate, (3) increasing convection, and (4) increasing liquid mass diffusivity. In an ingot the parabolic rate constant for solidification is so high, and the mass

Dr. C. M. Adams is Assistant Professor of Metallurgy, Massachusetts Institute of Technology.

diffusivity so low, that, in the absence of convection, little macrosegregation can take place. Convection is usually present to some extent in a killed-steel ingot because of temperature differences and residual stirring momentum from the tapping operation, so that some positive segregation will occur. Vigorous convection in a rimming ingot favors pronounced positive segregation.

When the mechanism of solidification is dendritic, the liquid-solid interface is no longer a plane, and its true area is much larger than its projected area viewed perpendicular to the mold wall. The ratio of projected to true liquid-solid interface areas may be regarded quantitatively as the sine of the angle between the principal growth direction and a plane parallel to the mold wall. When the liquid-solid interface is nearly planar, the ratio is unity, and the principal growth direction is perpendicular to the mold wall. When solidification is dendritic, the angle is very small, and growth and segregation are essentially "sideways." In the absence of mechanical influences, dendritic growth is accompanied by very little macrosegregation but pronounced microsegregation. Dendritic growth is favored by: (1) slow solidification, (2) steep liquidus, (3) high thermal conductivity, and (4) low superheat. It is important to realize that, in solidification of most steels, under conditions of parabolic growth from a metal mold wall, the direction of principal growth gradually shifts from nearly perpendicular to nearly parallel to the mold wall, and the segregation mechanism shifts accordingly.

The combination of dendritic solidification with certain of the mechanical influences indicated above can result in such effects as negative or inverse segregation in which the first regions to freeze are high in solute, or exaggerated positive segregation. For example, under conditions of dendritic freezing, solidification shrinkage can cause interdendritic solute-rich liquid to flow toward the mold wall; conversely, contraction of the completely solid layer

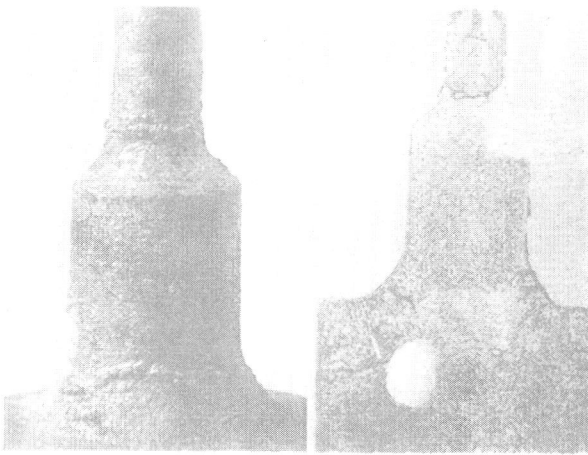

Fig. 1. Casting with sulfur print of cross section showing inverse segregation at fillet section. Mechanical restraint present during solidification due to mold design.

can move dendrites closer together, resulting in movement of interdendritic liquid away from the mold wall. In other words, the whole dendritic mass can behave like a sponge, accepting or rejecting liquid in conformity with the mechanical forces operating on it. One of the most drastic effects can result from hindered solid contraction of the first metal to freeze; under these circumstances, contraction in one region causes another, hotter area to

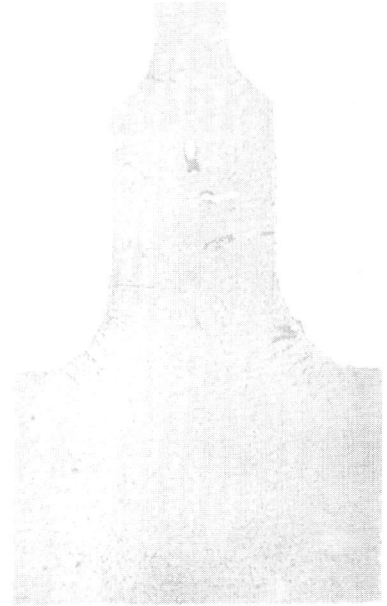

Fig. 2. Sulfur print of inverse segregation in casting. Mold design produced mechanical restraint during solidification.

Fig. 3. Sulfur print and etched surface of cross-sectioned area of casting showing planes of weakness as a result of mechanical stresses during solidification.

stretch. When solidification shrinkage is accompanied by plastic extension of the solid and separation of dendrites, the effects are additive, and very high local concentrations of solute are produced. Examples are shown in Figs. 1, 2, and 3, which are sulfur prints of sections of castings which were allowed to solidify under differential cooling conditions in which solid contraction was hindered. Although the metal was originally 0.04% sulfur, local concentrations can be high enough to induce cracking (Figs. 1, 2), with subsequent exudation of liquid, or to leave macroscopically invisible planes of weakness (Fig. 3). From this and other observations, it is apparent that mechanical influences are of determining importance in modifying simple chemical and thermal models used in consideration of segregation, and certainly the least important mechanism for solute transport in an ingot or casting is that of true diffusion.

Discussion

LARSEN asked whether severe restraint was necessary in order to set up the large stresses encountered. ADAMS answered that it was not particularly necessary. They were obtained by casting a flange on each end and using a sand mold which produced moderate stresses on cooling.

Section 9

Research Planning

The Conference convened for a short session to discuss the direction to be taken in research to meet the present and future needs of industrial steelmaking. A general discussion was held of the future direction of fundamental research and industrial needs for fundamental knowledge. Editor.

by F. D. Richardson

Summary of Needs for Fundamental Research

There are two prime objectives which we, who are concerned with fundamental research on the chemistry of iron and steelmaking, should have in view. It is our responsibility to ensure that fundamental knowledge is obtained where it is specially required to increase our understanding of existing processes. It is also our responsibility to devote a fair part of our research effort to projects which, though far more speculative, may well lay the foundations of entirely new processes. I shall consider the need for researches relating to current processes first; it is far easier to recognize what is required in this field than in the other, and I shall divide the topics on a scientific rather than on a process basis.

MOLTEN METAL SOLUTIONS

Early in this Conference we dealt with the activity of carbon at high concentrations in iron (1.5% to saturation) and the influence of carbon on the activity coefficient of oxygen. Now the position, in spite of the work reported, remains rather unsatisfactory. We can admittedly calculate the activity coefficients of carbon over the whole solution region; the values, however, are mainly based on uncertain interpolations, and I still think they should be accurately measured. The precise evaluation of the influence of carbon on the activity coefficient of oxygen still remains in doubt, partly on account of difficulties of analysis for oxygen in iron. The problem stands, indeed, on the very edge of analytical possibilities, and work on the carbon-oxygen interaction must go hand in hand with work on oxygen analysis. Fortunately, under the direction of Dr.

Kozakevitch, Dr. Cordier is tackling the problem, and as you know, IRSID has a very good analytical team. We can hope, therefore, that in a few years we shall have a full solution to this problem.

Accurate knowledge of the influence of carbon on the activity coefficient of oxygen could be helpful in the development of fundamental theory for solute-solute interactions in molten metals; there are, however, other systems in which it is easier to study these phenomena. I refer particularly to sulfur in metals and the effects of alloying elements on the behavior of sulfur at low concentrations. The solubility ranges, the magnitudes of the interactions, and the experimental techniques are all more favorable than they are in the case of the Fe—C—O system. On our side of the Atlantic, we in the Nuffield Group are studying the behavior of sulfur in copper and the effects of a whole range of alloying elements at different concentrations on the sulfur. By studying the activity coefficients as functions of temperature, we have hopes of expressing them in terms of partial molar heats and entropies. Thus far, the results are interesting, and they indicate that in ternary melts the activity coefficient of sulfur is influenced by the metal-metal as well as metal-sulfur interactions. It seems to me that the measurements which have been made on sulfur in iron and iron alloys at about 1600° C should be extended over a reasonable temperature range so that we can see how the partial heats and entropies of solution of the sulfur change in the presence of alloying elements. The measurements must, however, be made with the highest accuracy attainable at these temperatures; otherwise, the derived heats and entropies will have little or no significance. If, in addition, somebody would make similar studies

Dr. Richardson is Professor of Metallurgy, Imperial College, London, England.

of phosphorus and silicon in iron alloys, we would have a reasonable picture of the behavior of the four most important nonmetal solutes.

For simpler binary solutions, we ought to be more certain than we are about ideality in mixtures of iron with manganese, chromium, nickel, and cobalt. It appears, from studies of the partition of manganese between metal and slag, that manganese behaves ideally in molten iron. The latest work on the solid solutions of chromium and iron show that they are more or less ideal over the whole composition range. We not only have McCabe's Knudsen cell work on the vapor pressures of chromium over these alloys; there is also evidence which has been reported in Paris by Moreau of IRSID and which was obtained by an unusual type of study of the equilibria which can exist between chromium in iron, chromic oxide, hydrogen, and water vapor. It is interesting to note that if chromium is ideal in its solid α-alloys with iron it cannot be ideal in the liquid mixtures; provided the phase diagram is correct, it should show substantial deviations from Raoultian behavior. More knowledge on this simple binary system is important to the understanding of the iron-chromium-silicon and iron-chromium-carbon solutions. Let us hope that the vaporization technique for measuring activities which has been developed by Mr. Morris of the Bureau of Mines in Pittsburgh can be extended to cover the alloys I have mentioned. May I sound a word of warning concerning Mr. Morris' work on Fe—Si mixtures? In studying systems with silicon, there is a danger of having silicides as well as metals present in the *vapor*. Certainly platinum silicide is more volatile than platinum, and other silicides may exhibit similar properties.

LIQUID SLAGS

Next we may consider research on molten slags. One of the first things we should try to do is to resolve the discrepancies in the measured activity curves of silica in the lime-silica system. This binary system is of *prime* importance so far as all kind of slags are concerned, and the measurement of activities in simple binary mixtures is a step toward an understanding of activities in ternary and more complex slags. After the $CaO—SiO_2$ activities have been established, measurements for the magnesia-silica system should be attempted.

As I mentioned earlier in the Conference, the heat of fusion of silica should be properly established. If we knew it accurately, we would be able to interpret the depressions of the freezing point of silica caused by the addition of such oxides as Na_2O and MgO in terms of entropy and, hence, structural changes. We would thus be better able to understand the mechanism of breakdown of the silicate lattice on the addition of metal oxides. We need to know more about the structures, the heats of formation, and the heat capacities of silicate solid

solutions and of compounds formed between silicates; also, of the heats of fusion of the compounds. If we had adequate data for solid solutions, we could probably find useful relationships between the heats of formation of mixed silicates and their structures. The behavior of the solid solutions might also provide a basis for an understanding of the corresponding liquid mixtures. Attempts should be made to obtain the heats of mixing of the molten silicates themselves, but this, unfortunately, is no easy matter. One must first measure the heat contents of the melts in a dropping-type calorimeter; the mixtures have to be cooled quickly for such measurements, with the result that glasses and unstable silicate modifications may be formed in the process. The solidified silicate must therefore be subsequently dissolved in HF, and the further heat effect measured. I hope such things will some day be attempted in the calorimeters which have been built at IRSID and at Berkeley.

Better activity data for ternary silicate mixtures are needed so that the "ideal mixing" theory which I presented earlier to the Conference can be fully tested. We also need to know the effect of alumina on the activity coefficients of metal oxides as we substitute alumina for silica in ternary melts. Fortunately, a reasonable amount of work on silicates is now in progress. We remain, alas, appallingly ignorant about phosphate melts, except for the few special cases that were discussed in the Conference. It would be a good thing if in the next few years someone would begin to study phosphate systems seriously—first the binaries and then the more complex ternaries and quaternaries.

Sulfur partition between gas and slag seems to me to be an excellent means of obtaining an understanding not only of the way that sulfur is held in the slag but of other aspects of slag behavior and the general picture of slag structure. Sulfur capacities of phosphate and of silicate-phosphate mixtures should be tackled, and we at the Nuffield Research Group are planning to do this in the near future.

The solubilities of metals* in slags came into the discussions at our Conference. We are continuing these measurements at the Nuffield Research Group and are currently studying the solubility of lead in molten silicates at 1500° C. We have been lucky to find that an iridium container is not attacked either by lead or by the silicates. This effect is being studied as a function of temperature and slag composition for binary and ternary solutions. However, lead is not of particular interest to the iron and steel people, and someone should tackle the problem of the solubility of iron, chromium, and manganese in steelmaking slags. Chromium and manganese should be considerably more soluble than iron on account of their high vapor pressures. These two solubilities must be known if we are go-

---

* As distinct from metal oxides.

ing to interpret correctly the slag-metal equilibria in which there are only small quantities of metal and metal oxides dissolved in the slags. At the present time, one has to assume that all the chromium or all the manganese is present as the oxide, and this is probably not the case at low concentrations. The results might be revealing in other ways; they should shed some light on the amount of free space available in the silicate lattice, for the neutral metal atoms probably pack into holes in the network, and the solubilities may also correlate with other structurally dependent properties.

This brings us to the question of slag structure. We want a good structural background against which we can picture both the thermodynamic and kinetic aspects of slag behavior. In spite of the X-ray work which has so far been done on crystalline silicates and glasses and the density and viscosity measurements which Bockris and others have made on silicate melts, our present knowledge is only scanty. The first thing which would be of interest would be to reopen the X-ray studies on the structure of glasses. After talking with X-ray experts, I gather that there are now methods available by which it should be possible to determine whether there really exist in melts the large silicate groups or ions (as distinct from chains) which Bockris has proposed to account for his viscosity and density measurements. Coupled with this work, it would be desirable to have density measurements and measurements of the self-diffusion coefficients for cations in silicates as a function of composition. Bockris has suggested (on the basis of expansivity measurements) that in the early stages of disintegration of the silicate network caused by the addition of small amounts of metal oxide, the cations are locked in by surrounding silicon-oxygen links; he supposes this to happen up to 12 mole % of oxide and considers that there is a sudden change in structure at about this composition. Cation diffusion rates should therefore be very slow at from 0 to 12 mole % metal oxide and should increase sharply when the structure changes.

We ought also to know more about the structures of solid mixed silicates at elevated temperatures and establish whether the cations are randomly distributed with respect to one another in solid solutions. In cases where the cation distribution is random and the heats of formation of the solid silicate mixtures (of the type $XO \cdot aSiO_2 + YO \cdot aSiO_2$) are zero, the free energy of formation of the mixtures from the binary silicates $XO \cdot aSiO_2$, $YO \cdot aSiO_2$ can be taken as ideal; when the corresponding phase diagrams are accurately known (as should often be the case, thanks to the work at the Geophysical Laboratory in Washington), it should be possible to calculate activities in the melts with considerable accuracy.

SULFIDES

Sulfide melts have not been mentioned at this Conference. We ourselves studied some time ago Cu_2S—Na_2S at high copper-sulfide concentrations. With the present interest in decopperizing iron and in trying to get manganese out of various metallurgical systems, more should be known about the activity coefficients of copper sulfide, manganese sulfide, iron sulfide, and nickel sulfide at low concentrations in sodium sulfide. Unfortunately, measurements of these activities are not easy at high temperatures, because of the high volatility of some of the components. This is the reason why we ourselves could investigate the Cu_2S—Na_2S system only at high copper-sulfide concentrations. With a circulating gas system ($H_2 + H_2S$), we were limited to mixtures containing more than 50 mole % Cu_2S because of the volatility of Na_2S. It appears that facts of both fundamental and technological importance might emerge from such sulfide studies.

Oxide-sulfide systems have hardly been investigated at all and surely merit further study. One good piece of information which I can give you now is that we ourselves have measured activities in FeO—FeS mixtures. The system behaves ideally over the whole range of composition, so the results are in general agreement with Professor Chipman's 1948 calculations which were based on the heats of fusion and the somewhat uncertain phase diagram.

KINETICS

One notable thing about this Conference, which sets it apart from many others, is the interest that has been shown in the kinetics of ironmaking and steelmaking processes. We now need to digest some of the ideas which have been so well expressed by Professor Wagner and Dr. Darken, and then to decide what fundamental work will be most rewarding and most revealing. We must be careful not to fall into a trap. The advantage to the thermodynamic studies is that the results from the laboratory are independent of scale and inevitably apply to the works process insofar, of course, as equilibrium is attained or approached in both experiment and process. With the kinetic problems, we have, in my view, to exercise far more care in designing our experiments if they are to be at all relevant to actual processes. It is difficult enough to study true chemical kinetics; it is far more difficult to design useful diffusion-controlled experiments in which liquid-liquid boundaries are present. I must confess that I have no personal experience of kinetic studies involving liquids: there are others here, particularly from the Massachusetts Institute of Technology and the Carnegie Institute of Technology, who have conducted such investigations and can speak with more assurance. Nonetheless, I think that there is a great danger of making studies which are neither sufficiently well controlled to yield decisive fundamental data, nor of sufficient scale to have any bearing on either existing or conceivable full-scale processes.

In addition, one must steer clear of the tempta-tion to use the Δl term as an unexplained con-stant hiding many real uncertainties. Experiments should be designed which will reveal the character of this layer. In this matter, we can fortunately rely on Professor Wagner to steer us toward cer-tainty and away from "catastrophe" (and he uses that word with a remarkable sense of being able to avert catastrophe!), avoiding disaster with a proper choice of power terms and the various dimension-less numbers.

SOLID METALS

Another subject which we should consider briefly is the physical chemistry of solid metals, although solids have not actually been discussed at this Con-ference. First, somebody should attempt to clear up the question of the solubility of oxygen in solid iron, a question which is in a very unsatisfactory state. There are the negligibly small values ob-tained by Kitchener and co-workers for gamma iron, and the values of Seybolt for alpha iron; the latter, if extrapolated to the delta range, give solu-bilities which substantially exceed the values found for liquid iron. I personally think that the true values for alpha iron must be very much less than those suggested by Seybolt, even making allowance for the fact that oxygen may be like sulfur and thus is considerably more soluble in alpha and delta iron than in gamma iron.

The possibility of measuring activities in solid alloys by the Knudsen method is, as McCabe has shown, very attractive. Apart from iron-chromium alloys on which he has already made measurements, iron-manganese alloys would be worth checking; iron alloys with less volatile metals such as cobalt and vanadium might be studied, using radio-iron and collection methods. Activity measurements on all of these would be well worth while, and activities in the melts could then be calculated with the help of the phase diagrams. We could also tackle profit-ably activities in solid solutions of carbides and sul-fides and oxides, by measuring the compositions of the metal mixtures which can exist in equilibrium with the solid carbide, sulfide, or oxide phases, and making use of the activities of the various compo-nents of the metal phases.

NEW DIRECTIONS FOR RESEARCH

As I said at the beginning of this review, it is our responsibility to look beyond existing processes and to direct a fair share of our research effort in novel and visionary directions. The big criticism which I level at iron and steel research (I am talking among friends now!) is that it has been directed almost exclusively to revealing the mysteries of ex-isting processes and attempting to improve effi-ciencies by relatively small amounts. I know full well that it is extremely difficult to suggest lines of research which have any great likelihood of leading to novel processes of steelmaking; it is, however, our duty to force our minds to attempt this and to launch out on speculative projects. I think chlori-nation processes are worthy of consideration and that we may some day come to making iron via its chlorides, simultaneously, perhaps, turning out pure aluminum and silicon, and titanium! Such a proc-ess could be economic only as a whole, and the value of the nonferrous metals would have to be taken into account. It may not be as visionary as at first sight appears; many of you probably saw the paper which appeared in 1954 in Britain (*Journal of the Iron and Steel Institute*), in which Reeve of the United Steel Companies described a chlorination process for the concentration of lean British ores. The ferric oxide of the ore is converted to ferric chloride by means of gaseous HCl; the chloride is then reconverted by hydrolysis to a virtually pure oxide and the HCl used for further chlorination. Vanadium is a useful by-product, and the process as a whole comes near to being economic, if it is not actually so at the moment. I wonder if we ought not to look more closely into the physical chemistry of the chlorides of manganese, and chromium, and perhaps of other elements which appear in the iron- and steelmaking flow-sheet; most of these may not be of much economic interest in the pure state today, but in the long run they may become of con-siderable significance.

With this thought I must conclude these sugges-tions, hoping that some of you will be bold enough to take them up and helpfully improve upon them.

by P. P. Kozakevitch

Commentary on

The Direction
of IRSID'S Research Program
in Physical Chemistry

Because Prof. Richardson covered rather thoroughly the needs for fundamental research, Dr. Kozakevitch gave a short summary of IRSID's history and then outlined briefly the direction of research at IRSID. A summary of the comments follows. Editor.

A major program for studying slag-making reactions and metal-reduction reactions in the bosh and hearth of the blast furnace is to be undertaken in the near future. There is under way a sizable study as to methods for reducing iron ore by processes other than the blast furnace. The limited time available prevents giving details. The physical and chemical properties of blast-furnace slags are to be studied. The discussion on foams and emulsions is a precursor of this work.

Dr. Kozakevitch is Head of Physical Chemistry Department, Institut de Recherches de la Sidérurgie, Saint-Germain-en-Laye, France.

Thermodynamic studies are to be made of the activities in many liquid iron solutions: iron-carbon-phosphorus, iron–high-carbon alloys, and iron-phosphorus-silver. The high-temperature calorimetry will be continued. Studies on the viscosities of iron and iron alloys will be continued, and measurements of densities will also be undertaken.

There are many other plans under way, but he felt that those listed were of primary interest to the group.

by K. L. Fetters

Physical Chemistry of Steelmaking

and Plant Practice

It would be very satisfying, in presentation of a brief paper on the subject of the application of physicochemical ideas in plant practice, to be able to point to dramatic and spectacular new and improved processes which had sprung full-grown from the basic research in the field. That we cannot do so, in my opinion, does not in any way belittle the importance of such physicochemical research, but rather emphasizes the need for a realistic appraisal of what practical production gains we may expect from such research.

I am sure you are all familiar with the remarks of Mr. D. J. O. Brandt, the English metallurgist, who expressed some extremely caustic comments regarding the value of research in the iron and steel industry and who challenged the continuation of such research. Likewise, I suspect most of you have read a résumé of the defense of such research so well presented by Dr. Hermann Schenck in Tokyo in April, 1955. I do not propose to rehash these arguments, but I make reference to these papers to provide a statement on which I wish to elaborate. In written comments on Dr. Schenck's talk, Mr. D. J. O. Brandt writes: ". . . Dr. Schenck recommends the employment of intermediaries in steelworks to see that scientific research results are interpreted in steelmakers' language. It is obvious from a reading of the German Technical Press that such men are widely employed in the German industry. . . ."

It is my opinion that one of the most important "products" of physicochemical research is this trained manpower to serve as steelworks "intermediaries."

Dr. Fetters is Assistant Vice President, Youngstown Sheet and Tube Company, Youngstown, Ohio.

Study of a "sample" of 55 men who have done graduate or postgraduate research in the field of physical chemistry of steelmaking from two of our leading engineering schools, Massachusetts Institute of Technology and Carnegie Institute of Technology, since 1940 shows that 36% of these men are actively engaged in the United States iron and steel industry, that 38% have come from, and returned to work in, other countries, including Canada, and that 9% are engaged in reproducing their kind by teaching on the faculty of departments of metalurgy in the United States. If we correct our sample to consider only those originating in and subsequently employed in the United States, we find 59% employed in the iron and steel industry and 15% teaching. We have thus produced some very competent "intermediaries" as a direct product of sponsored physicochemical research in the colleges. In this study, no account was taken of the much larger number of undergraduates who have been better trained in the physicochemical field as a secondary by-product of the research projects and who also are very important steelworks intermediaries. Not only have we been producing the intermediaries, but we also have been producing the "primaries," or men who are in the direct supervisory line operating organization in charge of the steelworks. Many a steel plant superintendent and even some of the "top brass" in present steel plant operations have graduated from modern schools where the physical chemistry of steelmaking has been well taught on a high level, and many more earlier graduates have "caught up" by means of the technical journals and books written as the result of physical chemical research.

According to Gilbert Burok (on page 91 of For-

tune magazine for January, 1956), "One reason why steelmen have been less research-minded than many think they should be is that they haven't had to resort to laboratory research to become more efficient. Few operations, not even railroading, present more opportunities for spectacular increases in efficiency with relatively little outlay of money. Step up performance in one phase of the operation, and another cries to be improved; step it up there, and still another opportunity shows up."

Undoubtedly, much of what Mr. Burok says is true, but as we have improved efficiency, we have in some cases reached a point of diminishing returns where the scientific approach must be sought for further gains. A few instances to illustrate the point follow.

1. **Control of the Composition of Steel to Meet Specific Requirements.** One need only consider the drawing jobs regularly made with present-day drawing-quality steel or nonaging steels with their freedom from stretcher strains to see the progress that has been made in this field. Tinplate, welding rods, and many other specialty steels are now made according to what at one time would have been impossibly restrictive chemistry.

2. **Control of the Response of Steel to Heat Treatment.** In the early 1930's we learned how to control grain size by suitable deoxidation. Now grain-size control, along with hardenability control by balancing alloy compositions, has made mass-production heat treatment possible.

3. **Control of Steel Quality.** This is a rather paradoxical characteristic but is herein considered to be qualitatively definable in terms of steel cleanliness, soundness, and freedom from surface imperfections or defects. A good example of progress in steel quality control is found in cleanliness specifications to which we presently conform for severe forming or severe service requirements. Magnetic particle inspection and magnetic testing of present-day steels quickly reveal the improvements that have been effected in surface quality.

4. **Steelmaking Production Rates, Tons Per Hour.** Here we have a basic economic factor in keeping the relative cost of steel low as contrasted to greater increases in cost of many other commodities. Many gains in production rates may be attributed to mechanical improvements and the trend toward larger and larger production units. However, shorter heat times have been achieved by

slag control and working practices which have evolved from laboratory and plant slag-metal research.

Table 1 shows that in 1943 the average open-hearth furnace in the United States was tapping heats of 146.3 tons per heat at a tap-to-tap rate of 12.07 tons per hr. In 1954 the average open hearth had grown to where it tapped 182 tons per heat at an average rate of 15.81 tons per hr. Table 1 also shows that the average heat time in 1943 was 12 hr, 12 min and that this time had been reduced 42 min in 11 years to the 1954 average of 11 hr, 30 min heat time. Are we better or just bigger?

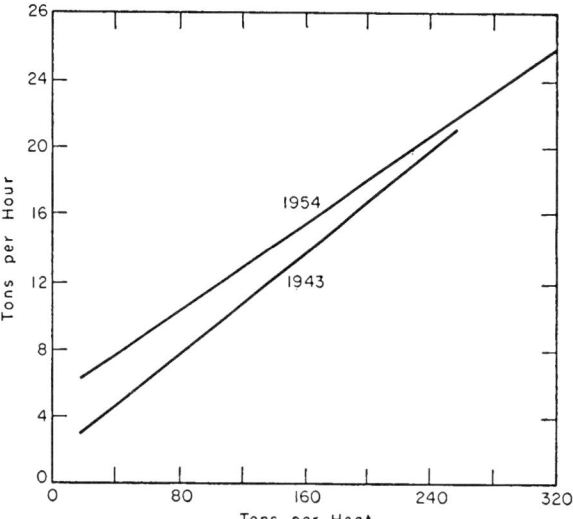

Fig. 1. Effect of size of heats on average production rates of basic open-hearth furnaces.

In Fig. 1 the data for most of the open-hearth shops in the industry are represented by the calculated regression lines for the tons per hr as a function of heat size. The lines represent 1943 and 1954. A line for 1948, not shown on the chart, indicates that for all practical purposes we did not gain in efficiency in the first half of the decade, but that we did a little better in the second half. Percentagewise we have increased our tons per hr by 31.0% in 11 yr with a 24.3% increase in heat size; $24.3 \div 31.0 = 78.4\%$, which we will round off to 80. That is a measure of the production increase that may be attributed to size, leaving 20% which may be claimed as "practice" improvement.

5. **Economies of Recovery of Residual Alloys and Economy of Yield of Additions.** From our knowledge of the effects of temperature, our ability to estimate better the degree of oxidation of the bath and our ability to predict more accurately the yield of deoxidizing and alloying additions, important economies have resulted. Not the least of these savings has been the lesser percentage of "off" heats. In addition to the contribution of basic physicochemical research, this situation has also been aided by the temperature-measuring devices

Table 1. Comparison of Open-Hearth Operations

	1943	1954
Tons/heat	146.3	182.0
Tons/hr*	12.07	15.81
Heat time*	12:12	11:30

* Calculated on tap-to-tap times.

and the rapid analytical methods which have partly evolved from the basic research; partly their development has been inspired when basic research showed the need for such controls.

6. **Ability to Meet Stringent Specification Standards Despite Decrease in Quality of Raw Materials.** Examples to illustrate this situation are numerous, such as the steady increase in the sulfur content of coking coals and our present ability to meet the same sulfur specifications as we did 20 years ago; and for certain products requiring special practices, even to keep below the sulfur levels of the past is ample illustration of progress.

Possibly one of the strongest claims that can be made for the contribution of physical chemistry to steelmaking progress has been the clarification of the "mysteries" surrounding the "qualities" of steels from different processes. We now understand rather well that the acid open-hearth steels, basic open-hearth steels, acid Bessemer steels, and basic Bessemer steels are all the same when their chemical constitution is the same and do not depend on some mysterious, intrinsic attribute of the process. The rapid acceptance of an open-hearth substitute for Bessemer-free machining steels is a good case in point where many American steel companies have discontinued Bessemer operations and gone to open hearths to produce these grades at a considerable savings in some instances. The ready acceptance of the "L-D," the "O-V," and other oxygen steelmaking methods illustrates the flexibility of modern practices with the unscientific prejudices removed by physical chemistry.

Fundamental steelmaking research is not spectacular and its application is evolutionary, rather than revolutionary, and the future benefits will be in proportion to the rate of research. This rate ultimately is primarily a dollars-and-cents matter, and we have in our control this rate by our appropriations for such research. We can only hope for a bright future if we maintain our fundamental research at a pace fast enough to keep us ahead of the declining quality of raw materials and the rising costs of materials and labor. Fundamental steelmaking research sponsored by the steel industry amounts to considerably less than one-half of one cent per ton of ingots produced. We can and must spend more.

by K. L. Fetters

Commentary on

Research Needs for

Process Development

At this point it seems appropriate to try to stand back a bit and survey this whole field of process development. Twenty-five years ago we could ill-define any of our fundamental processes of smelting and refining. We had a few figures and a few data with a few interpretations of reaction mechanisms. Twenty-five years later we now have good data for many of the reactions. We have developed a line of research and development by following certain definite lines and aspects which are continuing to produce worth-while data at a rapid rate.

I would like to refer again, to an over-all look at the whole program. Consider the entire field of raw materials for blast furnaces and steelmaking: the ore, the coke, the fluxes, the ferromanganese, manganese ores; the details of the blast furnace and what is involved there; the gas-solid reactions and gas-metal reactions on which there are now many data. At the Conference, there have been papers on desulfurization and dephosphorization of hot metal as it comes from the blast furnace; and, although desiliconizing has not been treated extensively, it is a very pertinent part of our present and future processes.

On the open-hearth problem, several reaction-rate studies have been made concerning the gas-solid reactions—that is, the reaction between combustion gases and the solid scrap. We have done some rate work on slag-metal reactions. In each of these cases I am calling attention to the fact that we are beginning to understand the kinetics which go with our equilibrium measurements.

Now it appears that the equilibrium studies of the gas-solid reactions and slag-metal reactions in the blast furnace and those of the slag-metal reac-

tions in the open hearth have both been carried to the point where we have a great many data today. We have studied the solute reactions in deoxidation to a considerable extent insofar as equilibrium is concerned, but we have not gone nearly so far where kinetics and rates of reaction are concerned. Phase separation—and by that I mean the formation of insoluble or separate phases as the result of deoxidation reactions—has not had nearly enough attention.

In the pouring of ingots and their solidification, we have again a continuation of the thinking inherent in the solute reactions—the sort of thing I referred to in the discussion of deoxidation and the deoxidation reaction. We have the reactions of freezing and the solidification of ingots. We also have the interaction with refractories. We should stand back and take a look at all these various aspects and determine where we may be far ahead in some of these fields of our development work but far behind in other fields. Thus we should see where we need to catch up. I am not talking from the standpoint of applied research in industry, which is actually industry's function and can be well taken care of by these intermediaries which we have been discussing. I am thinking of fundamental research because a lot more answers are needed in these fields.

I would like to refer to a speech given by Professor Dubridge of the California Institute of Technology. The title of it was "Things We Do Not Know." I think it would be good for us at this moment to take stock of our whole field of physical chemistry of steelmaking and tabulate the things which we do not know, as we have done with things we do know at this Conference. We ought to give greater thought to the transients—a word which I used in connection with Dr. Fischers's work—that is, some of the intermediate products of deoxidation. With our study of reaction rates and kinetics, we are thinking in terms of how quickly we get from here to there, and I do not believe we are giving sufficient thought to intermediate products which may be formed as well as the intermediate chemical steps. Fortunately at times, and perhaps unfortunately more times than not, those intermediate products end up in the final steel. I believe that there is quite a bit of fundamental information that we could well afford to get.

Now take an objective look at the tools which we have used in our physical chemistry research: we have been studying equilibrium and also the rates of reaction. Principally we have come in the last twenty years to where we have lots of glass apparatus, furnaces, and so on; crucible materials, temperature-measuring devices, methods of analysis; all of these are pretty well developed. It appears to me that with our great concentration on the number of data that we can get from these tools, we may not have reconsidered often enough some of the tools which are available but which the physical metallurgists have made more use of. We are starting to use the isotopic studies which should reveal a lot of useful information. The spectrograph is just another means of chemical analysis. I do not believe we have done enough in our research studies with such things as diffraction analysis, which would give us better identification and a few more answers in our physicochemical research—that is, X-ray diffraction and electron diffraction. We might also resort more frequently to the light microscope, the electron microscope, and also the petrographic microscope.

As the last point, I think that another field that we are going to have to start taking a look at is the chemistry of still higher temperatures than the 1600° C, as John Chipman's paper stated some years ago. I would like to say that I think there will be some of us living who in the future will be thinking of temperatures in the range 6000° C. These temperatures are certainly not unattainable by techniques we now know about. Future generations may talk about entirely different processes for refining of metals. I certainly do not recommend at the moment projects being started on the study of chemistry at 6000° C, but I think we ought to keep this field in mind so that some day we may be a little bit ahead of plant practice with our physical chemistry.

by B. M. Larsen

A Note on Physical Chemistry and the Practice of Steelmaking

In 1954 an English writer named Brandt,* in a brief polemic, likened the "Physical Chemistry of Steelmaking" to the sacred cow, who must be fed, even though she gives no milk. Though we could hardly be expected to accept this notion as essential truth, perhaps we might pretend that it has some validity, if only as an approach to a critical attitude that, in turn, might point to lines of study aimed toward developing the *science* to a state where it could have more effect on shaping the *practice* of steelmaking.

That the science has lagged behind the art, it would be unreasonable to argue against, and although more recent years have seen a very rapid development in the science, the art is now also changing more rapidly, and we still seem to lag behind the largely empirical development of actual steelmaking practices. A discussion of why this is so would perhaps be pointless, since we would probably agree, anyhow. Also, there are beginning to be a few cases where physical chemistry obviously has some directing influence, a notable instance being the work of Kalling in Sweden.

In my own experience, the applications of theory have in most cases been largely post-mortems, explaining a bit more plausibly from theory what we already knew by empirical experience. Two of our open-hearth shops once thought there should be some magical method of getting more of the bath sulfur to move from metal to slag. We were able to show that the furnaces were doing as well

as could possibly be expected with the existing conditions of furnace burdening, in fact, even somewhat better than some available laboratory data would predict. This is a typical example which saved some futile experimenting but was hardly as constructive as one would like. In general, the science has yet to give as much predictive power as we would like, and one is hard put to detect from the actual details of shop operation any very strong indication of the existence of physical chemistry.

Nearly all hot-metal open-hearth shops use run-off-slag practice, certainly to great practical advantage. Yet this practice is bad in many respects. Very wasteful of any manganese present in the charge, it also tends to cause an irregular iron loss. But it is bad mostly because it is difficult to make uniform and controllable. Heats cannot be scheduled properly, a shortcoming which results in overall plant inefficiencies, and in many obstacles to that approach toward almost completely automatic control that is becoming more and more urgent. Our science ought to be able to point toward a more efficient and controllable practice. But we are lacking in a clear-cut general theory of the open-hearth process, and when the operator points with pride to the amount of charge ore reduced to raise the yield, in run-off-slag practice, we can neither show him precisely how much this ore reduction costs him in speed and controllability, nor present him with a better method of operation.

If we attempt to outline an approximate overall theory of what determines open-hearth speed and control, we immediately run into the problem of how the required heat is transmitted through the slag layer into the metal. Regardless of what mechanism we may assume for this heat flow

* D. J. O. Brandt, *Iron & Coal Trades Rev.*, 1373 (Dec. 10, 1954).

Mr. Larsen is Assistant Director, Process Metallurgy, E. C. Bain Research Laboratory, United States Steel Corporation, Monroeville, Pennsylvania.

(which ultimately must decide how the process works at all), we immediately find for the needed heat conductivity of liquid slag an almost complete absence of any data in all of the literature. It could be argued that even a good approximate value for this specific conductivity would be of more practical use than reams of data on slag-metal equilibria and free energies of reaction.

There has been a general tendency to separate a process into parts for intensive study and then forget the sometimes critical connections between these parts. The fuel engineers have studied at length the flame development in the open-hearth combustion space and, generally, neglected the critical step of heat flow through the slag by means of the boils as well as the very important contribution of bath-boil gases to the available heat supply. Many physical chemists have relegated to the subconscious the critical rate of oxygen diffusion into the slag and have often tried vigorously to prove that slag and metal are at, or essentially at, equilibrium, without even stopping to consider that they were probably proving that the process cannot really work at all, this being apparently true without the critically important gradients and diffusion rates related to temperature and oxygen pressure. And both fuel engineers and chemists seem to neglect that twilight zone of top-slag film and adjacent film of gases above, that separates the combustion space from the slag-metal bath, a zone which is very important, indeed, to the whole process.

The open-hearth bath often goes into a foamy behavior that almost shuts off the critical heat flow. We have not apparently tried to produce such a foaming bath under controlled conditions to find out how it starts and how it relates to temperature, slag composition, viscosity, surface tension, and so on. In practice, we commonly burn many tons of limestone per heat underneath the bath. But we have not tried to bubble CO_2 through molten iron-carbon solution to get a general idea of the factors that may affect the rate of the resulting reaction. Are the rates of absorption and diffusion of oxygen into steelmaking slags critically dependent on viscosity or iron-oxide content? It would be extremely useful to have such data. Precisely what is the mechanism of production of the iron-oxide fume that plugs up furnace passages and helps to flux away refractories? Available data now published give a very puzzling picture of iron evaporation or fume production. Precisely how does the liquid slag eat into the dolomite banks, and how is rate of erosion related to temperature, boil mechanism and rate, and slag composition?

Does the film theory of reaction rate apply to heat flow, to the carbon reaction, to sulfur transfer? Has the carbon reaction really a heterogeneous mechanism at the metal-gas interface? If so, and if this process is also autocatalytic in the sense that it speeds itself up by its own stirring rate, what could we do to speed it up almost without limit? Does the boiling effect in open-hearth refining clean out particles in suspension from the liquid-metal bath, and could this somehow be made to give an almost perfect freedom from minute non-metallic particles in a refined bath at tap? If so, a gaseous deoxidation technique plus some means for getting the liquid metal into solid form without air oxidation should give steel of nearly perfect homogeneity.

These are questions which I mention primarily as examples of things that have seemed to be weak or missing or disregarded links in the general picture. I suspect that if we should give more effort to the development of general working theories about actual steelmaking practices this would point to such missing links where new or more precise data are needed. Perhaps we have been too circumscribed in the kind of problems mainly attacked in the laboratories. Merely to avoid misunderstanding, I should disavow any thought of disparaging any efforts to measure or theorize on any phase of this chemistry of high-temperature reactions. Yet, it is also interesting, and perhaps useful, to try to fit together what knowledge we do have into a picture of how an actual given complicated process works as a whole. And if we make the assumption that this is desirable, there is a point to be made. The point is simply that no matter how voluminous or exact are the data we do get, if a number of critical pieces of the picture puzzle are either missing or not known with sufficient exactness, all the rest may have too little effectiveness in showing just how to control and develop new methods in practice. If so, then these missing pieces must be sought, no matter how great may be the apparent difficulties involved.

by L. S. Darken

Commentary on

The Direction of Research for

Kinetic Studies

Perhaps some consideration should be given to what some people have referred to as "cross-effects," in other words, the phenomena occurring in sharp gradients such as the temperature gradient between slag and metal, or in the slag in the open-hearth bath. Such gradients, of course, are not limited to the open hearth. We could think about how these gradients could be produced in other processes and used to advantage. We know that in thermal gradients there are effects, such as the Soret effect, which are fairly well buried so far as steelmaking and high-temperature chemistry are concerned. A thermal gradient of 100° could be quite a substantial pump. That is, the thermal gradient itself could push things across the boundary. There is, in addition, the very high oxygen gradient in any process of oxidizing iron or carbon and iron, which produces an effect somewhat similar to a temperature gradient—for example, on elements such as sulfur that might want to move to the higher oxygen side or perhaps the lower oxygen side.

In this connection I would also like to suggest a problem. Supposing we have a thermal-conductivity experiment in gas and we are asked to pick out some gas having a high thermal conductivity. I would have suggested using hydrogen, but Hirshfelder of the University of Wisconsin has suggested a system which will be a hundred times better. It consists of a mixture of H_2 and I_2. At the low-temperature side, HI is formed, and at the high-temperature side, HI is decomposed. In this fashion there is a tremendous heat exchange that is a hundredfold times that obtained with hydrogen alone. The general problem thus suggested is that of vast changes in thermal conductivity associated with chemical reaction.

In summary, my suggestion is that we should consider these nonequilibrium conditions that can occur in gradients and then make use of them.

Dr. Darken is Assistant Director, Physical Chemistry, E. C. Bain Research Laboratory, United States Steel Corporation, Monroeville, Pennsylvania.

by M. Tenenbaum

Commentary on

Industrial Needs for

Fundamental Knowledge

Dr. Tenenbaum is very much concerned over the problem of the utilization of theoretical data in steelmaking operations, steelmaking development, and steelmaking industrial research. He charged us at this Conference with talking to ourselves. He said that the communication of information among the groups who are working in the theoretical end of the field and the industrial area who need to utilize the data is not sufficient today. That is, conferences, meetings, conventions, technical papers, and so on, do not fulfill this basic need. He even went so far as to say that the production of graduate students, men with doctoral degrees and the like, who ultimately go into industry, does not provide this means of communication. The reason for this is that they go into industry and in one or two years become highly oriented to their working environment and the problems that they are working on. Because of their intense devotion to these responsibilities, they become separated from the theoretical approach. He felt that one

This commentary was summarized and presented by Professor J.' F. Elliott. Dr. Tenenbaum is Superintendent, Quality Control Department, Inland Steel Company, East Chicago, Indiana.

answer to this problem would be to arrange for a person who is working on this fundamental field in some research activity to spend considerable time working in the industrial area where some of these theoretical considerations should be used. This may help to increase communication. He feels that the indictment that research is a sacred cow is to some extent true. This problem may be of our own making because of the extreme difficulty of communication as a result of the fact that the ideas and concepts are complex and are not exchanged back and forth easily. It is further complicated because one must always be careful to keep his mind oriented to the proper application of concepts.

Some subjects which he feels need study include bubble formation—that is, the reactions of bubbles with liquid and solid phases. Also, much more should be known about the behavior of oxygen and oxidizing systems not only with the oxygen in the slags, but reactions with gas phases. These would be the steelmaking systems. He feels that we do not understand the role of oxygen in steelmaking furnaces as well as we do the role of oxygen in the blast furnace.

Special Lecture

Kinetic Problems in Steelmaking

C. Wagner

by C. Wagner

Kinetic Problems in Steelmaking

GENERAL CHARACTERISTICS OF METAL-SLAG REACTIONS

A typical metal-slag reaction is the displacement reaction between iron and manganese,

$$[Mn] + Fe^{2+} \text{ (slag)} = Mn^{2+} \text{ (slag)} + [Fe] \quad (1)$$

This reaction involves necessarily the following principal steps:

1. Transport of the reactants from the bulk phases to the metal-slag interface.

2. The phase-boundary reaction at the metal-slag interface, which in turn may comprise several individual steps.

3. Transport of the reaction products from the metal-slag interface into the bulk phases.

Textbooks on chemical kinetics give a very adequate introduction to the evaluation of rate measurements on homogeneous inorganic or organic reactions in solutions or in the gas phase by using the classical concepts of first-order and second-order reactions according to van't Hoff, or more sophisticated concepts such as consecutive reactions and chain reactions. Textbooks, however, offer little direct help with regard to metal-slag reactions involving combinations of transport processes and phase-boundary reactions. Thus it is necessary to bridge the gap between available textbooks on chemical kinetics and the problems encountered in steelmaking. As a preliminary contribution, the present analysis has been made. This analysis is especially intended to be used as a framework for the planning of future research and for the evaluation and the co-ordination of experimental data.

We visualize immediately (1) transport processes

to and from the metal-slag interface, and (2) phase-boundary reactions as processes whose rate laws may be investigated separately.

Moreover, we may anticipate the following two principal limiting cases.

1. Equilibrium at the metal-slag interface may be virtually established and the rate of the over-all reaction may be determined by the rate of the various transport processes from and to the interface. Under more special conditions, the over-all rate may be determined by the transport rate of only one molecular species. Special cases have been discussed by Darken[1] and others.

2. Differences between concentrations at the metal-slag interface and bulk concentrations are insignificant, and the rate of the over-all reaction is determined essentially by the rate of the phase-boundary reaction which in turn may involve several consecutive steps, in particular electrochemical processes, as is discussed below.

KINETICS OF TRANSPORT-CONTROLLED REACTIONS

Principles of Reactions between Two Liquid Phases

If equilibrium of phase-boundary reactions is attained very rapidly, the rate of a heterogeneous reaction may be determined essentially by the rate of diffusion processes. A well-known example is the oxidation of a metal at elevated temperatures under conditions where the parabolic rate law is obeyed. In the case of metal-slag reactions, however, transport of reactants and reaction products to and from the metal-slag interface is effected not only by diffusion but also by convection. The rate of transport $j_{i(x)}$ in mole/cm²sec of a molecular species i in a liquid phase in the x direction may be written as

Dr. Wagner is Professor of Metallurgy, Massachusetts Institute of Technology, and is now Head of the Max-Planck Institut für physikalische Chemie, Göttingen, Germany.

Support of this study by the American Iron and Steel Institute is gratefully acknowledged.

$$j_{i(x)} = -D_i(\partial c_i/\partial x) + c_i u_x \qquad (2)$$

where D_i is the diffusion coefficient, c_i is the concentration of the molecular species i, and u_x is the flow velocity of the liquid in the x direction. The first term on the right-hand side of equation 2 accounts for diffusion, and the second term for transport due to fluid flow. Under most conditions, significant concentration differences are found only in the vicinity of an interface. According to Prandtl, considerations may therefore be confined to the concentration and velocity distribution in the so-called boundary layer, and concentration differences in the bulk liquid may be disregarded.

Under certain conditions, the concentration and the velocity distribution in a boundary layer may be calculated from first principles,[2] for example, for a fluid flowing along a flat plate, or a fluid in which a solid disk is rotated.

In many cases, however, details of the concentration distribution in a boundary layer are not known. If the rate of transport $j_{i(x)}$ to or from an interface has been determined and the diffusion coefficient is known, we may calculate from equation 2 the value of the concentration gradient directly at the interface $x = 0$ since $u_x = 0$ at $x = 0$ if a displacement reaction does not result in a volume change of the alloy or the slag. Hence

$$(\partial c_i/\partial x)_{x=0} = -j_{i(x)}/D_i \qquad (3)$$

A schematic concentration profile is shown in Fig. 1. Such a profile may be characterized by the

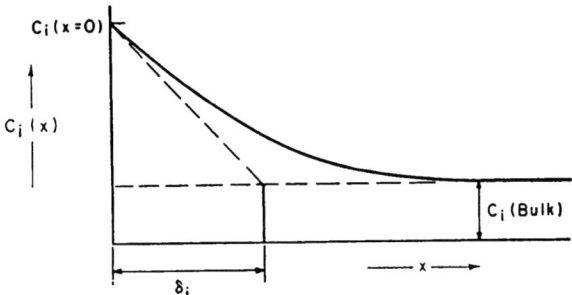

Fig. 1. Concentration profile of a boundary layer.

effective thickness δ_i of the diffusion boundary layer, which is defined by the relation

$$\delta_i = \frac{c_i(\text{bulk}) - c_i(x = 0)}{(\partial c_i/\partial x)_{x=0}} \qquad (4)$$

In view of equations 3 and 4 it follows that

$$j_{i(x)} = \frac{c_i(x = 0) - c_i(\text{bulk})}{\delta_i} D_i \qquad (5)$$

The value of δ_i may also be obtained as the distance of the intersection of the tangent line on the c_i versus x curve at $x = 0$ and the extrapolated plateau of the bulk concentration, as is shown in Fig. 1.

In the case of forced convection, the value of

$(\partial c_i/\partial x)_{x=0}$ is proportional to the concentration difference $c_i(x = 0) - c_i(\text{bulk})$ between interface and bulk liquid. Thus, in view of equation 4, the effective thickness of the boundary layer is independent of the values of $c_i(x = 0)$ and $c_i(\text{bulk})$.

In the case of a turbulent boundary layer, the velocity distribution in the boundary layer is not stationary but fluctuates with time. Under these conditions, the concentration profile also fluctuates with time, and accordingly, δ_i has to be defined as an average value.

The concept of the effective thickness of the diffusion boundary layer as formulated above does not involve approximations or premature simplifications. It must be recalled, however, that the geometry and flow conditions in the bulk liquid are decisive variables which determine the value of δ_i. In addition, it must be noticed that the concentration profiles for different molecular species involving different diffusion coefficients are not equal. This is shown schematically in Fig. 2. By

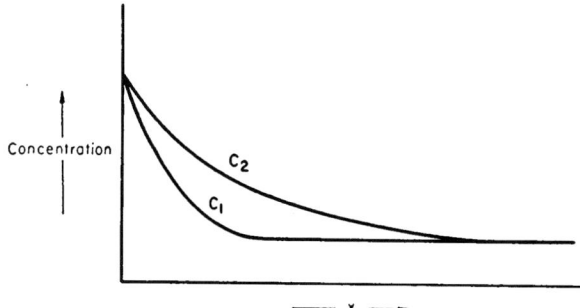

Fig. 2. Concentration profiles for species 1 and 2 in a boundary layer if $D_2 > D_1$.

and large, the greater the diffusion coefficient of a molecular species, the further extends the diffusion boundary layer; that is, the greater is the effective thickness. Consequently, for simultaneous diffusion of several molecular species, it is necessary to employ in equation 5 individual values of δ_i for each molecular species.

From a theoretical analysis of the concentration and the velocity distribution in the laminar boundary layer of a liquid flowing along a flat plate, it follows[2] that the value of δ_i is proportional to $D_i^{1/3}$ but exponents other than $1/3$ occur under other conditions.

A particularly simple reaction between constituents in two liquid phases is the redistribution of a solute between two virtually immiscible solvents which are mechanically stirred, for example,

$$\text{NH}_3 \text{ (in water)} = \text{NH}_3 \text{ (in isobutanol)} \qquad (6)$$

which has been investigated by Gordon and Sherwood.[3]

In contrast to reaction 1, only one molecular species is involved, but even in this case we have to consider two transport processes in series, namely, transport of ammonia from the bulk of the aqueous solution to the interface, and transport of ammonia from the interface to the bulk isobutanol solution.

According to the two-phase film theory,[3, 4, 5, 6] the rate of transfer j_i of solute i per unit area from phase I to phase II may be written in a form similar to equation 5:

$$j_i = \frac{c_i' - c_i'^*}{\delta_i'} D_i' = \frac{c_i''^* - c_i''}{\delta_i''} D_i'' \qquad (7)$$

where c_i' and c_i'', respectively, are the bulk concentrations of molecular species i in phases I and II, starred values are the respective concentrations next to the interface, δ_i' and δ_i'' are the effective thicknesses of the diffusion boundary layers, and D_i' and D_i'' are the diffusion coefficients of i in the phases I and II, respectively.

Equation 7 gives the transport rate per unit area for a particular area element of the interface involving the values δ_i' and δ_i''. Therefrom the total transport rate $\dot{n}_i$ from phase I to phase II is to be calculated as the integral $\int j_i dA$ over the total area A of the interface with due regard to local variations of δ_i' and δ_i'' corresponding to local variations of flow conditions. If the local variations of δ_i' and δ_i'' are unknown, the equation for the total transport rate $\dot{n}_i$ may be written as

$$\dot{n}_i = A \frac{c_i' - c_i^*}{\delta_i'} D_i' = A \frac{c_i''^* - c_i''}{\delta_i''} D_i'' \qquad (8)$$

where δ_i' and δ_i'' are appropriately defined average values. A rigorous definition of average values to be used in equation 8 is rather involved. In what follows, it is assumed that the local variations of δ_i' and δ_i'' are not excessive, and thus the mode of calculating average values is of minor importance.

At the interface, virtual establishment of distribution equilibrium is assumed. Thus

$$c_i''^* / c_i'^* = K \qquad (9)$$

where K is the distribution coefficient.

Upon combining equations 8 and 9 and eliminating concentrations at the interface, it follows that

$$\dot{n}_i = A \frac{D_i' D_i''}{\delta_i' \delta_i''} \frac{c_i' K - c_i''}{(D_i'/\delta_i') + (D_i''/\delta_i'')K} \qquad (10)$$

Under the conditions employed by Gordon and Sherwood,[3] the value of δ_i in each phase is supposed to be approximately proportional to $D_i^{1/2}$ for given stirring conditions. Then it is possible to calculate the individual values of δ_i' and δ_i'' from rate measurements for different solutes. Analogous calculations can be made for metal-slag reactions, as is discussed below.

Exchange Reactions Involving Radioactive Isotopes

As a particularly simple example of a displace-ment reaction, we consider exchange of a radioactive isotope between liquid steel and a slag:

$$^{59}[\text{Fe}] + {}^{56}\text{Fe}^{2+} (\text{slag}) = {}^{56}[\text{Fe}] + {}^{59}\text{Fe}^{2+} (\text{slag}) \qquad (11)$$

where ^{56}Fe is the normal atomic species of iron and ^{59}Fe is a radioactive isotope. This reaction has been investigated by Birchenall and Derge,[7] but the experimental results have not been evaluated quantitatively.

Convection is assumed to be provided by mechanical stirring or by induction heating. Moreover, chemical equilibrium between metal and slag for the abundant atomic species is assumed, and the change in the total number of the radioactive atoms due to decay is supposed to be negligible during an exchange experiment.

Exchange equilibrium between isotopes is supposed to be virtually established at the metal-slag interface. Accordingly the fraction of the radioactive isotope in the metal at the interface is supposed to be equal to the fraction of the radioactive isotope in the slag at the interface. In view of equation 8, the rate of transfer of the radioactive tracer, $\dot{n}_{i(\text{tr})}$, from metal (phase I) to slag (phase II) for any element $_i$ may be written as

$$\dot{n}_{i(\text{tr})} = A(D_i'/\delta_i')c_i'(\gamma_i' - \gamma_i^*)$$
$$= A(D_i''/\delta_i'')c_i''(\gamma_i^* - \gamma_i'') \qquad (12)$$

where γ_i', γ_i'', and γ_i^* are the fractions of the radioactive isotope in the metal, in the slag, and at the interface, respectively.

From equation 12 it follows that

$$\gamma_i^* = \frac{(D_i'/\delta_i')c_i'\gamma_i' + (D_i''/\delta_i'')c_i''\gamma_i''}{(D_i'/\delta_i')c_i' + (D_i''/\delta_i'')c_i''} \qquad (13)$$

Substitution of equation 13 in equation 12 yields

$$\dot{n}_{i(\text{tr})} = A \frac{(D_i'/\delta_i')c_i'(D_i''/\delta_i'')c_i''}{(D_i'/\delta_i')c_i' + (D_i''/\delta_i'')c_i''} (\gamma_i' - \gamma_i'') \qquad (14)$$

The changes in the concentration of the radioactive isotope with time t are given by the material balance

$$\dot{n}_{i(\text{tr})} = V'' d(c_i''\gamma_i'')/dt = -V' d(c_i'\gamma_i')/dt \qquad (15)$$

where V' and V'' are the volumes of alloy and slag, respectively.

The initial fraction of tracer in the metal is denoted by $\gamma_i^{(0)}$. Then we have the following material balance for any time,

$$V'c_i'\gamma_i'^{(0)} = V'c_i'\gamma_i' + V''c_i''\gamma_i'' \qquad (16)$$

whence

$$\gamma_i'' = (V'c_i'/V''c_i'')(\gamma_i'^{(0)} - \gamma_i') \qquad (17)$$

After equilibrium has been reached, the fractions of the radioactive isotope in the alloy and the slag are equal. Thus

$$\gamma_i' = \gamma_i'' = \gamma_i^{(eq)} \qquad \text{if } t \to \infty \qquad (18)$$

where $\gamma_i^{(eq)}$ denotes the equilibrium fraction of radio-active tracer.

Upon substitution of equation 18 in equation 16, it follows that

$$\gamma_i^{(eq)} = V'c_i'\gamma_i'^{(o)}/(V'c_i' + V''c_i'') \qquad (19)$$

From equations 17 and 19 we obtain

$$\gamma_i' - \gamma_i'' = \gamma_i' - (V'c_i'/V''c_i'')(\gamma_i'^{(o)} - \gamma_i')$$

$$= (\gamma_i' - \gamma_i^{(eq)})(V'c_i' + V''c_i'')/V_i'c_i' \qquad (20)$$

Substitution of equations 14 and 20 in equation 15 yields

$$d\gamma_i'/dt = -\kappa_i(\gamma_i' - \gamma_i^{(eq)}) \qquad (21)$$

where

$$\kappa_i = A \frac{(D_i'/\delta_i')c_i'(D_i''/\delta_i'')c_i''}{(D_i'/\delta_i')c_i' + (D_i''/\delta_i'')c_i''} \frac{V'c_i' + V''c_i''}{V'c_i'V''c_i''} \qquad (22)$$

Thus the change in tracer concentration is given by a first-order rate law. Integration yields

$$\gamma_i' = \gamma_i^{(eq)} + (\gamma_i'^{(o)} - \gamma_i^{(eq)})e^{-\kappa_i t} \qquad (23)$$

Similarly,

$$d\gamma_i''/dt = \kappa_i(\gamma_i^{(eq)} - \gamma_i'') \qquad (24)$$

$$\gamma_i'' = \gamma_i''^{(eq)}(1 - e^{-\kappa_i t}) \qquad (25)$$

The following special cases are noteworthy. From equation 22 it follows that

$$\kappa_i \cong AD_i'/\delta_i'V' \qquad \text{if} \qquad (D_i''/\delta_i'')c_i'' \gg (D_i'/\delta_i'c)_i'$$

and

$$c_i''V_i'' \gg c_i'V_i' \qquad (26)$$

and conversely,

$$\kappa_i \cong AD_i''/\delta_i''V'' \qquad \text{if} \qquad (D_i'/\delta_i')c_i' \gg (D_i''/\delta_i'')c_i''$$

and

$$c_i'V_i' \gg c_i''V_i'' \qquad (27)$$

From equations 26 and 27 we recognize that the rate constant κ_i is essentially determined by diffusion in the boundary layer within the metal if the concentration of the respective constituent is much greater in the slag than in the metal and if the volumes of alloy and slag do not differ widely. Conversely, the rate constant κ_i is essentially determined by diffusion in the boundary layer within the slag if the concentration of the respective constituent is much greater in the alloy than in the slag. If the diffusion coefficients are known, we may, therefore, calculate the effective thickness of the boundary layer within the slag from tracer transfer experiments with constituents which are preferentially in the alloy, for example, nickel in liquid steel. Conversely, we may calculate the effective thickness of the boundary layer within the alloy from tracer transfer experiments with constituents which are preferentially in the slag, such as manganese.

Since the viscosity of a slag is in many cases considerably greater than that of the alloy, we have to expect that

$$D_i'' < D_i' \qquad \text{and} \qquad \delta_i'' < \delta_i'$$

and, therefore,

$$D_i''/\delta_i'' < D_i'/\delta_i'$$

Consequently, diffusion in the boundary layer within the slag may be the controlling factor even at equal concentrations c_i' and c_i''. On the other hand, diffusion in the boundary layer within the alloy will be the controlling factor only if the concentration c_i' in the alloy is 10 or 100 times less than the concentration c_i'' in the slag.

Assume that from experiments with constituent α the value δ_α'' and from experiments with constituent β the value δ_β' are known. Then, according to Gordon and Sherwood,[3] we may use the approximations

$$\delta_i' = \delta_\beta'(D_i'/D_\beta')^{1/2} \qquad (28)$$

$$\delta_i'' = \delta_\alpha''(D_i''/D_\alpha'')^{1/2} \qquad (29)$$

in order to obtain the effective thickness of a boundary layer for other constituents i from the values of δ_α'' and δ_β'.

Substituting the values δ_i' and δ_i'' in equation 22, we may, therefore, predict the rate of transfer of radioactive tracer of any constituent i.

Instead of using the approximate equations 26 and 27, we may combine equation 22 for $i = \alpha$ and equation 22 for $i = \beta$ and substitute

$$\delta_\alpha' = \delta_\beta'(D_\alpha'/D_\beta')^{1/2} \qquad \delta_\beta'' = \delta_\alpha''(D_\beta''/D_\alpha'')^{1/2}$$

Thus we obtain two equations involving the unknowns δ_α'' and δ_β', which may be calculated. Subsequently, we may use equations 28 and 29 in order to obtain the values for other elements.

Displacement Reactions in Quasi-Binary Systems

Next we consider displacement reactions analogous to that formulated in equation 1. In a more general form, we rewrite equation 1 as

$$A \text{ (alloy)} + B^{2+} \text{ (slag)} = B \text{ (alloy)} + A^{2+} \text{ (slag)} \qquad (31)$$

In order to avoid complications resulting from the simultaneous occurrence of other displacement reactions, we confine considerations to a binary alloy consisting of metals A and B and a slag containing cations of A and B and other nonreactive ions, such as an iron-nickel alloy and a slag containing iron oxide, nickel oxide, lime, alumina, and silica. Convection is assumed to be provided by mechanical stirring, or by induction heating.

In addition to the phase-boundary reaction, there are necessarily the following transport processes:

a. Diffusion of A from the bulk alloy to the interface.

b. Diffusion of A^{2+} ions from the interface to the bulk slag.

c. Diffusion of B^{2+} ions from the bulk slag to the interface.

d. Diffusion of B from the interface to the bulk alloy.

As a limiting case, we assume that equilibrium at the interface is virtually established. Then we may formulate equations analogous to equation 8 for the transport rate of both A and B. In view of the stoichiometric relation in equation 31, the rate of transport of A from the alloy to the slag must be equal to the rate of transport of B from the slag to the alloy. The concentrations of A, B, A^{2+}, and B^{2+} at the interface, where equilibrium is assumed, are interrelated by the law of mass action. Thus we have the necessary number of equations in order to calculate the concentrations of A, B, A^{2+} ions, and B^{2+} ions at the interface, and finally the rate of the displacement reaction 31. Even in the case of equal valences of A and B as assumed in equation 31, there results a very involved equation.

Simple expressions are obtained if only one of the four transport processes listed controls the overall rate. Thus we assume that the concentrations of three constituents at the interface are virtually equal to the respective bulk concentrations, whereas the concentration of the fourth constituent differs substantially from its bulk concentration, and therefore, the transport of the latter constituent is the rate-determining step.

For further simplification, we assume the validity of the ideal law of mass action and, therefore, obtain for the interface the condition

$$\frac{c_B'^* c_A''^*}{c_A'^* c_B''^*} = K \qquad (32)$$

where concentrations are expressed in terms of mole/cm³ and K is the equilibrium constant.

Then we have the following four limiting cases.

1. The diffusion of A from the bulk alloy to the interface is the rate-determining step. Hence the rate $\dot{n}$ of reaction 31 is found to be

$$\dot{n} = A(D_A'/\delta_A')(c_A' - c_A'^*) \qquad (33)$$

In view of the general presupposition introduced above, it is assumed that the concentrations of B, A^{2+}, and B^{2+} at the interface can be replaced approximately by the respective bulk concentrations. Thus we may calculate $c_A'^*$ from equation 32 and obtain on substitution in equation 33

$$\dot{n} = A(D_A'/\delta_A')c_A'(1 - Q/K) \qquad (34)$$
where

$$Q = \frac{c_B' c_A''}{c_A' c_B''} \qquad (35)$$

is the "concentration quotient," which in the case of equilibrium becomes equal to the equilibrium constant K.

It can be shown that the rate $\dot{n}$ is less than the value calculated in equation 34 if the concentration of B^{2+} as the other reactant at the interface is less than its bulk concentration, and if the concentrations of the reaction products A^{2+} and B at the interface are greater than their bulk concentrations. Thus the rate calculated in equation 34 may be called the virtual maximum rate for the transport of A for given bulk concentrations of the reactants and reaction products.

2. The diffusion of A^{2+} ions from the interface to the bulk slag is the rate-determining step,

$$\dot{n} = A(D_A''/\delta_A'')(c_A''^* - c_A'')$$
$$= A(D_A''/\delta_A'')c_A''(K/Q - 1) \qquad (36)$$

3. The diffusion of B^{2+} ions from the bulk slag to the interface is the rate-determining step,

$$\dot{n} = A(D_B''/\delta_B'')(c_B'' - c_B''^*)$$
$$= A(D_B''/\delta_B'')c_B''(1 - Q/K) \qquad (37)$$

4. The diffusion of B from the interface to the bulk alloy is the rate-determining step,

$$\dot{n} = A(D_B'/\delta_B')(c_B'^* - c_B')$$
$$= A(D_B'/\delta_B')c_B'(K/Q - 1) \qquad (38)$$

Each of the rates calculated in equations 34 and 36 to 38 represents the virtual maximum rate for a particular transport process. From the preceding discussion of the chemical equation 31, it follows that all four transport processes are in series. By and large, if one of several processes in series has a virtual maximum rate which is much lower than those of the other processes, the rate of the over-all reaction is approximately equal to the virtual maximum rate of the former process. Consequently, if the bulk concentrations are given, if the value of the equilibrium constant is known, and if it is possible to estimate the order of magnitude of the values of (D_A'/δ_A'), and so on, then it is possible to find whether one of the foregoing limiting cases can be assumed. As a rule, the same order of magnitude of the values of (D_A'/δ_A'), and so on, may be assumed unless the slag is highly viscous.

In particular, in the vicinity of equilibrium, that is, if $Q \sim K$, the transport of the constituent involving the lowest concentration has the lowest virtual maximum transport rate, and accordingly, the transport of this constituent determines the rate of the over-all reaction.

On the other hand, if initially no B^{2+} ions are in the slag, the initial value of Q is zero, and thus the virtual maximum diffusion rates for transport of the reaction products away from the interface become infinite according to equations 36 and 38. Under these conditions, the rate of the over-all reaction is definitely controlled by the transport of the reactants to the interface, that is, either by transport of A in the metal, or transport of B^{2+} ions in the slag.

The following special reactions may be considered.

1. Steel containing 0.3 atomic % Mn reacts with a slag containing 10 mole % FeO. In view of the foregoing rules, the rate of the displacement reaction formulated in equation 1 is supposed to be determined by transport of manganese in the metallic phase toward the interface and, therefore, the value of (D_{Mn}'/δ_{Mn}') is obtainable. Measurements may be repeated with other concentrations of Mn and FeO. Within certain limits, the rate is supposed to be proportional to the manganese concentration and independent of the FeO concentration. If this is true, transport control is proved, since control by a phase-boundary reaction leads supposedly to a different dependence of the rate on the concentrations of the reactants, as is discussed below. Values of the effective thickness of the boundary layers in liquid steel under the same conditions of convective flow may subsequently be calculated with the help of equation 28.

2. Pure iron reacts with a slag containing 1 mole % NiO

$$[Fe] + Ni^{2+} \text{ (slag)} = [Ni] + Fe^{2+} \text{ (slag)} \quad (39)$$

Given the foregoing rules, the rate of reaction 39 is supposed to be determined by transport of nickel ions in the slag toward the interface. Thus the value of (D_{Ni}''/δ_{Ni}'') is obtainable. Values of the effective thickness of the diffusion boundary layer in the slag under the same conditions of convective flow may be calculated with the help of equation 29.

These examples show how values of the thickness of the diffusion boundary layer can be obtained from measurements of the initial rate of displacement reactions in a rather straightforward way, which may be even simpler than the evaluation of the results of exchange reactions involving radioactive isotopes.

It is possible that the rate of the over-all reaction is controlled by different steps at the beginning and at the final stage during approach to equilibrium. As an example, we consider the reaction between initially pure iron and a manganese-bearing slag initially free of iron. Then the rate of the reaction

$$[Fe] + Mn^{2+} = [Mn] + Fe^{2+} \quad (40)$$

is controlled initially by diffusion of Mn^{2+} ions as reactant toward the interface. In view of the large value of the equilibrium constant of reaction 40, the manganese concentration in the slag will decrease only slightly unless the volume of the alloy is very much greater than that of the slag. If the volume of the metal is somewhat but not excessively greater than that of the slag, the concentration of iron ions in the slag will increase more

rapidly than the concentration of manganese in the metal, and the manganese concentration in the metal during the final approach to equilibrium will be lower than the concentrations of the other constituents. Consequently, the rate of the reaction 40 will be controlled by transport of manganese in the metal during the final stage, in contradistinction to control by transport of manganese ions in the slag during the initial stage. It is obvious that under these conditions the rate cannot be represented by one of the classical rate laws for a first-order or a second-order reaction.

In general, however, somewhat simpler conditions can be expected even if none of the foregoing four limiting cases is realized. In particular, it can be shown that in the vicinity of equilibrium the rate of approach to equilibrium of any reaction may be represented by a first-order rate law, with the deviation of the concentration of one constituent from its equilibrium value as a measure of the driving force.

To supplement the treatment of the four specified limiting cases, we consider conditions where the virtual maximum rates for transport of A in the metal and transport of A^{2+} ions in the slag are much greater than the corresponding rates for B and B^{2+} ions, and the latter values are of the same order of magnitude. Then the rate of the over-all reaction is controlled by both transport of B in the metal and transport of B^{2+} ions in the slag. Hence we have the rate equation

$$\dot{n} = A(D_B''/\delta_B'')(c_B'' - c_B''^*)$$
$$= A(D_B'/\delta_B')(c_B'^* - c_B') \quad (41)$$

Replacing the concentrations of A and A^{2+} ions at the interface in equation 32 by bulk concentrations, we have

$$c_B'^* c_A''/c_A' c_B''^* = K \quad (42)$$

Upon combining equations 41 and 42 and eliminating the concentrations of B and B^{2+} ions at the interface, we obtain

$$\dot{n} = A \frac{D_B' D_B''}{\delta_B' \delta_B''} \frac{c_B'' - c_B' c_A''/K c_A'}{(D_B'/\delta_B') + (D_B''/\delta_B'')c_A''/K c_A'} \quad (43)$$

The rate of change in the concentration of B^{2+} ions in the slag is given by the relation

$$\dot{n} = -V''(\partial c_B''/\partial t) \quad (44)$$

In addition, we have the material balances

$$V'c_A'^{(o)} + V''c_A''^{(o)} = V'c_A' + V''c_A'' \quad (45a)$$
$$V'c_B'^{(o)} + V''c_B''^{(o)} = V'c_B' + V''c_B'' \quad (45b)$$

where the initial concentrations of the various constituents are denoted by $c_A'^{(o)}$, and so forth.

Finally, we have the condition that the number of equivalents of A^{2+} ions and B^{2+} ions in the slag is the same at any time:

$$V''c_A''^{(o)} + V''c_B''^{(o)} = V''c_A'' + V''c_B'' \quad (46)$$

Substituting equations 44 and 46 in equation 43, we may obtain the standard form of a rate law which gives dc_B''/dt as a function of the instantaneous concentration c_B'' and the various initial concentrations.

The assumption that the rate of the over-all reaction is controlled essentially only by diffusion of B and B^{2+} ions but not by diffusion of A and A^{2+} ions is satisfied if the concentrations of A and A^{2+} ions are much greater than the concentrations of B and B^{2+} ions. Thus the change in the concentrations of A and A^{2+} ions with time may be ignored. Letting $c_A' \cong c_A'^{(o)}$ and $c_A'' \cong c_A''^{(o)}$ and substituting equations 44 and 46 in equation 43, we obtain a first-order rate law,

$$-(dc_B''/dt) = \kappa(c_B'' - c_B''^{(eq)}) \tag{47}$$

where

$$c_B''^{(eq)} = \frac{V'c_B'^{(o)} + V''c_B''^{(o)}c_A''^{(o)}/Kc_A'^{(o)}}{V' + V''c_A''^{(o)}/Kc_A^{(o)}} \tag{48}$$

is the final equilibrium concentration of B^{2+} ions in the slag, and

$$\kappa = A \frac{D_B'D_B''}{\delta_B'\delta_B''}$$
$$\times \frac{V' + V''c_A''^{(o)}/Kc_A'^{(o)}}{V'V''[(D_B'/\delta_B') + (D_B''/\delta_B'')c_A''^{(o)}/Kc_A'^{(o)}]} \tag{49}$$

is a rate constant.

Consequently, if the rate is controlled in part by transport of B^{2+} ions in the slag and in part by B in the metal, rate measurements give only a combination of the values of (D_B'/δ_B') and (D_B''/δ_B''). The individual values may, in principle, be calculated on combining the results of measurements for different initial concentrations of the reactants and reaction products. Such a procedure, however, is not recommended whenever these values are directly obtainable from measurements corresponding to limiting cases 3 and 4.

The foregoing discussion has been confined to displacement reactions involving divalent metals. If the valences of the metals participating in a displacement reaction are different from each other, the valences appear as factors but the resulting equations are essentially of the same type as in the preceding examples.

Simultaneous Displacement Reactions in Multicomponent Systems

In multicomponent systems, several displacement reactions may take place simultaneously. As an example, we consider the reactions occurring between an iron-aluminum alloy and a nickel-oxide-bearing slag,

$$[Fe] + Ni^{2+} \text{ (slag)} = [Ni] + Fe^{2+} \text{ (slag)} \tag{50a}$$

$$2[Al] + 3Ni^{2+} \text{ (slag)} = 3[Ni] + 2Al^{3+} \text{ (slag)} \tag{50b}$$

We assume that the concentration of nickel oxide in the slag is much greater than the concentration of aluminum in the alloy. For instance, we may have a slag containing 10 mole % NiO, and liquid steel containing 0.2 atomic % aluminum. Under these conditions, the virtual maximum rate of transport of aluminum toward the interface is much less than that of nickel ions. Thus initially only a few nickel ions arriving at the interface have the chance to participate in reaction 50b, and most nickel ions will be consumed by reaction 50a. Thereby iron ions accumulate in the slag, and accordingly, another displacement reaction may take place,

$$2[Al] + 3Fe^{2+} \text{ (slag)} = 3[Fe] + 2Al^{3+} \text{ (slag)} \tag{51}$$

which may formally be obtained by combining equations 50a and 50b.

If the volume of the alloy is much greater than the volume of the slag and especially the initial amount of aluminum in the alloy is much greater than the amount of nickel oxide in the slag, the final state of the system can be described essentially as a displacement of nickel ions by aluminum ions in the slag according to reaction 51 with an iron-ion concentration which is lower than the transient concentration reached during the initial stage. This example illustrates transient overshooting of a final equilibrium concentration due to occurrence of simultaneous displacement reactions.

The foregoing discussion shows that this phenomenon may definitely occur under conditions involving exclusive transport control. Observations of overshooting are available especially for systems involving transfer of sulfur[8, 9, 10] where at least partial control by phase-boundary reactions has been ascertained.[10]

In the case of transport control, rate equations analogous to equations 33 and 41 may be formulated. From these equations the concentrations at the interface may be eliminated with the help of the pertinent equilibrium conditions. Upon combining rate equations and material balances, equations for the rate of change of the concentrations of all constituents may be deduced. These equations involve values of (D_i'/δ_i') and (D_i''/δ_i'') for the rate-determining transport processes. In principle, these values may be obtained from a careful analysis of measured rates of change in the concentrations of the various constituents in systems involving simultaneous displacement reactions. In view of the complexity of the conditions, however, such a procedure is not recommended. Instead, values of (D_i'/δ_i') and (D_i''/δ_i'') should be determined from measurements on quasi-binary systems with the help of methods suggested above. Then it is possible to calculate the rate of change in the concentrations in systems involving several simultaneous displacement reactions and to compare these values with experimental values in order to check the consistency of the theoretical analysis.

The Dependence of the Effective Thickness of the Boundary Layer on the Rate of Gas Evolution

Philbrook and Kirkbride[11] have investigated the rate of the reduction of iron oxide in a slag by graphite-saturated iron,

$$[C] + Fe^{2+} \text{ (slag)} + O^{2-} \text{ (slag)} = [Fe] + CO \text{ (gas)} \tag{52a}$$

Since the concentration of carbon in the alloy is much greater than the concentration of iron oxide in the slag, we may assume that diffusion of iron ions from the bulk slag to the alloy-slag interface determines the rate. Since the equilibrium concentration of iron oxide in the slag is much lower than the initial concentration except for the end of the reaction, we may use the approximation

$$\dot{n} = \frac{A D_{Fe}'' c_{Fe}''}{\delta_{Fe}''} \tag{52b}$$

If the thickness of the boundary layer is constant, equation 52b yields a first-order rate law. Actually, however, Philbrook and Kirkbride have found that the rate $\dot{n}$ is proportional to the square of the iron-oxide concentration in the slag. This suggests that the thickness of the boundary layer decreases with increasing rate of gas evolution causing more convection. Tentatively, we may assume that

$$\delta_{Fe}'' = b(\dot{n}/A)^{-\beta} \tag{52c}$$

where b and β are constants.

Substituting equation 52c in equation 52b and solving for $\dot{n}/A$, we obtain

$$\frac{\dot{n}}{A} = \left(\frac{D_{Fe}'' c_{Fe}''}{b} \right)^{1/(1-\beta)} \tag{52d}$$

Hence

$$\frac{\dot{n}}{A} = \left(\frac{D_{Fe}'' c_{Fe}''}{b} \right)^2 \quad \text{if} \quad \beta = \tfrac{1}{2} \tag{52e}$$

in accord with experimental data.

The assumption that δ_{Fe}'' is inversely proportional to a fractional power of the rate of gas evolution is supported by similar results obtained for the rate of dissolution of magnesium rods in hydrochloric acid with transport of HCl to the magnesium surface as the rate-determining step.[12] More recently, Green and Robinson[13] have determined the rate of reduction of $Hg(CN)_2$ from a solution containing a large excess of NaOH on a mercury electrode at which hydrogen was evolved and have found that the rate of $Hg(CN)_2$ reduction was essentially proportional to the square root of the rate of hydrogen evolution for current densities ranging from 0.1 to 1 amp/cm^2.

Electrochemical Measurements

It is well known that the rates of transport of an ionic or molecular species to a metal-electrolyte interface can be determined from measurements of the saturation current of a polarized electrode as it is done in polarography. Such measurements are much less laborious than the methods suggested above, which require counts of radioactivity, or chemical analyses. A cell for the determination of a saturation current involves necessarily a counter electrode, which preferably should be of greater size than the working electrode at which the saturation current is to be measured. This imposes rather serious limitations on the geometry of the working electrode and the resulting hydrodynamic conditions. Results obtained with such a cell will, therefore, hardly be representative for conditions usually employed in the study of displacement reactions. Consequently, measurements of saturation currents do not promise to give pertinent information which can be used in conjunction with other investigations, especially measurements of the rate of displacement reactions.

Discussion

It has been shown above that values of (D_i'/δ_i') and (D_i''/δ_i'') can be obtained either from studies of the exchange of radioactive isotopes or from measurements of the rate of displacement reactions. If equations 28 and 29 for the relation between the effective thicknesses of diffusion boundary layers and diffusion coefficients are adopted and the diffusion coefficients of the constituent elements are known or can be estimated with sufficient confidence, it is necessary only to determine the values of (D_i'/δ_i') for a single constituent in the metal from which values for the other constituents can be calculated with the aid of equation 28. In addition, the value of (D_i''/δ_i'') for the same or another constituent in the slag is to be determined. Therefrom values for other constituent ions in the slag may be calculated with the aid of equation 29. Consequently, the determination of values of (D_i'/δ_i') and (D_i''/δ_i'') for all constituents of a metal-slag system for fixed conditions of convective flow is a manageable problem. These values in turn may be used in order to calculate the rate of displacement reactions which have not been investigated directly.

A considerable simplification is possible by using the same value of the diffusion coefficient for different atomic or ionic species as an approximation. By and large, diffusion coefficients of different molecular species in liquid phases do not differ widely. Moreover, in view of equations 28 and 29, the values of the quotients (D_i'/δ_i') and (D_i''/δ_i'') are proportional to the square root of D_i' and D_i'', respectively. Therefore, a difference of 20% in the individual diffusion coefficients corresponds to a difference of only 10% in the rates of the respective transport processes under comparable conditions.

It must be recalled, however, that the values of δ_i' and δ_i'' depend on the conditions of convective flow. If gas is passed through a bed containing a catalyst in the form of small grains, the dependence of the ratio of the effective thickness of the diffusion boundary layer to the diameter of the grains of the catalyst may be represented as a function of two dimensionless groups, that is, Reynolds number and Schmidt number. This procedure does not lend itself to an analogous treatment of metal-slag reactions because, in the latter case, conditions of convective flow are much more complex and so far it has been impossible to characterize flow conditions by dimensionless groups. At present, it is, therefore, necessary to determine values of (D_i'/δ_i') and (D_i''/δ_i'') for each set of flow conditions by special experiments. Results obtained in one laboratory can, therefore, be compared to results in another laboratory only if the setups are virtually identical. The same is true for a comparison of observations made in different industrial furnaces. By the same token, results obtained in laboratory experiments cannot be applied to operations in industrial furnaces.

This is a rather pessimistic conclusion. Nevertheless, rate studies seem to be of considerable scientific value in order to differentiate between control by transport processes and control by phase-boundary reactions. In particular, if the observed rate of an over-all reaction is considerably lower than the rate calculated from values of (D_i'/δ_i') and (D_i''/δ_i'') for exclusive transport control, it can be concluded that the assumption of virtual establishment of equilibrium at the interface does not hold and that, therefore the rate of the respective phase-boundary reaction can be and should be investigated. In this case, a knowledge of the values of (D_i'/δ_i') and (D_i''/δ_i'') is needed in order to set up experiments under conditions where the rate of the over-all reaction is determined essentially only by the rate of the phase-boundary reaction. Therefore, a direct determination of the rate law of the phase-boundary reaction is possible.

In this paper, no attempt is made to discuss the structure of the boundary layers under conditions which are characteristic of steelmaking, or pertinent laboratory experiments. Possibly, numerical values of δ_i' and δ_i'' may give clues on flow conditions in the boundary layers for a future semi-empirical treatment. In particular, it seems of interest to determine the ratio δ_i'/δ_i'' of the effective thicknesses of the boundary layer in the alloy and the slag first, for induction-heated systems with supply of stirring energy only to the metallic phase, and second, for large ratios of the viscosities of the phases involved, that is, a metallic phase having a low viscosity and a slag having a high viscosity.

KINETICS OF PHASE-BOUNDARY REACTIONS

Alternative Mechanisms

A displacement reaction between iron and manganese as considered in equation 1 may take place by a direct exchange of the ions or atoms involved in a single step similar to certain types of organic reactions in aqueous or nonaqueous solutions; for example,

$$Cl^- + CH_2I \cdot COOH = I^- + CH_2Cl \cdot COOH \quad (52)$$

If the elements involved in a displacement reaction have unequal valences, a more involved situation is encountered. For instance, in the case of the displacement reaction between aluminum and iron,

$$2[Al] + 3Fe^{2+} = 3[Fe] + 2Al^{3+} \quad (53)$$

a single-step reaction would require simultaneous rearrangement of five atoms. Such a mechanism is extremely unlikely.

Instead, displacement reactions involving different metals of equal or unequal valences may take place by virtue of consecutive electrochemical cathodic and anodic processes. Thus the mechanism of the phase-boundary reaction of displacement reactions is supposed to be analogous to the mechanism of corrosion reactions which take place in aqueous solutions at room temperature.[14, 15] For instance, for the displacement reaction between iron and manganese we may consider the sequence

$$[Mn] = Mn^{2+} + 2e^- \quad (54a)$$

$$Fe^{2+} + 2e^- = [Fe] \quad (54b)$$

and similarly for reaction 53

$$2\{[Al] = Al^{3+} + 3e^-\} \quad (55a)$$

$$3\{Fe^{2+} + 2e^- = [Fe]\} \quad (55b)$$

When alloys corrode in aqueous solutions, consecutive anodic and cathodic processes occur frequently at distinctly different sites corresponding to different phases of a metallic structure. This is called local cell action. Consecutive anodic and cathodic processes, however, may also be operative on a uniform metal-electrolyte interface such as a liquid metal-slag interface. Under these conditions, anodic and cathodic processes take place at random;[16] that is, the instant and the site of an individual cathodic process are supposedly not correlated to the instant and the site of a proceeding anodic process, in view of the large reservoir of conduction electrons in a metal. However, the rates of anodic and cathodic processes taken as average values depend on the electrode potential and, therefore, are interrelated by the electrode potential.

It is known that electrochemical processes definitely occur at a metal-slag interface. The electrochemical mechanism suggested in equations 54a to 55b is, therefore, not only a possibility but a necessity. It is open to question, however, whether electrochemical processes alone account for the observable over-all rate of a metal-slag reaction, or whether the rate of a direct exchange of two atoms or ions between metal and slag is appreciable or even predominant in comparison with the rate of consecutive electrochemical processes. To answer this question, one may determine the rate of the individual electrochemical processes with the help of electrochemical measurements for a comparison with measurements of the rate of displacement reactions such as formulated in equations 1 and 53. Investigations of this kind, however, are more difficult than analogous investigations[16] for reactions between alloys and aqueous solutions, as is discussed below.

The rate of anodic dissolution of a metal i, $\dot{n}_i$, is equal to the product of the metal-electrolyte interface area A and the current density J_i of this particular process divided by the product of valence z_i and the Faraday constant F,

$$\dot{n}_i = AJ_i/z_iF \qquad (56)$$

This equation also accounts for cathodic deposition of metal i with a negative value of $\dot{n}_i$ for transfer from slag to metal and a negative value for the current density J_i of the cathodic process.

Equations interrelating current density and electrode potential have been established by Erdey-Grúz and Volmer[17] and others.[18] If metal i in the alloy and in the slag is present either in a dilute solution obeying Henry's law or as a solvent whose concentration is virtually constant, we may use the equation[19]

$$J_i = k_i'c_i'^* \exp\left[(1 - \alpha_i)z_iEF/RT\right]$$
$$- k_i''c_i''^* \exp\left[-\alpha_iz_iEF/RT\right] \quad (57)$$

which gives the current density as the difference of two terms for anodic dissolution of metal i and its cathodic deposition as individual electrochemical phase-boundary reactions, where E is the potential difference between the electrode under consideration and a reference electrode measured without IR drop in the electrolyte, k_i' and k_i'' are the rate constants for the anodic and the cathodic process, respectively, $c_i'^*$ is the interface concentration of the metal to be oxidized anodically, $c_i''^*$ is the interface concentration of the metal ions to be reduced cathodically, and α_i is a parameter whose values lies between zero and unity.

When a metal is deposited from an aqueous solution involving complex ions, a homogeneous reaction may precede electrodeposition. For instance, it has been shown[20, 21, 22, 23] that cadmium deposition from a solution containing $Cd(CN)_4^{2-}$ ions and free cyanide ions takes place mostly from $Cd(CN)_3^-$ ions which are formed by dissociation of $Cd(CN)_4^{2-}$ ions. Similarly, if silicon is deposited

from a slag, it is conceivable that silicon in the slag occurs in the form of different complexes corresponding to different silicate ions, different rate constants have to be assigned to silicon deposition from different silicate ions, depletion of the rarer types of silicate ions at the metal-slag interface may occur, and the rate constants for the mutual transformation of the various silicate ions must be taken into account. In what follows, complications of this kind are disregarded.

In general, metal-slag reactions take place without net flow of electrical current. Thus we have

$$\Sigma J_i = 0 \qquad (58)$$

that is, an equivalence of anodic and cathodic processes. From equations 56 and 58 it follows that

$$\Sigma z_i\dot{n}_i = 0 \qquad (59)$$

In addition, we may note equations analogous to equation 41 for the rate of transport of all constituents i. Upon combining these transport equations with equations 56 and 57 for the rate of individual electrochemical processes with equation 59, we have a set of equations from which may be calculated the concentrations of all constituents in the metal and in the slag at the interface, the single electrode potential E, and the rate of transfer for each constituent i from metal to slag or in the reverse direction. For these calculations it is necessary to know the constants (D_i'/δ_i'), (D_i''/δ_i''), k_i', k_i'', α_i, and the bulk concentrations.

For the following discussion we confine our considerations to conditions where the over-all rate is much less than the rate for exclusive diffusion control, and therefore, concentrations of the various constituents next to the interface are virtually equal to the respective bulk concentrations. Thus equation 57 becomes

$$J_i = k_i'c_i' \exp\left[(1 - \alpha_i)z_iEF/RT\right]$$
$$- k_i''c_i'' \exp\left[-\alpha_iz_iEF/RT\right]$$
$$\text{if} \qquad c_i'^* \cong c_i' \qquad \text{and} \qquad c_i''^* \cong c_i'' \qquad (60)$$

In some applications of equation 60, it is convenient to write the electrode potential E as the sum of the equilibrium potential $E_{i(eq)}$ and the overpotential η

$$E = E_{i(eq)} + \eta \qquad (61)$$

Letting $J_i = 0$ and $E = E_{i(eq)}$ in equation 60 and solving for $E_{i(eq)}$, we obtain

$$E_{i(eq)} = (RT/z_iF) \ln\left(k_i''c_i''/k_i'c_i'\right) \qquad (62)$$

Upon substitution of equations 61 and 62 in equation 60, it follows that

$$J_i = J_{i(o)} \{\exp\left[(1 - \alpha_i)z_i\eta F/RT\right]$$
$$- \exp\left[-\alpha_iz_i\eta F/RT\right]\} \quad (63)$$

where

$$J_{i(o)} = (k_i'c_i')^{\alpha_i}(k_i''c_i'')^{1-\alpha_i} \qquad (64)$$

is the so-called exchange-current density for specified concentrations of species i in the metal and the slag.

If the overpotential is small, that is, the exponents in equation 63 are small in comparison to unity, we may use series expansions for the exponential functions and obtain a linear relation between current density and overpotential,

$$J_i = J_{i(o)} z_i \eta F / RT \qquad (65)$$

It must be recalled that equations 60 to 65 hold only if $c_i'^* \cong c_i'$ and $c_i''^* \cong c_i''$; that is, concentration polarization is negligible. This can be checked by calculating the differences $(c_i' - c_i'^*)$ and $(c_i'' - c_i'')$ from equation 8 for given experimental conditions.

At present, it seems likely that most metal-slag reactions are transport-controlled under the conditions usually employed in laboratory experiments or in metallurgical practice. This is in accord with results of polarization measurements on electrodes in molten salts, which indicate that, in general, concentration polarization exceeds activation polarization. Some metal-slag reactions, however, exhibit rather low rates, and therefore, the underlying electrochemical phase-boundary reactions are expected to involve a significant activation polarization. This is especially true for the reduction of silica to silicon in liquid steel and, moreover, for the evolution of CO. These reactions are discussed in detail below.

Exchange Reactions Involving Radioactive Isotopes

Diffusion-controlled exchange of radioactive isotopes between liquid steel and a slag has been considered earlier. Now we consider the same reaction under conditions involving control by a phase-boundary reaction. As a special example we consider exchange of radioactive silicon

$$^{31}[\text{Si}] + {}^{28}\text{Si}^{4+} (\text{slag}) = {}^{28}[\text{Si}] + {}^{31}\text{Si}^{4+} (\text{slag}) \quad (66)$$

Chemical equilibrium between alloy and slag for the abundant atomic species is assumed; that is, the net exchange rate is zero. Thus the overpotential η is equal to zero, and it follows from equations 56 and 63 that the number of gram atoms of species i passing per unit area per unit time from the alloy to the slag or in the opposite direction is equal to $J_{i(o)} A / z_i F$. If the mole fractions of radioactive isotope in the alloy and the slag are γ_i' and γ_i'', respectively, the net rate of transfer of radioactive isotope from alloy to slag is found to be

$$\dot{n}_{i(\text{tr})} = (J_{i(o)} A / z_i F)(\gamma_i' - \gamma_i'') \qquad (67)$$

Equation 67 is of the same type as equation 14 for a transport-controlled exchange reaction. Using the auxiliary equations 15 to 20, we obtain

from equation 67 a first-order rate law as formulated in equation 21 with the rate constant

$$\kappa_i = \frac{J_{i(o)} A}{z_i F} \frac{V' c_i + V'' c_i''}{V' c_i' V'' c_i''} \qquad (68)$$

In view of equation 64, equation 68 may be rewritten as

$$\kappa_i = \frac{A (k_i' c_i')^{\alpha_i} (k_i'' c_i'')^{1-\alpha_i}}{z_i F} \frac{V' c_i + V'' c_i''}{V' c_i' V'' c_i''} \qquad (69)$$

On the other hand, if we assume a single-step exchange reaction corresponding to a bimolecular reaction as formulated in equation 66, we have

$$\dot{n}_{i(\text{tr})} = A k_{i(\text{ex})} c_i' c_i'' (\gamma_i' - \gamma_i'') \qquad (70)$$

Under these conditions, the rate constant of the first-order rate law in equation 21 is found to be

$$\kappa_i = A k_{i(\text{ex})} (c_i'/V'' + c_i''/V') \qquad (71)$$

Upon doubling the concentrations of species i in both metal and slag, the rate constant κ_i remains the same if equation 22 or equation 69 holds, but κ_i increases by a factor of 2 if equation 71 holds. Thus, if control by phase-boundary reactions has been ascertained and, therefore, only equations 69 and 71 have to be considered, the dependence of κ_i on concentration provides a means to differentiate between the electrochemical mechanism underlying equation 67 and a single-step exchange underlying equation 71.

Evolution of Carbon Monoxide

Liquid steel saturated with graphite reacts readily with a slag containing iron oxide under evolution of carbon monoxide. The over-all reaction

$$[\text{C}] + \text{FeO} (\text{slag}) = [\text{Fe}] + \text{CO} (\text{g}) \qquad (72)$$

may take place as a single-step reaction, or by virtue of consecutive electrochemical processes, namely,

$$[\text{C}] + \text{O}^{2-} (\text{slag}) = \text{CO} (\text{g}) + 2e^- \qquad (73)$$

$$\text{Fe}^{2+} (\text{slag}) + 2e^- = [\text{Fe}] \qquad (74)$$

In contradistinction to conditions prevailing in the investigation by Philbrook and Kirkbride,[11] we assume vigorous forced convection so that the reaction is controlled by phase-boundary reactions rather than transport processes. In accord with a recent analysis of experiments by Ramachandran, King, and Grant,[10] the anodic evolution of CO is assumed to be the particular step which determines the rate. Thus the electrode potential is essentially given by the equilibrium for reaction 74, that is, by Nernst's formula

$$E = E_0 + (RT/2F) \ln c_{\text{Fe}}'' \qquad (75)$$

where E_0 is the equilibrium potential of an iron electrode extrapolated to unit concentration of iron ions in the slag.

Tentatively, we assume that the anodic evolution is a single-step reaction involving an oxygen ion in the slag and a carbon atom in the metal. Thus we have to use equation 60 in a slightly modified form and obtain for the rate $\dot{n}_{CO}$ of CO evolution

$$\dot{n}_{CO} = \{(A/2F)k_{CO}'c_C'c_O'' \exp [2(1 - \alpha_{CO})EF/RT]$$
$$- k_{CO}''p_{CO} \exp [-2\alpha_{CO}EF/RT]\} \quad (76)$$

Substitution of equation 75 in equation 76 yields

$$\dot{n}_{CO} = (A/2F)$$
$$\{k_{CO}'c_C'c_O''(c_{Fe}'')^{1-\alpha_{CO}} \exp [2(1 - \alpha_{CO})E_0F/RT]$$
$$- k_{CO}''p_{CO}(c_{Fe}'')^{-\alpha_{CO}} \exp [-2\alpha_{CO}E_0F/RT]\} \quad (77)$$

The carbon concentration in the metal may be kept constant by using a graphite crucible which provides liquid iron saturated with graphite. If the concentration of iron ions in the slag is much greater than the relatively low equilibrium concentration corresponding to about 0.03 to 0.1% FeO according to Hatch and Chipman,[24] we may disregard the second term in equation 77, that is, the backward reaction, and obtain

$$\dot{n}_{CO} = \kappa_{CO}(c_{Fe}'')^{1-\alpha_{CO}} \quad (78)$$

where

$$\kappa_{CO} = (A/2F)k_{CO}'c_C'c_O'' \exp [2(1 - \alpha_{CO})E_0/RT] \quad (79)$$

If $\alpha_{CO} = 0.5$, the rate of CO evolution becomes proportional to the square root of the concentration of iron ions in the slag. By and large, a fractional power of the concentration of a reactant is characteristic of rate control by an electrochemical process. For instance, Brönsted and Kane[25] have found that the rate of the reaction

$$Na \text{ (in Hg)} + H^+ \text{ (aq)} = Na \text{ (aq)} + \tfrac{1}{2}H_2 \text{ (g)} \quad (80)$$

is essentially proportional to the square root of the sodium concentration in mercury. Frumkin[26] has shown that this result is due to consecutive electrochemical processes with cathodic evolution of hydrogen as the rate-determining step.

In contradistinction to equation 78, a single-step reaction between FeO or iron ions and oxygen ions in the slag with carbon dissolved in iron according to equation 72 gives a rate which is proportional to the first power of iron ions in the slag if only the forward reaction is considered. This difference for the dependence of the rate of CO evolution on the concentration of iron ions in the slag enables us to decide experimentally whether the electrochemical mechanism plays a decisive part.

Anodic evolution of CO may also occur in conjunction with other cathodic processes, for example, sulfur transfer from metal to slag,

$$[S] + 2e^- = S^{2-} \text{ (slag)} \quad (81)$$

Upon adding corresponding sides of equations 73 and 81, we have the over-all reaction

$$[S] + [C] + O^{2-} \text{ (slag)} = S^{2-} \text{ (slag)} + CO \text{ (g)} \quad (82)$$

This reaction, however, does not occur alone because, in addition to the electrochemical processes in equations 73 and 81, either iron ions are reduced cathodically according to equation 74, or conversely iron is dissolved anodically,

$$[Fe] = Fe^{2+} \text{ (slag)} + 2e^- \quad (83)$$

Ramachandran, King, and Grant[10] have recently studied the simultaneous reactions which take place when sulfur-bearing iron saturated with graphite is allowed to react with a slag essentially free of iron oxide and sulfur. Assuming consecutive electrochemical processes, we may interpret the experimental results as follows. Initially, sulfur transfer according to equation 81 as cathodic reaction is accompanied by CO evolution according to equation 73 and iron dissolution according to equation 83 as anodic processes, whereas in the later stage the cathodic processes are sulfur transfer according to equation 81 and reduction of iron ions according to equation 74 accompanied by evolution of CO according to equation 73 as anodic process.

Reduction of Silica

Fulton, Grant, and Chipman[27, 28] have studied the reaction

$$[C] + SiO_2 \text{ (slag)} = [Si] + 2CO \text{ (g)} \quad (84)$$

and have found that the equilibrium of this reaction is attained very slowly.

Reaction 84 may take place as a bimolecular reaction between silica and in the slag and carbon dissolved in iron. As an alternative possibility, we may assume consecutive electrochemical processes.

In this context, it is profitable to describe the composition of a slag in terms of the number of the ions of the chemical elements which are present. These ions may be designated as "component ions" in order to indicate that these ions are the particles from which a slag can be built up in compliance with the principle of electrical neutrality, but no assumption regarding the association of the various ions is made. Thus equation 84 may be rewritten as

$$[C] + Si^{4+} \text{ (slag)} + 2O^{2-} \text{ (slag)} = [Si] + 2CO \text{ (g)} \quad (85)$$

presumably involving the following electrochemical processes

$$2\{[C] + O^{2-} \text{ (slag)} = CO \text{ (g)} + 2e^-\} \quad (86)$$
$$Si^{4+} \text{ (slag)} + 4e^- = [Si] \quad (87)$$

In view of available information, especially results reported by Ramachandran, Grant, and

King,[10] it may be assumed that in this case, equilibrium for reaction 86 is virtually established and reaction 87 is the rate-determining step. For liquid iron saturated with graphite and CO of atmospheric pressure, the electrode potential E has, therefore, a constant value. In view of equation 60, the rate of silicon transfer $\dot{n}_{Si}$ from metal to slag is found to be

$$\dot{n}_{Si} = (A/4F)\{k_{Si}'c_{Si}' \exp[4(1-\alpha_{Si})EF/RT]$$
$$- k_{Si}''c_{Si}'' \exp[-4\alpha_{Si}EF/RT]\} \quad (88)$$

When equilibrium between metal and slag has been reached, we have the equilibrium concentrations $c_{Si}'^{(eq)}$ and $c_{Si}''^{(eq)}$ in the metal and the slag, respectively, and the reaction rate is equal to zero. Thus equation 88 becomes

$$0 = (A/4F)\{k_{Si}'c_{Si}'^{(eq)} \exp[4(1-\alpha_{Si})EF/RT]$$
$$- k_{Si}''c_{Si}''^{(eq)} \exp[-4\alpha_{Si}EF/RT]\} \quad (89)$$

whence

$$E = (RT/4F) \ln[k_{Si}''c_{Si}''^{(eq)}/k_{Si}'c_{Si}'^{(eq)}] \quad (90)$$

Substituting equation 90 in equation 89 and using equation 64, we obtain

$$\dot{n}_{Si} = \frac{J_{Si(o)}A}{4F}\left[\frac{c_{Si}' - c_{Si}'^{(eq)}}{c_{Si}'^{(eq)}} - \frac{c_{Si}'' - c_{Si}''^{(eq)}}{c_{Si}''^{(eq)}}\right] \quad (91)$$

Equation 91 also applies to reduction of silica corresponding to a transfer of silicon from slag to alloy with a negative value of $\dot{n}_{Si}$, $c_{Si}' < c_{Si}'^{(eq)}$, and $c_{Si}'' > c_{Si}''^{(eq)}$.

In view of equations 68 and 91, the exchange current density $J_{Si(o)}$ may be calculated from (1) measurements of the rate of transfer of radioactive silicon between alloy and slag, and (2) measurements of the rate of silica reduction by liquid iron saturated with graphite. It must be noticed, however, that according to equation 64 the exchange current density depends on the concentrations of silicon in the alloy and of silica in the slag according to equation 64. Thus, for the aforementioned comparison, it is necessary to measure the rate of transfer of radioactive silicon with a system containing the equilibrium concentrations attained in a reduction experiment.

Electrochemical Measurements

Most of the usual experimental methods for the determination of direct-current density-potential curves at electrodes in aqueous solutions do not readily lend themselves to investigations on metal-slag systems at about 1500° C for these reasons.

A separation between cathode and anode compartment is difficult to accomplish because of the limited choice of materials for the construction of vessels for electrolytic cells.

For the same reason, it is hardly possible to measure single electrode potentials during flow of current with the help of a reference electrode and a Luggin capillary whose opening is placed next to the working electrode, although measurements in molten salts have been extended recently by Piontelli and his associates[29] up to 530° C

Alternating-current measurements on cells involving molten salts have been made by Randles,[30] Laitinen and Osteryoung,[31] and Berzins and Delahay.[32]

It may be most promising to apply the so-called interruptor method to a cell with three alloy electrodes of equal composition and a slag as electrolyte. Reference is made to theoretical calculations by Berzins and Delahay.[32] Chemical equilibrium between alloy and slag is assumed if no current is passed. Accordingly, there is no potential difference among the three electrodes. After passing a current pulse between electrode 1 as cathode and electrode 2 as anode, the potential difference between electrodes 1 or 2 and electrode 3 as reference electrode may be measured as a function of time with the help of an oscilloscope, according to Hickling.[33] Extrapolation to the time when the current was interrupted gives the potential difference between electrode 1 or 2 and electrode 3 corresponding to the applied current without IR drop in the electrolyte, that is, the overpotential η.

A rather serious difficulty may be that in many cases the applied current is used not only for one but several electrode reactions. As a special case, we consider an iron-silicon alloy as electrode in contact with a silicate melt. In view of the displacement reaction

$$2[Fe] + Si^{4+}(slag) = [Si] + 2Fe^{2+}(slag) \quad (92)$$

the slag contains necessarily a small concentration of iron ions in addition to silicon ions. Thus when the alloy is polarized cathodically, the reactions

$$Fe^{2+}(slag) + 2e^- = [Fe] \quad (93)$$

$$Si^{4+}(slag) + 4e^- = [Si] \quad (94)$$

take place simultaneously. When a short pulse is applied, reduction of iron ions will prevail since concentration polarization becomes appreciable only after a longer time and activation polarization for discharge of divalent iron ions is presumably much less than that for discharge of tetravalent silicon ions. However, when a pulse of appropriate length is applied, iron ions in the vicinity of the cathode will be depleted in view of their low bulk concentration, and therefore, discharge of silicon ions according to equation 94 will predominate at the end of the pulse provided that other cations such as calcium and magnesium ions are not discharged.

Neglect of discharge of iron ions is possible only if the concentration of iron ions at the metal-slag interface is much lower than the bulk concentration and, accordingly $-\eta \gg 2RT/F$. Thus only the second term on the right-hand side of equation 63 is significant.

Letting $J_{Si} \cong J$ in equation 63, we obtain

$$\log J \cong \log J_{Si(o)} + 2.30 \times 4\alpha_{Si} |\eta| F/RT$$

$$\text{if } |\eta| \gg RT/2F \qquad (95)$$

In view of equation 95, a plot $\log |J|$ versus $|\eta|$ is expected to give a straight line within the limits stated above. From such a plot the value of $\log J_{Si(o)}$ is obtained as the ordinate intercept. Upon comparing the value of $J_{Si(o)}$ deduced from electrochemical measurements with the value of $J_{Si(o)}$ deduced from measurements of the exchange of radioactive silicon according to equation 68, the hypothesis of an electrochemical mechanism may be tested most directly.

Local Cell Action

When a metal-slag reaction is investigated in a graphite crucible, it is possible that a cathodic reaction takes place at the metal-slag interface and an anodic reaction at the graphite-slag interface, with electrons migrating in the graphite and the metal ions migrating in the slag. This is illustrated schematically in Fig. 3 for sulfur transfer from

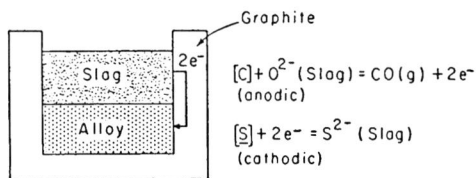

Fig. 3. Local cell mechanism for the reaction $[C] + [S] + O^{2-}(slag) = S^{2-}(slag) + CO(g)$.

metal to slag and evolution of CO as the over-all reaction. The occurrence of a local current has been shown by Baak and King[34] in model investigations on sulfur transfer between liquid silver and a borate melt in a graphite crucible. Additional investigations are needed in order to determine the significance of local cell reactions in comparison to electrochemical reactions at the metal-slag interface alone without participation of a second interface.

CONCLUDING REMARKS

Since metal-slag reactions involve necessarily several steps, rate equations are involved unless only one step controls the rate. In order to obtain pertinent basic information, it is, therefore, most desirable to select conditions under which only one step controls the rate. Thus, simple limiting cases have been stressed, but for a comparison, more complex situations have also been considered, for example, in equations 22 and 40 to 49, and other cases have been discussed in a qualitative fashion as, for instance, the occurrence of simultaneous displacement reactions in multicomponent systems.

A study of the rate of both transport processes and phase-boundary conditions is needed. In some respects, investigations on transport processes are simpler and, therefore, should precede investigations on phase-boundary reactions.

For the evaluation of measurements of the rates of transport processes, a better knowledge of diffusion coefficients in both alloys and slags is needed. A discussion of appropriate experimental methods, however, is beyond the scope of this paper.

From a fundamental point of view, investigations on phase-boundary reactions are considered to be most interesting because there are alternative mechanisms between which an experimental decision is needed.

Although there may be only a few reactions whose rate is controlled by a phase-boundary reaction, a study of these reactions is also of special practical interest. By and large, transport-controlled processes are relatively fast, and equilibrium may be approached within a reasonable time, whereas processes controlled by a phase-boundary reaction are slow, and accordingly, equilibrium may not be reached in metallurgical practice.

References

1. L. S. Darken, "Kinetics of Metallurgical Processes" in *Basic Open Hearth Steelmaking*, Second Edition, pp. 592 ff., Physical Chemistry of Steelmaking Committee, A.I.M.E., New York, 1951; L. S. Darken and W. R. Gurry, *Physical Chemistry of Metals*, pp. 465 ff., McGraw-Hill Book Co., New York, 1953.
2. E. R. G. Eckert, *Introduction to the Transfer of Heat and Mass*, McGraw-Hill Book Co., New York, 1950.
3. K. F. Gordon and T. K. Sherwood, *Chem. Eng. Prog. Symposium Ser.*, No. 10, 15 (1954).
4. W. G. Whitman, *Chem. & Met. Eng.*, 29, No. 4, 146 (1923).
5. W. K. Lewis and W. G. Whitman, *Ind. Eng. Chem.*, 16, 1215 (1924).
6. W. G. Whitman and D. S. Davis, *Ind. Eng. Chem.*, 16, 1233 (1924).
7. C. E. Birchenall and G. Derge, *Trans. A.I.M.E.*, 197, 1648 (1953).
8. N. J. Grant, U. Kalling, and J. Chipman, *Trans. A.I.M.E.*, 191, 666 (1951).
9. N. J. Grant, J. W. Dauding, and R. J. Murphy, *Trans. A.I.M.E.*, 191, 1451 (1951).
10. S. Ramachandran, T. B. King, and N. J. Grant, *Trans. A.I.M.E.*, 206, *J. Metals*, 8, 1549 (1956).
11. W. O. Philbrook and L. D. Kirkbride, *Trans. A.I.M.E.*, 206, 351 (1956).
12. B. Roald and W. Beck, *J. Electrochem. Soc.*, 98, 278 (1951)
13. M. Green and P. H. Robinson, *Trans. Faraday Soc.*, in press.
14. U. R. Evans, *An Introduction to Metallic Corrosion*, pp. 26 ff., Edward Arnold and Co., London, 1948.
15. H. H. Uhlig, *Corrosion Handbook*, pp. 3 ff., John Wiley and Sons, New York, 1948.
16. C. Wagner and W. Traud, *Z. Elektrochem.*, 44, 391 (1938).
17. T. Erdey-Grúz and M. Volmer, *Z. physik. Chem. Leipzig*, A, 150, 203 (1930).

18. J. O'M. Bockris, "Electrode Kinetics" in *Modern Aspects of Electrochemistry*, edited by J. O'M. Bockris and B. E. Conway, pp. 180 ff., Academic Press, New York, 1954.
19. K. Vetter, *Z. Elektrochem.*, **59**, 596 (1955).
20. L. Gierst and A. Juliard, *Compt. rend. C.I.T.C.E.*, **1**, 197 (1950); *J. Phys. Chem.*, **57**, 701 (1953).
21. H. Gerischer, *Z. physik. Chem. Leipzig*, **202**, 292 (1953).
22. H. Gerischer, *Z. physik. Chem. Frankfurt*, **2**, 79 (1954).
23. P. Delahay and T. Berzins, *J. Am. Chem. Soc.*, **75**, 2486 (1953).
24. G. G. Hatch and J. Chipman, *Trans. A.I.M.E.*, **185**, 274 (1949).
25. J. N. Brönsted and N. L. R. Kane, *J. Am. Chem. Soc.*, **53**, 3624 (1931).
26. A. Frumkin, *Z. physik. Chem. Leipzig, A*, **160**, 116 (1932).
27. J. C. Fulton, N. J. Grant, and J. Chipman, *Trans. A.I.M.E.*, **197**, 185 (1953).
28. J. C. Fulton and J. Chipman, *Trans. A.I.M.E.*, **200**, 1136 (1954).
29. R. Piontelli, *Z. Elektrochem.*, **59**, 778 (1955); R. Piontelli and G. Montanelli, *J. Chem. Phys.*, **22**, 1781 (1954); R. Piontelli and G. Sternheim, *J. Chem. Phys.*, **23**, 1358, 1971 (1955); R. Piontelli, G. Sternheim, and N. Francini, *J. Chem. Phys.*, **24**, 1113 (1956).
30. J. E. B. Randles and W. White, *Z. Elektrochem.*, **59**, 666 (1955).
31. H. W. Laitinen and R. A. Osteryoung, *J. Electrochem. Soc.*, **102**, 598 (1955).
32. T. Berzins and P. Delahay, *J. Am. Chem. Soc.*, **77**, 6448 (1955); *Z. Elektrochem.*, **59**, 792 (1955).
33. A. Hickling, *Trans. Faraday Soc.*, **33**, 1540 (1937); A. Hickling and F. W. Salt, *Trans. Faraday Soc.*, **36**, 1226 (1940).
34. T. Baak and T. B. King, private communication.

Index